Methods in Plant Biochemistry
Volume 10
Molecular Biology

METHODS IN PLANT BIOCHEMISTRY

Series Editors
P. M. DEY
Department of Biochemistry, Royal Holloway and Bedford New College, UK
J. B. HARBORNE
Plant Science Laboratories, University of Reading, UK

1. Plant Phenolics: J. B. HARBORNE
2. Carbohydrates: P. M. DEY
3. Enzymes of Primary Metabolism: P. J. LEA
4. Lipids, Membranes and Aspects of Photobiology: J. L. HARWOOD and J.R. BOWYER
5. Amino Acids, Proteins and Nucleic Acids: L. ROGERS
6. Assays for Bioactivity: K. HOSTETTMANN
7. Terpenoids: B. V. CHARLWOOD and D. V. BANTHORPE
8. Alkaloids and Sulphur Compounds: P. G. WATERMAN
9. Enzymes of Secondary Metabolism: P. J. LEA
10. Molecular Biology: J. BRYANT

Methods in Plant Biochemistry

Series editors
P. M. DEY and J. B. HARBORNE

Volume 10
Molecular Biology

Edited by

J. BRYANT

Department of Biological Sciences, University of Exeter, UK

ACADEMIC PRESS
Harcourt Brace & Company, Publishers
London   San Diego   New York
Boston   Sydney   Tokyo   Toronto

ACADEMIC PRESS LIMITED
24–28 Oval Road
London NW1 7DX

US edition published by
ACADEMIC PRESS INC
San Diego, CA 92101

Copyright © 1993, by
ACADEMIC PRESS LIMITED

All Rights Reserved
No part of this book may be reproduced in any form, by photostat, microfilm or any other means, without written permission from the publishers

This book is printed on acid-free paper

A catalogue record for this book is available from the British Library

ISBN 0-12-461020-X

Typeset by Colset Private Ltd, Singapore
Printed in Great Britain by The University Press, Cambridge

Contents

Contributors	vii
Series Preface	ix
Preface	xi

1	RNA Extraction and Fractionation *J. Speirs and T. Longhurst*	1
2	*In Vitro* Translation of Plant Messenger RNA *J. Speirs*	33
3	cDNA Cloning and Screening *R. P. Hodge and R. J. Scott*	57
4	Nucleic Acid Blotting and Hybridisation *P. A. Sabelli and P. R. Shewry*	79
5	Applications of Protein Blotting in Plant Biochemistry and Molecular Biology *R. J. Fido, A. S. Tatham and P. R. Shewry*	101
6	The Polymerase Chain Reaction *C. J. R. Thomas*	117
7	Non-radioactive *In Situ* RNA Hybridisation Using Digoxigenin and an Application for Co-localisation Studies with Radioisotopes *E. Y. Tanimoto and T. L. Rost*	141
8	Immunolocalisation of Antigens in Plants with Light and Transmission Electron Microscopy *J. A. Jernstedt, T. J. Jones and T. L. Rost*	159
9	Protoplast Fusion *R. Marchant, M. R. Davey and J. B. Power*	187

10 Import of *In Vitro* Synthesised Proteins into Intact Chloroplasts and Isolated Thylakoids from Higher Plants   207
  C. Robinson

11 Seed Development   219
  A. C. Cuming

12 Molecular and Genetic Analysis of Tomato Fruit Development and Ripening   251
  J. J. Giovannoni

Index   287

Contributors

A. C. Cuming, Department of Genetics, University of Leeds, Leeds LS2 9JT, UK
M. R. Davey, Plant Genetic Manipulation Group, Department of Life Science, University of Nottingham, Nottingham NG7 2RD, UK
R. J. Fido, Department of Agricultural Sciences, University of Bristol, AFRC Institute of Arable Crops Research, Long Ashton Research Station, Long Ashton, Bristol BS18 9AF, UK
J. J. Giovannoni, Department of Horticultural Sciences, Texas A & M University, College Station, TX 77843, USA
R. P. Hodge, Department of Botany, University of Leicester, Leicester LE1 7RH, UK
J. A. Jernstedt, Department of Agronomy and Range Science, University of California, Davis, CA 95616, USA
T. J. Jones, E. I. DuPont de Nemours and Co., DuPont Agricultural Products, Experimental Station, Wilmington, DE 19880-0402, USA
T. Longhurst, Department of Clinical Oncology, Royal North Shore Hospital, St Leonards, NSW 2065, Australia
R. Marchant, Plant Genetic Manipulation Group, Department of Life Science, University of Nottingham, Nottingham NG7 2RD, UK
J. B. Power, Plant Genetic Manipulation Group, Department of Life Science, University of Nottingham, Nottingham NG7 2RD, UK
C. Robinson, Department of Biological Sciences, University of Warwick, Coventry CV4 7AL, UK
T. L. Rost, Section of Botany, University of California, Davis, CA 95616-8537, USA
P. A. Sabelli, Department of Agricultural Sciences, University of Bristol, AFRC Institute of Arable Crops Research, Long Ashton Research Station, Bristol BS18 9AF, UK
R. J. Scott, Department of Botany, University of Leicester, Leicester LE1 7RH, UK
P. R. Shewry, Department of Agricultural Sciences, University of Bristol, AFRC Institute of Arable Crops Research, Long Ashton Research Station, Bristol BS18 9AF, UK
J. Speirs, CSIRO Division of Horticulture, PO Box 52, North Ryde, Sydney, NSW 2113, Australia
E. Y. Tanimoto, Section of Botany, University of California, Davis, CA 95616-8537, USA

CONTRIBUTORS

A. S. Tatham, Department of Agricultural Sciences, University of Bristol, AFRC Institute of Arable Crops Research, Long Ashton Research Station, Long Ashton, Bristol BS18 9AF, UK

C. J. R. Thomas, Advanced Technologies (Cambridge) Ltd, Cambridge Science Park, Cambridge CB4 4WA, UK

Preface to the Series

Scientific progress hinges on the continual discovery and extension of new laboratory methods and nowhere is this more evident than in the subject of biochemistry. The application in recent decades of novel techniques for fractionating cellular constituents, for isolating enzymes, for electrophoretically separating nucleic acids and proteins and for chromatographically identifying the intermediates and products of cellular metabolism has revolutionised our knowledge of the biochemical processes of life.

While there are many books and series of books on biochemical methods, volumes specifically catering for the plant biochemist have been few and far between. This is particularly unfortunate in that the isolation of DNA, enzymes or metabolites from plant tissues can often pose special problems not encountered by the animal biochemist. For a long time, the Springer series *Modern Methods in Plant Analysis*, which first appeared in the 1950s, provided the only comprehensive guide to experimental techniques for the investigation of plant metabolism and plant enzymology. This series, however, has never been completely updated; a second series has recently appeared but this is organised on a techniques basis and thus does not provide the comprehensive coverage of the first series. One of us (JBH) wrote a short guide to modern techniques of plant analysis *Phytochemical Methods* in 1976 (second edition, 1984) which showed the need for an expanded comprehensive treatment, but which by its very nature could only provide an outline of available methodology.

The time therefore seemed ripe to us to produce an entirely new multi-volume series on methods of plant biochemical analysis, which would be both thoroughly up-to-date and comprehensive. The success of *The Biochemistry of Plants*, edited by P. K. Stumpf and E. E. Conn and published by Academic Press, was an added stimulus to produce a complementary series on the methodology of the subject. With these thoughts in mind, we planned individual volumes covering: phenolics, carbohydrates, amino acids, proteins and nucleic acids, terpenoids, nitrogen and sulphur compounds, lipids, membranes and light receptors, enzymes of primary and secondary metabolism, plant molecular biology and biological techniques in plant biochemistry. Thus we have tried to cover all the major areas of current endeavour in phytochemistry and plant biochemistry.

The main aim of the series is to introduce to the scientist current knowledge of techniques in various fields of biochemically-related topics in plant research. It is also intended to present the historical background to each topic, to give experimental details of methods and analyses and appraisal of them, pointing out those methods that are most suitable for immediate application. Wherever possible illustrations and structures have been used and one or more case treatments presented. The compilation of known data and properties, where appropriate, is included in many chapters. In addition, the reader is directed to relevant references for further details. However, for the sake of clarity and completeness of individual reviews, some overlap between chapters of volumes has been allowed.

Finally, we extend our warmest thanks to our volume editors for undertaking the important task of organising each volume and cooperating in preparing the contents lists. Our special thanks go to the staff of Academic Press and to the many colleagues who have made this project a success.

P. M. DEY
J. B. HARBORNE

Preface to Volume 10

In 1976, I published a book entitled *Molecular Aspects of Gene Expression in Plants*.[1] I had wished originally to entitle it Plant Molecular Biology. However, I was advised by the Series Editor for the Academic Press Botanical Monographs Series, the late Professor James Sutcliffe, that the term 'molecular biology' would not necessarily be understood in the way I intended by all members of the plant science community. At that time, his advice was correct, but how different things are today. One of the most spectacular features of research in the plant sciences over the past 15 years has been the very rapid expansion of plant molecular biology, and I am sure that all who read the term 'molecular biology' know that it is about 'genes and their function as seen at the molecular level'.[1]

This expansion has followed the advent in 1973 of recombinant DNA technology — the ability to manipulate DNA *in vitro* — coupled with an amazing range of associated technical advances undreamt of in the late 1970s. When I published *Molecular Aspects of Gene Expression in Plants*,[1] recombinant DNA techniques had barely entered the plant science world, but I predicted that 'During the next decade we shall undoubtedly see . . . direct efforts to modify the genetic content of plant cells and attempts to regulate gene expression artificially'.[1] Both of those predictions have been amply fulfilled, but what I had failed to see was that recombinant DNA techniques would have an enormous impact on our understanding of plant genes themselves and of the ways in which the genes work. It is in this latter area that the expansion of our knowledge has been most spectacular, whilst the applications in agriculture and horticulture, although very important, have proceeded more slowly.

So, it is this flowering of plant molecular biology which has led to the production of this volume and the next in this series. In each of the two volumes, three different themes are discernible: (1) techniques for working with plant genes (Chapters 1-8 in this volume); (2) specific techniques for modifying the genetic contents of plant cells (Chapter 9 in this volume); and (3) applications of molecular biology techniques to understanding how plants work (Chapters 10-12 in this volume). Assembling the two volumes has not been easy. Firstly, had I included all possible applications of plant molecular biology, particularly in theme 3, the size of the two volumes would have been unmanageable. It has therefore been necessary to be selective, but I hope that the selection made serves to illustrate how molecular biology may be applied to different research areas right across the plant sciences. Secondly, practitioners of plant molecular biology are very busy people, active in their research and fired up by the general enthusiasm (and funding opportunities!) in this exciting field. It is a significant sacrifice to take time away from the bench to write a chapter for a book such as this. Some have, entirely justifiably, been unwilling to do so. However, those authors who did respond to my invitation to contribute have done so enthusiastically, and I am very grateful to them. I am also grateful to the series editors, Professor J. B. Harborne and Dr P. M. Day for their help in the early stages of preparing this book, and to the staff of Academic Press, not only for their help and support, but also for their cheerfully administered 'prods' at times when it seemed that things were moving rather slowly.

JOHN A. BRYANT

1. Bryant, J. A. ed. (1976). *Molecular Aspects of Gene Expression in Plants.* Academic Press, London & New York.

1 RNA Extraction and Fractionation

JIM SPEIRS[1] and TERRY LONGHURST[2]

[1]CSIRO Division of Horticulture, PO Box 52, North Ryde, Sydney, NSW 2113, Australia

[2]Department of Clinical Oncology, Royal North Shore Hospital, St Leonards, NSW2065 Australia

I.	Introduction	2
II.	RNA isolation	2
	A. General background	2
	B. SDS/phenol method	3
	C. Hot phenol method	7
	D. Detergent — no phenol method	8
	E. Differential solubility method	10
	F. Extraction using proteinase K	11
	G. Use of chaotropic agents and CsCl centrifugation	13
	H. Use of sodium perchlorate	15
	I. Use of cetyltrimethyl ammonium bromide	16
III.	Purification of RNA	21
	A. Measurement of RNA purity	21
	B. Removal of protein, DNA and carbohydrate	21
	C. Isolation of poly(A)$^+$ RNA	22
IV.	RNA fractionation	24
	A. General background	24
	B. Sucrose gradient fractionation	25
	C. Denaturing sucrose gradient fractionation	25
	D. Non-denaturing gel electrophoresis	26
	E. Glyoxal and dimethylsulphoxide denaturation and gel electrophoresis	26
	F. Formamide/formaldehyde denaturation and electrophoresis	28
V.	Conclusions	30
	References	30

I. INTRODUCTION

Of the numerous types of RNA found in all cells, the ribosomal RNAs (rRNAs) and transfer RNAs (tRNAs) are by far the most abundant. As a consequence, the rRNAs and tRNAs have been relatively easy to isolate and fractionate, and have been extensively characterised (c.f. Noller, 1984; Bjork et al., 1987; Sugiura and Wakasugi, 1989; Chapman and Buchheim, 1991; and references therein). Of the less abundant classes, messenger RNA (mRNA), which constitutes some 1-5% of the total RNA, is of particular interest currently as it is an important route by which information on gene structure and regulation can be obtained. The tissue-specific nature of many mRNAs, however, often leads to the need to isolate RNA, in good yields and condition, from difficult tissues, and the literature contains a multiplicity of isolation methods developed for individual plant tissues. In this chapter we review these methods, characterise them in terms of applicability to various tissues, ease of use and efficiency, and detail a number of representative methods.

Methods for the fractionation of isolated RNA have also developed markedly in recent years with the change of emphasis away from the ribosomal RNAs to the less abundant species. We look at a number of methods for the fractionation of total RNA and for the isolation and fractionation of mRNA.

II. RNA ISOLATION

A. General Background

It is probably true to say that isolation of RNA from plant tissues is generally more difficult than from animal tissues. There are a variety of reasons for this. Plant cells are bounded by a cell wall composed of cross-linked complex carbohydrates. These need to be mechanically disrupted in order to release the cell contents. In some tissues, such as ripening fruit, partial solubilisation of cell wall polysaccharides has occurred. These carbohydrates tend to co-purify with the RNA and, having very similar chemical properties, are difficult to remove. Cells of many plant tissues contain large vacuoles, and have correspondingly high ratios of H_2O to solutes. In these tissues, particularly if the cells are not actively dividing, the abundance of RNA relative to tissue fresh weight is very low, and RNA yields are correspondingly low. Ribonucleases (RNases) are ever present and can be very abundant in some tissues such as seed endosperm. The use of detergents for the disruption of tissue releases ribonucleases from the lysozomes, and appropriate steps have to be taken to inactivate the enzyme. A list of some of the protein denaturants, peptidases or specific RNase inhibitors used to protect RNA from hydrolytic attack during isolation, is shown in Table 1.1. Polyphenolic compounds and their quinone oxidation products are also often abundant and will interfere with the purification of intact RNAs if not counteracted.

As with the isolation and purification of RNA from any source, a number of basic precautions should be observed to minimise the danger of exposing the RNA to ribonuclease from sources other than the tissue itself. Ribonucleases are very stable enzymes, may be capable of regaining activity after denaturation, for example with detergent, and should be considered present in any solution or piece of equipment not

TABLE 1.1. RNase inhibitors used in some RNA isolation methods.

Inhibitor	Mode of action	Reference
Sodium dodecyl sulphate		
Sodium sarkosyl	Denaturation	Noll and Stutz (1968); Parish (1972)
Guanidine hydrochloride	Denaturation	Cox (1968)
Guanidine thiocyanate	Denaturation	Jencks (1969)
Phenol	Denaturation	Kirby (1968); Parish (1972)
Heparin	Binds RNase	Parish (1972); Palmiter (1974)
RNasin	Binds RNase	Scheele and Blackburn (1979)
Diethylpyrocarbonate	Denaturation	Fedorcsak and Ehrenberg (1966)
β-mercaptoethanol/dithiothreitol	Reduces disulphide bridges in RNase	Sela et al. (1957)
Vanadyl-ribonucleoside complexes	Binds RNase	Berger and Birkenmeier (1979)
Iodoacetate	Denaturant	Parish (1972)
Cetyltrimethyl ammonium bromide (CTAB)	Binds RNA	Bellamy and Ralph (1968)
Cetylpyridinium bromide	Binds nucleic acid	Geck and Nasz (1983)
Sodium perchlorate	Denatures RNase	cf. Newbury and Possingham (1979)
Proteinase K	Hydrolyses RNase	Wiegers and Hilz (1971)

specifically treated for their removal. Precautions should include autoclaving of all solutions where possible (but not solutions containing volatile organic compounds or detergents), autoclaving, if possible, all plasticware and pipettes, and heat sterilisation of glassware at 180°C for 2 h. In laboratories where RNase is frequently used, it is advisable to soak equipment in 0.1% diethylpyrocarbonate solution in H_2O (v/v) for 30 min followed by drying, in place of or as well as autoclaving, as autoclaving alone does not completely inactivate RNase. Hands are sources of RNase and contact with solutions, spatula, inside surfaces of containers, etc. must be avoided. Disposable gloves can be worn as an extra precaution.

Tissues from which RNA is to be isolated are best used without storing, although good quality RNA preparations can be made from tissue stored over liquid N_2. RNA can also be prepared from tissue stored at $-80°C$ after initial freezing in liquid N_2. Freeze-dried tissue can be used provided the tissue is initially frozen in liquid N_2, is stored at $-80°C$ before and after drying, and does not thaw during the drying process.

Extracted RNA in aqueous solution can be stored at $-80°C$ or lower, although our experience is that RNA preparations stored at $-80°C$ will show deterioration over a 6-month period. Long-term storage of RNA is best as an ethanol precipitate at $-80°C$.

B. SDS/Phenol Method

One of the most common methods of isolating RNA from plant tissue is to disrupt the ground tissue in a buffer containing a detergent, usually sodium dodecyl sulphate (SDS), and to deproteinise the solution by extraction with phenol, or a phenol/chloroform mixture. A number of buffer compositions with varying pH have been reported for use on a number of different tissues. It is likely that most of these variations result only in minor differences in quality and quantity of the extracted RNAs. Palmiter (1974), however, recommends the use of an acid pH buffer because it minimises RNase activity (Brown, 1967), allows ethylenediaminetetraacetic acid (EDTA) to be used (at an alkaline pH EDTA precipitates in ethanol), promotes the rapid removal of interface material, and minimises DNA recovery. He also points out that potassium should not be used, as it precipitates SDS.

We routinely add sodium acetate at pH 5.0 to our extracts prior to precipitation with ethanol. This satisfies the requirement for salt to promote precipitation of the RNA and creates an acid pH during precipitation, thus preventing EDTA from precipitating.

Mg^{2+} is used in some buffers in place of EDTA when it is important to maintain the integrity of the ribosomal RNAs, some of which are nicked and fragment in the absence of magnesium (Speirs and Grierson, 1978). For most isolations, magnesium is replaced with EDTA, which helps to break down and disperse protein aggregates and chelates metal ions required for RNase activity.

The method given below is a general method used in our laboratory with good results for many plant tissues.

1. Isolation of RNA from wheat leaf

The method is as described by Speirs and Brady (1981).

Extraction buffer

Tris/HCl	50 mM, pH 8.5 (4°C)
magnesium acetate	5 mM
SDS	0.5%

Chloroform/phenol. (1:1; v/v) chloroform : [phenol, *m*-cresol, 8-hydroxyquinoline (500:70:0.5; w/v)]

Method

- For total leaf RNA, freeze 2 g of leaf under liquid N_2 and grind to a fine powder in a mortar and pestle.
- Transfer the powder to 2 ml of extraction buffer in a separate mortar and pestle at 4°C and grind to a smooth paste.
- Add a further 8 ml of extraction buffer and continue to grind to disperse the paste in the buffer.
- Add 10 ml of chloroform/phenol and grind briefly to mix the organic and aqueous phases. Transfer to a sealable centrifuge tube and shake vigorously for 2–3 min.
- Recover the aqueous phase by centrifugation and re-extract by shaking with an equal volume of chloroform/phenol and recentrifuging.
- Precipitate the RNA from the aqueous phase by adding a 0.1 vol of 20% sodium acetate (w/v) pH 6.0 and 2.5 vols of ethanol and storing overnight at $-20°C$.
- Pellet the RNA by centrifugation and wash the pellet in 80% ethanol for a few minutes.

1. RNA EXTRACTION AND FRACTIONATION

- Re-pellet the RNA and dry under vacuum.
- Redissolve the RNA in a small volume (c. 100–200 µl) of 50 mM Tris/HCl, pH 7.4, 5 mM magnesium acetate, adjust to 2% sodium acetate and reprecipitate by the addition of 2.5 vol of ethanol and storage overnight at $-20°C$.
- This fairly crude RNA preparation will contain DNA which can be removed by redissolving the preparation in 50 mM Tris/HCl pH 7.4, 5 mM magnesium acetate, and treating with 20 µg ml^{-1} deoxyribonuclease 1 (RNase-free) at 37°C for 30 min.
- Extract with chloroform/phenol as above and reprecipitate.

An alternative and preferable method of purification is described in Section III.B, for use when dealing with the isolation of poly(A)$^+$ RNA.

2. Isolation of RNA from wheat leaf chloroplast

Wheat leaf chloroplasts can be isolated essentially by the method of Hartley and Ellis (1973) as modified by Speirs and Brady (1981).

Isolation buffer

sucrose	350–450 mM*
HEPES/NaOH	25 mM, pH 7.6 (4°C)
EDTA	2 mM
sodium isoascorbate	2 mM
β-mercaptoethanol	30 mM

*Sucrose concentration varies depending on the age of the leaf tissue (see Speirs and Brady, 1981).

Resuspension buffer

N-tris(hydroxymethyl)methyl glycine/KOH	50 mM, pH 8.3 (4°C)
KCl	200 mM
MgCl$_2$	6.6 mM

Method

- Homogenise 10 g batches of leaves for 15 s in 200 ml of isolation buffer (used as a slurry — near freezing) in a Waring blender or equivalent.
- Strain the homogenate through 8 layers of muslin and centrifuge at 2500 × g for 1 min.
- Discard the supernatant and gently resuspend the chloroplast pellet in 2 ml of resuspension buffer.
- Re-pellet the chloroplasts by centrifuging at 2500 × g for 1 min and discard the supernatant.
- Drain the tube and suspend the chloroplast pellet immediately in 5 ml of extraction buffer (see total RNA above).
- Continue as for total leaf RNA.

Note: We use phenol mixed with chloroform 1:1 (v/v) for deproteinising our RNA extractions, as the mixture separates better with centrifugation and gives a better-defined interface which can be avoided. Added advantages of using the phenol/chloroform mixture are reported by Palmiter (1974). Phenol alone retains about 10–15% aqueous phase which will lead to equivalent losses of RNA during extractions, whereas mixing with chloroform prevents the aqueous retention (Palmiter, 1974). Messenger RNA will preferentially fractionate into a phenol phase under certain conditions (Smith *et al.*, 1970; Lee *et al.*, 1971), but this loss can be prevented by the addition of chloroform (Penman, 1966; Perry *et al.*, 1972; Palmiter, 1973).

Palmiter (1974) recommends that the RNA extracts be shaken with phenol alone, before the addition of chloroform, otherwise 'substantial losses of RNA are incurred, probably due to the formation of insoluble protein RNA aggregates'. With most tissues we have found this not to be necessary but, with peach fruit tissue for example (see below), we have found the two-step extraction to be profitable.

3. Isolation of RNA from tomato fruit

Simple isolation techniques based on SDS/phenol/chloroform extractions do not work well on tomato fruit tissues for a number of reasons. Because the fruit tissues contain a high proportion of water, the abundance of RNA per g fresh weight is low and the RNA is difficult to purify without major losses. Tomato fruit, like many other fruit, soften as they ripen and the cell wall carbohydrate becomes more soluble, co-purifying with nucleic acids and forming gels which are difficult to separate from the RNA. Again, like many fruit, the tomato is highly acidic and extraction media must be well buffered to compensate.

Our laboratory uses an extraction procedure based on the solubilisation of polysomes and their subsequent extraction with detergent and phenol/chloroform (Speirs *et al.*, 1984). The initial solubilisation of polysomes is based on the method of Davies *et al.* (1972).

Extraction buffer

Tris/HCl	400 mM, pH 8.5
sucrose	200 mM
magnesium acetate	30 mM
KCl	60 mM
polyvinylpyrrolidone* (M_r over 40 000)	1% (w/v)
Triton X–100	0.1% (v/v)

Autoclave before the addition of:

β-mercaptoethanol	60 mM

*Inactivates phenolic compounds.

Chloroform/phenol. 1:1 (v/v) chloroform : phenol saturated with TE (10 mM Tris/HCl, 1 mM EDTA, pH 8.0)

Method

- Slice the tomato pericarp tissue into liquid N_2 and store at $-80°C$.
- For each extract, powder frozen tissue in a mortar and pestle or in a cooled coffee grinder.
- Sprinkle 5 g of the powdered tissue into 10 ml of extraction buffer in a mortar and pestle at 4°C and grind to form a smooth suspension while thawing.
- Filter the suspension through one layer of autoclaved Miracloth (Calbiochem) and centrifuge the filtrate at 23 500 × g, 4°C for 10 min.
- Decant the supernatant through two layers of autoclaved Miracloth (Calbiochem) and make the cleared extract to 0.5% with SDS.
- Add an equal volume of chloroform/phenol and shake vigorously for 2 min.
- Separate the phases by centrifugation at 13 000 × g for 10 min, 4°C.
- Remove the aqueous phase, make to 2% (w/v) with sodium acetate, pH 5, and precipitate the RNA by the addition of 2.5 vols of cold ethanol and overnight storage at $-20°C$.
- Pellet the precipitated nucleic acids by centifugation at 16 000 × g for 10 min at 4°C and wash by resuspension in 80% ethanol.
- Recentrifuge, drain the pellet and partially dry under vacuum.

Note: The nucleic acid pellet so obtained will contain, in addition to RNA, contaminating DNA and carbohydrate. Removal of these contaminants is dealt with in Section III.B.

C. Hot Phenol Method

A number of RNA isolation methods utilising hot phenol extractions are to be found in the literature. They are similar to the SDS/phenol method above but protein denaturation and extraction is undertaken at 50–60°C, presumably improving tissue disruption and nuclease inhibition. The method below was developed for the isolation of RNA from chicken embryo tissue (Benveniste *et al.*, 1973). Inclusion of a sedimentation step through CsCl (Boedtker *et al.*, 1974, 1976) improved the quality of the RNA and made the technique suitable for use on tissues rich in carbohydrate. The method is applicable to most plant tissues and produces good quality RNA, although yields are variable (50–100 μg g^{-1} for leaf tissue, 5–10 μg g^{-1} for fruit tissue; E. Lin and R. C. Gardner, personal communication).

Extraction buffer

NaCl	0.25 M
sodium acetate	25 mM, pH 8.0
EDTA	5 mM
SDS	5% (w/v)

Sarkosyl buffer

Tris/HCl	50 mM, pH 7.5
EDTA	10 mM
sarkosyl	1% (w/v)

Method

- Prepare a 250 ml flask containing 10 ml extraction buffer, 30 ml phenol/chloroform, and 1% β-mercaptoethanol (added just before the powdered tissue), and bring to 60°C in a water bath, in a fume hood.
- Grind 10 g of tissue in liquid N_2 in a cooled mortar and pestle or in a coffee grinder, and transfer the powered tissue to the hot extraction mixture.
- Shake vigorously for 4 min, keeping the flask at 60°C as much as possible.
- Transfer to centrifuge tubes and centrifuge at 12 000 × g, 20°C for 10 min.
- Remove aqueous phase and keep on ice.
- Re-extract the organic phase by shaking with 10 ml fresh extraction buffer, 60°C, 4 min.
- Centrifuge at 12 000 × g, 20°C, 10 min.
- Pool aqueous phases and shake with 30 ml phenol/chloroform at 60°C, for 4 min.
- Centrifuge at 12 000 × g, 20°C, for 10 min.
- Precipitate nucleic acids by the addition of 2.5 vols of ethanol and storage for 2 h at −20°C.
- Pellet nucleic acids by centrifuging at 12 000 × g, 4°C, 30 min.
- Air dry the pellet and redissolve in 5–7 ml sarkosyl buffer (can heat to 60°C to help dissolve the pellet).
- Add 0.5 g CsCl per ml.
- Carefully layer the nucleic acid solution over a 2.5 ml CsCl cushion (6.6 g CsCl in 4 ml 0.01 M EDTA) in a Beckman SW 41 tube, or equivalent.
- Carefully overlay the solutions with 0.5 ml sarkosyl buffer and top up the tube with paraffin oil.
- If necessary prepare balance tubes.
- Centrifuge at 27 500 × g, (av.) for 17 h at 4°C.

RNA will pellet through the CsCl cushion. DNA, protein and carbohydrate will band or float upwards. Traces of denatured RNase may be present in the upper CsCl layer, and care must be taken to avoid contamination of the pellet with this layer as RNase can renature and become active again. A simple method to avoid contamination is described below (but see also Section F).

- Remove the supernatant using a pipette.
- Invert the tube and drain the pellet thoroughly.
- Cut the tube about 1 cm from the bottom.
- Redissolve the RNA pellet in a small volume of H_2O.
- Reprecipitate the RNA by adding 0.1 vol of 3 M sodium acetate, pH 6.0, and 2.5 vols of ethanol.
- Store RNA under ethanol at −20°C or −80°C until required.

D. Detergent – No Phenol Method

This is a modification of the method of Dellaporta *et al.* (1983) for the rapid preparation of DNA, using SDS and sequential potassium acetate and isopropanol precipitation steps. The method has been used successfully for isolation of RNA from cotton

1. RNA EXTRACTION AND FRACTIONATION

leaf, pollen and young embryos, embryos of several *Gossypium* species and leaves of tobacco and *Arabidopsis* (Hughes and Galau, 1988) and, with modifications, from grape berries (Tesniere and Vayda, 1991).

The method of Hughes and Galau (1988) is described below.

Extraction buffer

Tris/HCl	200 mM, pH 8.5
lithium dodecylsulphate	1.5% (w/v)
LiCl	300 mM
EDTA	10 mM
sodium deoxycholate	1%
NP 40 (Nonidet)	1%

Autoclave, then add as solids

thiourea	5 mM
aurintricarboxylic acid	1 mM
dithiothreitol	10 mM

Method

- Grind 0.8 g of fresh leaf tissue in 10 ml of ice-cold extraction buffer (in a 30 ml Duall tissue grinder — Kontes, Vineland, NJ).
- Transfer to a 15 ml plastic centrifuge tube and allow to freeze slowly at −80°C.
- Incubate the homogenate in a 37°C water bath until just thawed.
- Immediately chill in wet ice and add 1/3 vol of 8.5 M potassium acetate, pH 6.5, mix thoroughly by inversion and incubate on ice for 15 min.
- Remove precipitate by centrifugation at 5000 × g for 20 min, 4°C.
- Filter the supernatant through one layer of autoclaved Miracloth (Calbiochem) into a fresh centrifuge tube, in ice.
- Add 1/9 vol of 3.3 M sodium acetate, pH 6.1, and 1/2 (final) vol of cold isopropanol.
- Mix and incubate at −20°C for at least 1 h.
- Pellet the isopropanol precipitate by centrifugation at 5000 × g for 30 min.
- Discard the supernatant and drain the pellet well.
- Resuspend the pellet in 800 µl TE (10 mM Tris/HCl, pH 8.0, 1 mM EDTA), and transfer to an Eppendorf tube.
- Add 1/4 vol of 10 M LiCl (to 2 M) and incubate on ice for 5–12 h.
- Pellet the LiCl precipitate by centrifugation at 10 000 × g for 30 min.
- Discard the supernatant, drain the pellet and redissolve in 200 µl TE.
- Add 1.5 vols of 5 M potassium acetate (pH not adjusted), mix and incubate on ice for 3–5 h.
- Pellet the precipitate by centrifugation at 10 000 × g for 30 min.
- Discard the supernatant, drain the pellet and resuspend in 100 µl TE by incubating in ice for 1 h, with occasional vortexing.
- Centrifuge at 10 000 × g for 30 min, and remove the supernatant to a fresh Eppendorf tube.

- Precipitate the RNA by adding 1/9 vol of 3.3 M sodium acetate, pH 6.1, and 2 vols of ethanol, and incubating at −20°C for at least 1 h.

The RNA can be stored under ethanol at −20°C or −80°C until required.

Notes: The authors, Hughes and Galau (1988) emphasise the importance of using an extraction buffer with high pH and high ionic strength and containing all three detergents, for the success of this extraction protocol. They also stress the importance of using potassium acetate adjusted to pH 6.5, in the initial steps of the extraction. An optional extraction procedure with phenol/chloroform followed by chloroform on its own can be introduced prior to the final precipitation of purified RNA.

Tesniere and Vayda (1991) made use of a similar procedure for extraction of RNA from grape berry but found a carry-over of small amounts of contaminating material, in the final RNA pellet, which caused gel formation. An additional purification step, involving slow addition of 95% ethanol to an aqueous solution of the RNA, to a final ethanol concentration of 30% (v/v), precipitated the contaminants which could then be removed by centrifugation

E. Differential Solubility Method

Manning (1991) has recently published a method for the purification of nucleic acids which is based on differences in their solubilities in 2-butoxyethanol from those of carbohydrates and polyphenolic compounds. The method allows a one-step separation of nucleic acids from contaminants. Once purified, the different solubilities of RNA and DNA in LiCl solutions allows their separation from each other. The method has been proved on tissues previously noted for difficulty, and appears to be applicable to a variety of tissues.

1. Isolation of RNA by differential precipitation

This method has been successfully applied to the following tissues by Manning (1991): dried seed of carrot (*Daucus carota* L. cv.); mature leaf, mature green and fully ripe fruit of cherry (*Prunus avium* L.); fruit pericarp from pear (*Pyrus communis* L.); perimedulla tissue from the tuber of potato (*Solanum tuberosum* L.); roots and mature leaves from tomato (*Lycopersicon esculentum* Mill.); and achenes (white stage) and receptical tissue (white and fully ripe stages) of fruit of strawberry (*Fragaria* x *ananassa* Duch.). The protocol outlined below is a general one applicable to any of these tissues.

Extraction buffer

boric acid/Tris	0.2 M, pH 7.6
EDTA	10 mM
SDS	0.5% (w/v)
β-mercaptoethanol	0.3 M

Phenol/chloroform: phenol (saturated with H_2O) : chloroform (1:1)

Method

- Freeze the tissue under liquid N_2, and powder either in a mortar and pestle or in a coffee grinder.
- Sprinkle onto extraction buffer (2.5–10 ml g^{-1}) in a prechilled mortar and pestle and grind to a smooth suspension.
- Bring to room temperature and shake vigorously with an equal volume of phenol/chloroform.
- Separate the phases by centrifugation at 20 000 × g for 10 min at 4°C and re-extract the interface, organic phase and pellet by shaking with a second volume of extraction buffer, and centrifuging.
- Combine the two aqueous phases and extract again by shaking with an equal volume of phenol/chloroform and centrifuging.
- To the aqueous phase, add 1 M sodium acetate/acetic acid, pH 4.5, to bring the Na$^+$ concentration to 80 mM (including the contribution of Na$_2$ EDTA in the extraction buffer).

Precipitation of carbohydrates and polyphenolic compounds

- Add 0.4 vol 2-butoxyethanol and incubate on ice for 30 min.
- Centrifuge at 20 000 × g for 10 min, 4°C, and collect the supernatant, taking care to avoid the gel-like pellet containing carbohydrate.
- Add 1 vol of 2-butoxyethanol to the supernatant and incubate in ice for 30 min to precipitate the nucleic acids.

Precipitation of nucleic acids

- Pellet the nucleic acids by centrifugation at 20 000 × g for 10 min at 4°C.
- Wash the pellet with 1:1 mixture of 0.2 M boric acid/Tris, 10 mM Na$_2$ EDTA (pH 7.6) : 2-butoxyethanol, to remove traces of polyphenolic compounds.
- Wash with 70% ethanol containing 0.1 M potassium acetate.
- Wash with 100% ethanol.
- Drain and dry under vacuum.

Separation of RNA and DNA

- Redissolve the nucleic acid pellet in H$_2$O to a concentration of > 1 mg ml^{-1} and adjust to 3 M LiCl by adding 1/3 vol of 12 M LiCl.
- Incubate on ice for 60 min and pellet the precipitated RNA by centrifugation at 11 600 × g for 10 min in a microfuge.
- Wash the pelleted RNA twice with 1 ml per wash of 3 M LiCl.
- Wash pellet with 70% ethanol.
- Wash pellet with 100% ethanol, drain and dry under vacuum.

F. Extraction using Proteinase K

Proteinase K (available from Boehringer, Mannheim) is a highly active peptidase which does not exhibit pronounced cleavage specificity. It is not inactivated by chelating agents (e.g EDTA) or sulphhydryl reagents (β-mercaptoethanol or dithiothreitol). It is stable over a wide pH range (Ebeling et al., 1974) and its activity is stimulated as much as seven-fold by the presence of SDS (Orth, 1976). Its properties make it a highly effective inactivator of RNases and DNases (Wiegers and Hilz, 1971, 1972) and make it particularly suitable for isolation of native RNA and DNA from tissues or cell lines.

The method described below makes use of proteinase K for RNA extraction from peach fruit. It is a slight modification of the method developed by Callahan et al. (1989) which, in turn, was developed from a method of Mozer et al. (1980) for the isolation of RNA from barley aleurone layers.

Callahan et al. (1989) report having attempted a number of methods for isolating peach fruit RNA, including both hot and cold phenol extractions according to the protocols of Galau et al. (1981), Morgens et al. (1985) and Morgens et al. (1987), mixed alkyltrimethylammonium bromide (CTAB) extraction (Taylor and Powell, 1982) and two guanidinium isothiocyanate methods (Maniatis et al., 1982). Only the method utilizing proteinase K resulted in the isolation of RNA translatable in a cell-free protein synthesising system, and omission of proteinase K from the tissue lysate resulted in the isolation of non-translatable RNA which did not convert to translatable upon subsequent proteinase K treatment.

LiCl precipitation and ethanol precipitations from sodium and potassium salts are used to remove carbohydrate from the isolated RNA.

Extraction buffer

Tris/HCl	100 mM, pH 9.0
NaCl	100 mM
SDS	1% (w/v)
polyvinylpyrrolidone* (M_r over 40 000)	1% (w/v)

Autoclave, then add just before use

β-mercaptoethanol	1% (v/v)
Proteinase K	100 μg ml^{-1}

*Inactivates phenolic compounds.

Method

- Freeze tissue in liquid N_2 and lyophilise.
- Grind freeze-dried tissue under liquid N_2 in a mortar and pestle or in a coffee grinder.
- Transfer 2 g (peach tissue) to 5 ml extraction buffer in a mortar and pestle at 4°C.
- Grind tissue to smooth paste and add a further 5 ml extraction buffer.
- Continue grinding, add a final 10 ml extraction buffer and mix to a smooth suspension.
- Let stand for 5 min.
- Filter through one layer of autoclaved Miracloth (Calbiochem), squeezing gently to recover most of the liquid phase.
- Centrifuge at 12 000 × g for 10 min at 4°C.

- Vigorously shake the supernatant for 2 min with 20 ml of phenol (saturated with 50 mM Tris/HCl pH 7.5).
- Add 5 ml chloroform/isoamyl alcohol (24:1;v/v) and continue shaking for 2 min.
- Centrifuge at 12 000 × g for 10 min at 4°C.
- Re-extract the aqueous phase with 20 ml phenol/chloroform, and centrifuge as before.
- Extract aqueous phase with 20 ml chloroform/isoamyl alcohol (24:1; v/v) and centrifuge as before.
- To the aqueous phase add:
  3 M sodium acetate, pH 4.8 to bring the extract to pH 5-6 (1/10 vol of aqueous phase for peach)
  SDS (10% w/v) to a final concentration of 0.1%
  NaCl (5 M) to a final concentration of 0.5 M
- Incubate in ice for 2 h.
- Centrifuge (preferably in a swing-out rotor) at 16 000 × g for 30 min at 4°C.
- Decant supernatant and add 6 M LiCl to a final concentration of 2 M.
- Incubate in ice (in a cold room) overnight.
- Centrifuge at 12 000 × g for 45 min at 4°C.
- Drain pellet and redissolve in 0.9 ml H_2O.
- Add 1/10 vol 1 M NaCl and 2.5 vols ethanol.
- Precipitate RNA at −20°C overnight.
- Collect precipitate by centrifugation, drain and redissolve in 0.9 ml H_2O.
- Repeat ethanol precipitation but from 0.2 M KCl at −20°C overnight.
- Repeat KCl/ethanol precipitation.
- Collect final precipitate by centrifugation, drain and redissolve RNA in 200 µl H_2O.
- Quantitate spectrophotometrically.

We routinely obtain yields of 0.9–1.0 mg RNA per g freeze-dried mesocarp and pericarp tissue for peach fruit just leaving second stage of development and 500–600 µg per g freeze-dried mesocarp and pericarp tissue for soft, ripe fruit. Callahan et al. (1989) reported yields of 5 mg per g freeze-dried tissue for whole, early stage 1 fruit, and 100 µg per g freeze-dried mesocarp and pericarp tissue from soft, ripe fruit. They also used this method to obtain high quality RNA from freeze-dried peach ovules (with embryos) and from freeze-dried peach leaf.

G. Use of Chaotropic Agents and CsCl Centrifugation

One approach to minimise enzymic degradation of RNA during its isolation is to rapidly denature all the cell proteins including RNase during the procedure. Early attempts to isolate intact RNA using different concentrations of the chaotropic agent guanidinium chloride to denature the cell proteins had mixed success (Grinnan and Mosher, 1951; Volkin and Carter, 1951). It was subsequently shown (Sela et al., 1957) that concentrations of guanidinium chloride above 4 M are required to inactivate RNase effectively, and moreover, that the inactivation is reversible unless the structure of the RNase is modified by, for example, complete reduction of its four disulphide bridges. An efficient method for RNA isolation from plant tissues, in particular from potato tuber, using guanidinium chloride solubilisation coupled with phenol chloroform extraction, is described by Logemann et al. (1987).

Chirgwin et al. (1979; see also Ullrich et al., 1977), improved the use of chaotropic agents for RNA isolation by replacing guanidinium chloride with guanidinium thiocyanate and including high concentrations of the reducing agent mercaptoethanol in the isolation buffer. The thiocyanate salt of guanidinium has the advantage over the chloride salt in that both cation and anion are strong chaotropic agents (Jencks, 1969). The combined presence of these, together with the mercaptoethanol, will maximise the rate of protein denaturation during tissue disruption and RNA isolation.

In addition to refining the isolation step, Chirgwin and colleagues used the CsCl centrifugation procedure of Glisin et al. (1974) to recover the isolated RNA relatively free from contaminants including, importantly, any residual RNase.

Chirgwin et al. (1979) developed their method for the isolation of high molecular weight, biologically active RNA from rat pancreas, a tissue rich in RNase A (Beintema et al., 1973). The method is applicable to most plant tissues, including those rich in RNase. The RNA will be recovered essentially free of contaminating protein, carbohydrate and DNA. A typical application to plant tissue, is given below.

The method given is that of Colbert et al. (1983) for the isolation of RNA from seedlings of *Avena sativa*.

Extraction buffer

guanidinium thiocyanate	4 M
Tris/HCl	50 mM, pH 7.6
sodium lauroyl sarcosine	2% (w/v)
β-mercaptoethanol	1% (v/v)
vanadyl ribonucleoside complex*	10 mM

*Vanadyl ribonucleoside complex, a potent inhibitor of ribonuclease (see Table 1.1) is prepared as a 200 mM stock according to Berger and Birkenmeier (1979).

Tris/EDTA/SDS

Tris/HCl	10 mM, pH 7.4
EDTA	5 mM
SDS	1% (w/v)

Method

- Tissue is frozen under liquid N_2 and ground to a powder either in a mortar and pestle or in a coffee grinder.
- 1 g of frozen, powdered tissue, is added to 5 ml of extraction buffer, at room temperature, and thoroughly homogenised (Colbert et al., 1983, used an Ultra Turrax blender) but vigorous grinding in a mortar and pestle can be used.
- Centrifuge the homogenate at 3000 × g for 10 min to remove coarse cell debris.
- Add CsCl to the supernatant to a final concentration of 2.4 M.
- Centrifuge at 30 000 × g for 10 min.

- Layer the resulting supernatant onto a 2 ml pad of 5.7 M CsCl in 0.1 M EDTA (pH 7.5) in a Beckman SW 41 tube (or equivalent).
- Centrifuge at 209 000 × g for 18 h at 7°C.

RNA has a high buoyant density in CsCl and will pellet through the 5.7 M CsCl pad. Most of the DNA, protein and carbohydrate will float upwards and will be discarded with the CsCl solution at the end of centrifugation.

Recovery of clean RNA from the pellet can be accomplished in a number of ways (cf. Maniatis et al., 1982; Slater, 1991). Both methods include precautions to avoid any possibility of contamination of the RNA with residual ribonuclease. The procedure outlined in Maniatis et al. (1982) is outlined below.

- Discard the supernatant at the end of centrifugation and carefully dry the walls of the centrifuge tube.
- Redissolve the pellet in Tris/EDTA/SDS. This will take time.
- Extract the dissolved RNA by vortexing with an equal vol of chloroform/1-butanol (4:1; v/v).
- Centrifuge and recover the aqueous phase.
- Re-extract the aqueous phase with an equal vol of chloroform/1-butanol, centrifuge and combine the two aqueous phases.
- Add 0.1 vol of 3 M sodium acetate (pH 5.2) and 2.5 vols of cold ethanol and precipitate the RNA by storing at −20°C for a minimum of 2 h.
- Recover the RNA by centrifugation, wash the pellet with 80% ethanol, drain and dry.
- Dissolve the pellet in a small volume of H_2O, add 0.1 vol of 3 M sodium acetate (pH 5.2) and re-precipitate by adding 2.5 vols of ethanol.
- Store the RNA under ethanol at −70°C.

A typical yield for *Avena sativa* seedlings was 0.5 mg RNA per g fresh weight of tissue (Colbert et al., 1983)

Note: This procedure is probably inappropriate for tissues containing large quantities of water, such as fruit tissue. It may be applicable to freeze-dried fruit tissues, but this has not been tested in our laboratory. A variation of the method of Chirgwin et al. (1979), involving homogenization of tissue in 5 M guanidine monothiocyanate and direct precipitation of the RNA from the guanidinium by 4 M LiCl, has been described by Cathala et al. (1983). The method may be applicable to plant tissues, but has not been tried in our laboratory.

H. Use of Sodium Perchlorate

Another chaotropic agent, sodium perchlorate, has been used to separate protein/nucleic acid complexes (Marmur, 1961) and to isolate nucleic acids from plant tissues (Wilcockson, 1973) and from viruses (Wilcockson and Hull, 1974).

Some plant tissues contain an abundance of substances such as quinones and phenols which interfere with nucleic acid isolation, severely reducing quantity and quality of yield. In extreme cases, such as in grapevine, interfering substances almost totally prevent the isolation of ribosomal RNA from leaf tissue by standard detergent/phenol methods (Rezaian and Krake, 1987). To overcome problems with these 'difficult' tissues, a number of methods have been tried, including the use of sodium perchlorate as a dissociating and denaturing agent (Newbury and Possingham, 1979; Rezaian and Krake, 1987).

The simple method of Rezaian and Krake (1987), is outlined below.

Extraction buffer

Na-perchlorate	5 M
Tris/HCl	0.2 M, pH 7.5
SDS	5% (w/v)
polyvinylpyrrholidone	8.5% (w/v)
β-mercaptoethanol	0.1% (v/v)

Method

- Grind young expanding leaves (of grapevine, *Vitis vinifera*) to powder under liquid N_2.
- Suspend the powdered leaf tissue in 4 vols of extraction buffer and stir at room temperature for 30 min.
- Centrifuge the suspension at 2000 r.p.m. in a bench centrifuge for 5 min in a cone (Amicon centriflo) plugged with glass wool approximately 15 mm high.
- Discard the cone and further clarify the filtrate by centrifugation at $27\,000 \times g$ for 15 min.
- Discard the resulting small pellet and precipitate the nucleic acids from the clear supernatant by the addition of an equal volume of ethanol.
- Centrifuge at $10\,000 \times g$ for 10 min and discard the supernatant.
- Wash the pellet with 70% ethanol, dry briefly and resuspend in TE (10 mM Tris/HCl, pH 7.5, 1 mM EDTA).

Note: Both RNA and DNA are recovered in this method. DNA recovery is significantly reduced by increasing the concentration of Na-perchlorate in the extraction buffer to 7.5 M (Rezaian and Krake, 1987). For the preparation of highly purified RNA an additional purification step should be employed such as pelleting through CsCl (Glisin *et al.*, 1974; see also Section II.E) or washing with high salt (Palmiter, 1974; see also Section III.A).

I. Use of Cetyltrimethyl Ammonium Bromide

The principle of this method is the disruption of plant tissue in the presence of the cationic detergent cetyltrimethyl ammonium bromide (CTAB) and the reducing agent β-mercaptoethanol, both of which protect the nucleic acids from the actions of nucleases. Cell wall debris, denatured proteins and most polysaccharides are removed by extraction of the suspension with chloroform : isoamyl alcohol followed by precipitation of nucleic acid/CTAB complexes by lowering the salt concentration of the solution. Residual protein (including RNases) and carbohydrates remain soluble under these conditions. CTAB and any remaining carbohydrate and protein contaminants are removed by pelleting the redissolved RNA (and simultaneously banding the DNA) by centrifugation through a CsCl cushion.

Jones (1951, 1953) first demonstrated the precipitation of nucleic acids from dilute salt solutions, by the use of CTAB, and methods for the separation of DNA and RNA based on differential solubilities of their CTA salts in sodium chloride were developed by Dutta et al. (1953) and Jones (1963).

Combining the phenol extraction procedure of Kirby (1956) with CTAB precipitation of nucleic acids from the aqueous isolate, Bellamy and Ralph (1968) developed a method for the isolation and purification of nucleic acids from a variety of tissues. Murray and Thompson (1980) described a refined method using CTAB for the isolation of high molecular weight plant DNA, and an extension of their method (Taylor and Powell, 1982) allows the simultaneous isolation from lyophilised or fresh tissue of good yields of good quality plant DNA and RNA. The Taylor and Powell (1982) method is described below.

2 × Extraction buffer

cetyltrimethyl ammonium bromide (CTAB)*	2% (w/v)
Tris/HCl	100 mM, pH 8.0
EDTA	20 mM
NaCl	1.4 M

Precipitation buffer

CTAB*	1% (w/v)
Tris/HCl	50 mM, pH 8.0
EDTA	10 mM
β-mercaptoethanol	1% (v/v)

*It is important to use the Cetyl-polymer as opposed to the mixture of polymers that some products contain. We find that the Sigma product, catalogue number H-5882, works well.

1 M CsCl/salt solution

Tris/HCl	50 mM, pH 8.0
EDTA	5 mM
NaCl	50 mM
CsCl	1.0 M

5.7 M CsCl/salt solution

Tris/HCl	50 mM, pH 8.0
EDTA	5 mM
NaCl	50 mM
CsCl	5.7 M

Method

- Freeze 10–50 g of fresh tissue in liquid N_2 and grind to a powder either in a precooled mortar and pestle or blender or coffee grinder.
- Transfer to a precooled container.

TABLE 1.2. RNA extraction methods and approximate yields for various plant tissues.

Method	Tissue		Yield (µg g^{-1} F.W.)		Reference
			Total nucleic acid	RNA	
SDS/phenol	Wheat leaf	– full expansion	1330		Lee and Speirs (unpubl.)
		– senescent	260		Lee and Speirs (unpubl.)
	Tomato	– fruit (pericarp)			
		mature green		90	Biggs et al. (1986)
		ripe		30	Biggs et al. (1986)
	Pea	– seedlings	3000		Cashmore (1982)
Hot phenol	Unspecified fruit			10	Lin and Gardner (pers. comm.)
	Unspecified leaf			100	Lin and Gardner (pers. comm.)
Detergent – no phenol	Cotton	– leaf		400	Hughes and Galau (1988)
		– pollen		900	Hughes and Galau (1988)
	Grape	– berry		30	Tesniere and Vayda (1991)
Differential precipitation	Strawberry	– achene (white)	1800		Manning (1991)
		– receptacle (white)	117		Manning (1991)
		– receptacle (ripe)	83		Manning (1991)
	Cherry	– fruit (mature green)	88^a		Manning (1991)
		– fruit (ripe)	28^a		Manning (1991)
		– leaves	1750		Manning (1991)
	Carrot	– seed	355		Manning (1991)
	Pear	– fruit (ripe)	51		Manning (1991)
	Potato	– tuber	45		Manning (1991)
	Tomato	– roots	334		Manning (1991)

Proteinase K	Peach	– fruit (young)	500[a]	Callahan et al. (1989)
	Peach	– fruit (mature)	10[a]	Callahan et al. (1989)
		– fruit (mid-dev.)	100[a]	Speirs (unpubl.)
		– fruit (ripe)	60[a]	Speirs (unpubl.)
	Barley	– seed	n.a.	Mozer (1980)
	Avocado	– fruit (harvest)	56	Christoffersen et al. (1982)
		– fruit (preclimacteric)	60	Christoffersen et al. (1982)
		– fruit (climacteric)	60	Christoffersen et al. (1982)
		– fruit (postclimacteric)	28	Christoffersen et al. (1982)
Guanidine HCL	Sweet potato	– tuber	60	Yeh et al. (1991)
	Rice	– seedling	100	Yeh et al. (1991)
Guanidine isothiocyanate	Oat	– seedling	500	Colbert et al. (1983)
Sodium perchlorate	Grapevine	– leaf	200	Rezaian and Krake (1987) and (pers. comm.)
CTAB	Unspecified	– leaf	400	Taylor and Powell (1982)

Yields are presented as maximum reported. Varietal differences and, in some instances, differences in tissue ages, are not noted. Total nucleic acids are RNA + DNA.

[a] Where necessary, fresh weight yields have been approximated by dividing yields from freeze-dried tissue by a factor of 10.

n.a., data not available.

- Add β-mercaptoethanol to 2% of the initial volume.
- Add an equal volume (10–50 ml) of boiling 2 × extraction buffer, mix buffer and powder well (using a spatula if necessary), and transfer immediately to a water bath at 65°C.
- Stir the mixture manually until it reaches a temperature of 50°C.
- Cool the mixture to room temperature and extract by gently shaking with an equal volume of chloroform : isoamyl alcohol (24:1) in a sealable centrifuge bottle or tube, until the phases are mixed. (It is important to keep the temperature of the extract above 15°C from now on, as lower temperatures will result in premature precipitation of CTAB/nucleic acid complexes.)
- Centrifuge at 13 000 × g for 10 min at room temperature.
- Using a wide bore pipette, gently transfer the upper aqueous phase to a fresh bottle or tube and repeat the chloroform : isoamyl alcohol extraction.
- Again, gently transfer the aqueous phase to a fresh container and add an equal volume of precipitation buffer. The reduction in salt concentration will result in the precipitation of CTAB/nucleic acid complexes.
- Mix well and allow to stand at room temperature for 30 min.
- Centrifuge at low speed (4000 × g) for 5 min and remove the supernatant. (Centrifugation at higher speed may result in a compacted nucleic acid pellet which is difficult to redissolve.)
- Dissolve the pellet in 1 M CsCl/salt solution, approximately 4 ml per 50 g starting material — more for leaves, less for callus).
- Layer over a 2 ml cushion of 5.7 M CsCl/salt solution in a centrifuge tube for a Beckman SW 50.1, or similar.
- Centrifuge at 120 000 × g for 12 h at 20°C.

The DNA will band at or just below the interface between the 1 M and 5.7 M CsCl solutions and the RNA will pellet in the bottom of the tube. DNA can be recovered from the interface with a Pasteur pipette and can be further purified by ethidium bromide/CsCl isopycnic centrifugation, as described by Taylor and Powell (1982).

- To recover the RNA, carefully drain and wipe dry the centrifuge tube, dissolve the RNA pellet in 10 mM Tris/HCl, pH 7.6, 1 mM EDTA, 1% (v/v) β-mercaptoethanol and precipitate the RNA by the addition of 2.5 vols ethanol and overnight storage at −20°C.

The RNA, when dissolved, is suitable for fractionation on oligo(dT) cellulose into poly(A)$^+$ and poly(A)$^-$ RNA. Yields of 150–400 μg total RNA per g fresh weight of leaf tissue are reported by Taylor and Powell (1982).

Note: The method can be used for lyophilised tissue, using the same volumes as for fresh tissue but approximately 1/10 weight of tissue (i.e. 1 g of lyophilised tissue should be extracted in 10 ml of 2 × extraction buffer, and so on throughout).

A list of the methods outlined in this section, and including, where possible, approximate yields of RNA obtained using the methods, is set out in Table 1.2. The list is by no means exhaustive and is included only as a guide to help selection of appropriate methods available for the extraction of various tissues.

III. PURIFICATION OF RNA

A. Measurement of RNA Purity

Depending on the tissue and extraction procedure, RNA preparations often contain impurities of DNA, protein or carbohydrate. DNA contamination of RNA samples is not always immediately obvious, although difficulty in dissolving the RNA, and stickiness of the dissolved pellet are indicators. Also a high molecular weight smear may be seen during electrophoresis.

The concentration of RNA in solution is most commonly determined spectrophotometrically by reading the absorbance of a diluted sample at 260 nm. An A_{260} of 1, for a 1 cm path length is equivalent to an RNA concentration of $c.$ 40 μg ml^{-1}. It is recommended that a full spectrum of the RNA, between 320 nm and 220 nm be obtained in addition to the normal A_{260} reading, as both nucleic acids, RNA and DNA, have a characteristic spectrum, shown in Fig.1.1, deviation from which is an indication of some degree of impurity. For a pure sample of RNA in water, the A_{260}/A_{280} ratio should be 1.8–2.0 (Sambrook et al., 1989). The presence of protein as a contaminant, for example, will increase the A_{280} absorbance, causing a decrease in A_{260}/A_{280} ratio below 1.8. Some, but not all, carbohydrate contaminants will also cause skewing of the spectrum, as will phenols and other non-desirables. There are a number of methods for removing DNA, protein and carbohydrate contaminants.

B. Removal of Protein, DNA and Carbohydrate

Protein should have been removed in the main during RNA isolation and purification, and further extraction with phenol/chloroform (see Section II.B) followed by ethanol precipitation, should remove any residual, if necessary. Proteinase K treatment can also be used (see Section II.D and Slater, 1991).

DNA can be removed by digestion with ribonuclease free deoxyribonuclease I (see Section II.B) followed by phenol/chloroform extraction and ethanol precipitation, if one has confidence in the purity of the DNase. DNA, carbohydrate and protein contaminants can be removed by pelleting the RNA through CsCl, according to the method of Glisin (1974), as described in Sections II.C and II.F. An alternative, which

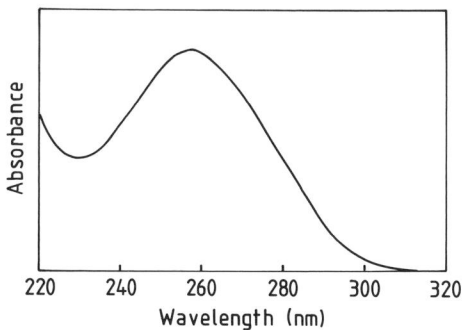

FIG. 1.1. Spectrum of purified RNA.

removes DNA, carbohydrate and low molecular weight RNA, is a method based on differential solubilities of the various solutes in high salt, used by Palmiter (1973). The method is described below.

Method

- Precipitate the RNA preparation under ethanol and pellet by centrifugation.
- Wash the pellet in 80% ethanol and semi-dry under vacuum.
- DNA, carbohydrate and low molecular weight RNA are selectively dissolved by addition of 3 M sodium acetate pH 6.0 (3.5 ml per 0.5–2 mg RNA) and frequent vortexing over a 1-h period, with storage on ice between mixes. If the pellet is slow to break up, a flame-sterilised curved glass rod can be used to break up the pellet mechanically.
- Pellet the undissolved RNA by centrifugation at 23 500 × g for 10 min, 4°C.
- Discard the supernatant, drain the tubes for 1 min and remove any remaining supernatant from the lip of the tube.
- Dissolve the RNA pellet in 1 ml 0.15 M sodium acetate, pH 4.8, and 0.5% (w/v) SDS at room temperature.
- Transfer the solution to a clean tube and precipitate the RNA by the addition of 2.5 ml cold ethanol.
- Pellet the RNA by centrifugation, wash the pellet in 80% ethanol and dry.
- Store the RNA either dry or under ethanol at −20°C or −80°C.

C. Isolation of poly(A)$^+$ RNA

Messenger RNAs with poly adenylated 3′-tails (poly(A)$^+$ RNA) can be isolated from preparations of total cell RNA by specific binding of the poly(A) tail. Affinity chromatography on oligo(dT)-cellulose columns (Aviv and Leder, 1972) is the most commonly used method, but other methods are available. These are generally commercial developments which utilise binding of the mRNA poly(A) tail with oligo(dT) or oligo(dU) immobilised on a variety of alternative solid supports such as nitrocellulose or nylon membranes (e.g. messenger affinity paper, Hybond-mAP, Amersham). Also available commercially (PolyATtract™, Promega Corp) is a method of binding poly(A)$^+$ RNA to paramagnetic particles via biotinylated oligo(dT) and streptavidin coupling. Messenger RNA selected in this way can be isolated very simply by magnetic attraction of the particles.

Affinity chromatography on oligo(dT)-cellulose (Aviv and Leder, 1972) is described below. We routinely include an additional purification step prior to mRNA isolation, which is designed to remove any residual carbohydrate present as a contaminant in the RNA preparations, as this can interfere with the chromatography. For this purpose, the RNA is passed through a column of cellulose (Mozer, 1980) before fractionation on the oligo(dT)-cellulose.

1. RNA EXTRACTION AND FRACTIONATION

Materials

2 × Binding buffer

NaCl	1 M
Tris/HCl	20 mM
EDTA	2 mM
SDS	0.02 % (w/v)

Adjust to pH 7.5 and autoclave.

1 × Binding buffer

NaCl	0.5 M
Tris/HCl	10 mM
EDTA	1 mM
SDS	0.01 % (w/v)

Adjust to pH 7.5 and autoclave.

Elution buffer

Tris/HCl	10 mM
EDTA	1 mM
SDS	0.01 % (w/v)

Adjust to pH 7.5 and autoclave.

Cellulose. Cellulose powder (Sigmacell type 50) is treated prior to use by autoclaving in 0.25 M NaOH and repeatedly washing with ribonuclease-free water until all trace of dark coloration is removed. The treated cellulose is re-autoclaved in H_2O before being packed in a 1 ml bed volume column and equilibrated with 1 × binding buffer.

Oligo(dT)-cellulose. A 0.3 ml bed volume column of oligo(dT)-cellulose (Collaborative Research Inc., Type 2) is prepared for use by washing sequentially in 2 × 10 ml elution buffer, 1 × 10 ml 0.5 M NaOH and 2 × 10 ml of 1 × binding buffer.

1. Removal of carbohydrate

Method

- Dissolve a pellet of total RNA (1 mg) in 200 µl H_2O and incubate at 65°C for 5 min followed by rapid cooling on ice.
- Add 200 µl 2 × binding buffer and 600 µl 1 × binding buffer.
- Pass the total RNA through the column of pretreated cellulose which has been equilibrated in 1 × binding buffer.
- Collect the run-through and pass through the column twice more.
- Wash the RNA from the column with 2 × 1 ml of binding buffer and combine the eluate fractions.
- Incubate the RNA fraction at 65°C for 5 min and rapidly cool on ice.

2. Oligo(dT)-cellulose fractionation

- Load the RNA onto the oligo(dT)-cellulose column equilibrated in 1 × binding buffer.
- Collect the eluate and wash the column with 1 ml of 1 × binding buffer.
- Combine the eluates and incubate at 65°C for 5 min.
- Cool the solution on ice and reload onto the oligo(dT) column.
- Wash the column with 1 ml of 1 × binding buffer. The eluate, containing poly(A)$^-$ RNA, can be discarded or the RNA can be precipitated by the addition of 2.5 vols of ethanol and storage at $-20°C$.
- Wash the column with a further 5 ml of 1 × binding buffer and discard the eluate.
- Elute the poly(A)$^+$ RNA from the column with elution buffer and collect 1 ml fractions.
- Measure the absorbance of the fractions at 260 nm and pool fractions containing RNA (poly(A)$^+$ RNA), usually fractions 1 and 2. (Note: acid-washed cuvettes should be used to avoid possible contamination with RNase.)
- Calculate the yield of poly(A)$^+$ RNA. The yield will depend on the poly(A)$^+$ content of the tissue from which total RNA was extracted, but should be greater than 1% of the amount of total RNA loaded onto the column.
- Precipitate the poly(A)$^+$ RNA by the addition of 0.1 vol 3 M sodium acetate, pH 4.8, and 2.5 vols of ethanol, followed by overnight storage at $-20°C$.
- Pellet the poly(A)$^+$ RNA by centrifugation, wash the pellet in 80% ethanol and dry under vacuum before dissolving in 200 μl H$_2$O and transferring to a microfuge tube.
- Precipitate the poly(A)$^+$ RNA under ethanol, as above, and store under ethanol at $-20°C$ or $-80°C$ until required.

IV. RNA FRACTIONATION

A. General Background

Total cell RNA is comprised of a number of different RNA types including ribosomal, transfer and messenger (see Slater, 1991) which vary in size from approximately 100 bases for tRNAs to greater than 5000 bases for mRNA precursors. Under native conditions, different RNA species can be fractionated according to size only to a limited extent. For example the cytoplasmic ribosomal RNAs (27S and 17S) can be separated from each other on sucrose gradients but cannot be fully separated from their chloroplastic counterparts (26S and 16S rRNAs) by the same method. Secondary structures within RNA molecules and aggregation between them prevent these molecules from behaving in a strictly linear manner, and lead to smearing and trapping during fractionation. Denaturing the RNA to eliminate secondary structures and aggregates allows fractionation more closely related to size, but leads in many cases to fragmentation of the ribosomal RNAs (cf. Speirs and Grierson, 1978). Hence, the method of fractionation employed will depend on several factors, including the species of RNA to be examined and the use to which that RNA will be put. Some examples of methods for fractionating RNA, under native and denaturing conditions, are outlined below.

1. RNA EXTRACTION AND FRACTIONATION

B. Sucrose Gradient Fractionation

Fractionation of native RNA on sucrose gradients is most applicable to isolation and purification of the abundant ribosomal, transfer and 5S RNAs. Non-denaturing sucrose gradients can be used to enrich mRNA fractions, but resolution will be poor and fractionation under denaturing conditions would be preferable.

Method

- Prepare a gradient of 5–40% sucrose in TE buffer (10 mM Tris/HCl, pH 8.0, 1 mM EDTA), in a Beckman SW 41 tube (or equivalent).
- Dissolve the RNA (100 µg) in 100 µl TE, and carefully layer on top of the gradient.
- Centrifuge for 16 h at 80 000 × g and 4°C.
- Pierce the bottom of the centrifuge tube with a hypodermic needle and collect 0.5 ml fractions.
- Dilute aliquots of each fraction and determine the absorbance at 260 nm.
- Dilute fractions containing RNA with an equal volume of H_2O, add 0.1 vol 3 M sodium acetate, pH 6.0, and 2.5 vols ethanol and precipitate the RNA overnight at −20°C.
- Pellet the precipitated RNA by centrifugation and wash the pellet with 80% ethanol to remove traces of sucrose.
- Store the RNA under ethanol at −20°C or −80°C until used.

C. Denaturing Sucrose Gradient Fractionation

Sucrose gradient fractionation of mRNA under non-denaturing conditions results in poor resolution of the RNAs due to the formation of secondary structures and aggregates. Where poly(A)$^+$ RNA can be obtained from a particular tissue, in large amounts, denaturing sucrose gradients can be used to obtain reasonably discrete fractions of mRNA. The technique allows preparation of RNA fractions that can be precipitated from the sucrose, analysed by *in vitro* translation or by hybridisation, and used as templates for cDNA synthesis and cloning. The method described below is that of Mozer (1980) and has been used to fractionate the abundant mRNA for α-amylase from barley aleurone layers.

Dimethylsulphoxide (DMSO) buffer

DMSO	95% (v/v)
deionised formamide	4% (v/v)
1 M Tris/HCl, pH 7.4 *	
1 M LiCl	1% (v/v)
100 mM EDTA	

*Made separately as a stock solution.

Gradient solutions

$\left.\begin{array}{l} 5\% \text{ sucrose} \\ 10\% \text{ sucrose} \\ 15\% \text{ sucrose} \\ 20\% \text{ sucrose} \end{array}\right\}$ made up in DMSO buffer

RNA. Poly(A)$^+$ RNA, *c.* 10 μg (more or less can be used depending on availability and method of analysis after fractionation). Dissolved in 100 μl deionised H$_2$O.

Method

- Prepare a step gradient of sucrose/DMSO — 5%, 10%, 15% and 20% — in a Beckman SW41 centrifuge tube, or equivalent, and allow to equilibrate overnight at room temperature.
- To the RNA dissolved in 100 μl of H$_2$O, add 400 μl of DMSO buffer.
- Heat at 60°C for 5 min, cool and carefully layer on top of the sucrose step gradient.
- Centrifuge at 195 700 × g at 28°C for 48 h.
- Fractionate the gradient either by pumping through a UV monitor or by collecting fractions from the base.
- Precipitate the RNA by adding 2.5 vols ethanol and storing overnight at −20°C.
- Wash the pellets 3 × with 70% ethanol, 0.3 M ammonia acetate.
- Store under ethanol at −20°C or −80°C until used.

D. Non-denaturing Gel Electrophoresis

Fractionation of RNA through polyacrylamide gels in the presence of SDS provides better resolution of the molecules than can be obtained on native sucrose gradients, without the problem of rRNA fragmentation associated with fully denaturing conditions (cf. Speirs and Grierson, 1978). RNA fractionated on polyacrylamide tube gels or slab gels can be visualised by scanning the gel in the UV wavelength in spectrophotometers modified for this purpose. Radiolabelled RNAs can be monitored by slicing the gel, drying the slices, and determining radioactivity in a scintillation counter. The use of polyacrylamide tube gels and slab gels is well described in Grierson (1982) and Slater (1983; 1991).

Non-denaturing fractionation can also be achieved using the agarose gel system described by McMaster and Carmichael (1977) (see also Section IV.E).

E. Glyoxal and Dimethylsulphoxide Denaturation and Gel Electrophoresis

There are a number of published methods for denaturing RNA prior to gel fractionation. Those include the use of methyl mercury (Bailey and Davidson, 1976), glyoxal and dimethylsulphoxide (McMaster and Carmichael, 1977) and formamide/formaldehyde (Fourney *et al.*, 1988). The glyoxal/dimethylsulphoxide and formamide/formaldehyde methods are outlined in this chapter. Although the methyl mercury method produces good fractionation and allows relatively straightforward transfer of fractionated RNA to membranes for hybridisation, we have not included the method here because of the toxic nature of the denaturant.

Denaturation of RNA with glyoxal and dimethylsulphoxide (DMSO) prior to fractionation by agarose gel electrophoresis has been described by McMaster and Carmichael (1977). When performed properly, this method gives good RNA fractionation, but care is required in order to obtain consistently good results. The running buffer used is of relatively weak ionic strength and must be recirculated during gel running to prevent an H^+ gradient forming between the electrodes which, if formed, reduces the quality of the fractionation. The following method can be used for non-denaturing RNA electrophoresis if glyoxal and DMSO are left out of the sample preparation.

Gel buffer and running buffer

        sodium phosphate   10 mM, pH 7.0

Loading buffer

        sodium phosphate   10 mM pH 7.0
        glycerol   50% (v/v)
        bromophenol blue   0.25% (w/v)
        xylene cyanol FF   0.25% (w/v)

Glyoxal. Glyoxal is either obtained as 40% (6 M) or dissolved to 6 M for use. It is important to deionise the glyoxal by passage through a mixed-bed resin (Bio-Rad AG 501-X8) before use. Deionised glyoxal should be stored in small aliquots at $-20°C$ to prevent oxidation.

Method

- Prepare a 1.0–1.5% agarose gel. Gels are prepared in a 10 mM sodium phosphate buffer, pH 7.0. The agarose is melted in the buffer, cooled and, before pouring, sodium iodoacetate can be added to 10 mM to inhibit ribonucleases. The gel is allowed to set for at least 30 min before electrophoresis.
- RNA (up to 30 μg of total RNA or 3 μg of poly(A)$^+$ RNA) is prepared by dissolving in 3.7 μl H$_2$O and adding 2.7 μl 6 M glyoxal, 8.0 μl DMSO and 3.0 μl 0.1 M sodium phosphate, pH 7.0, in a microfuge tube. The RNA is incubated at 50°C for 60 min followed by cooling on ice.
- Add 4 μl of loading buffer and load onto gel.
- The gel should be run with recirculating buffer or the buffer can be changed every 30 min. Electrophoresis should be at 3–4 V cm^{-1} until the bromophenol blue has run approximately 60% of the gel length.
- Stain the gel by placing in 200 ml of 30 μg ml^{-1} acridine orange in 10 mM phosphate buffer, pH 7.0, for 30 min.
- Destain in 10 mM phosphate buffer, pH 7.0, for 1 h before photographing on a UV transilluminator.
- RNA can be transferred to a nylon membrane (e.g. Zeta-Probe, Bio-Rad) by overnight capillary transfer with 50 mM Tris (unneutralised). Rinse the membrane in 2 × SSC before use.

F. Formamide/Formaldehyde Denaturation and Electrophoresis

This method, as described by Fourney *et al.* (1988), is the method of choice in our laboratory. It is quick and reliable, easily monitored by following ethidium-stained RNA bands, and allows efficient subsequent transfer of the RNA to membrane for hybridisation.

10 × MOPS/EDTA buffer

MOPS [3-(*N*-morpholino) propansulphonic acid]	0.2 M
sodium acetate	50 mM
EDTA	10 mM

Adjust to pH 7.0 and autoclave.

Electrophoresis sample buffer

formamide (deionised)	0.75 ml
10 × MOPS/EDTA buffer	0.15 ml
formaldehyde (37%)	0.24 ml
deionised, RNase-free H_2O	0.1 ml
glycerol	0.1 ml
10% (w/v) bromophenol blue	0.08 ml

Make fresh just before use, or store at $-20°C$ in small aliquots.

Electrophoresis buffer. 1 × MOPS/EDTA buffer.

Formaldehyde. Analytical reagent grade, 37%.

Ethidium bromide. 1.0 mg ml^{-1} in RNase-free deionised H_2O.

10 × SSC

NaCl	1.5 M
Na_3 citrate	0.15 M

Adjust to pH 7.0 and autoclave.

Method

All glassware is sterilised by heating at 180°C for 2 h. Where indicated, buffers are sterilised by autoclaving. Plasticware is sterilised by autoclaving (if appropriate) or by soaking in 0.1% diethylpyrocarbonate (DEPC) in sterilised, RNase-free, H_2O for 30 min.

Sample preparation

- Dissolve RNA in 25 mM EDTA, 0.1% SDS.
- Transfer 1–3 µg poly(A)$^+$ RNA, or 10–30 µg total RNA, to a microcentrifuge tube and make the volume up to 5 µl with sterile H_2O.

- Add 25 µl electrophoresis sample buffer and incubate at 65°C for 15 min.
- Add 1 µl ethidium bromide solution and mix thoroughly.

Sample is now ready for loading onto gel.

Gel preparation. RNA is fractionated on agarose gels containing between 1.0% and 1.5% agarose, 1 × MOPS/EDTA buffer, and 0.66 M formaldehyde. The general procedure is listed below.

- For a 60 ml — 1.2% gel, add 0.72 g agarose to 51 ml sterilised H_2O, and dissolve by heating.
- Cool agarose solution to about 50°C and add 6 ml 10 × MOPS/EDTA buffer.
- In a fume hood, add 3 ml 37% formaldehyde, and mix gently.
- Pour gel, insert comb and allow to set for at least 1 h before use.

Electrophoresis

- Flush sample wells with buffer and load samples.
- If very pure poly(A)$^+$ RNA is being fractionated, an extra track containing total RNA should be run, to provide molecular weight markers in the form of visible bands of ribosomal RNAs. We usually find, with poly(A)$^+$ RNA purified once through oligo(dT)-cellulose, that sufficient ribosomal RNA remains in the sample to be visible.
- Electrophorese at 3–4 V cm^{-1}.
- The migration of the ribosomal RNAs on the gel can be visualised by transferring the gel to a short-wave transilluminator (302 or 254 nm). Electrophoresis can be continued if necessary, until the ribosomal bands are approximately halfway down the gel.
- Photograph the gel on the transilluminator, through a yellow or orange filter.

Transfer of RNA to membrane

- We routinely soak RNA gels, with gentle agitation, in 0.05 M NaOH made up in 1 × SSC, for 15–20 min prior to transfer. This improves transfer efficiency, but is not essential and can be omitted for thin gels and/or low molecular weight RNAs.
- Soak the gel, with gentle agitation, for two 20-min periods in an amount of 10 × SSC sufficient to cover the gel.
- Pre-wet a charged nylon membrane, such as Zeta Probe (BioRad), in water followed by 10 × SSC, for a few minutes.
- Make use of capillary action to transfer the RNA in 10 × SSC from the gel to the membrane. The procedure is described in Fourney *et al.* (1988) and is essentially the same as that described by Southern (1975) for DNA transfer (see also Maniatis *et al.*, 1982).
- Fix the RNA to the membrane by baking for 2 h at 80°C prior to hybridisation.

V. CONCLUSIONS

This chapter has outlined a selection of procedures with application to the isolation, purification and fractionation of plant RNAs, with emphasis on plant mRNAs. Space (and time) constraints preclude an exhausive coverage of available methods. For some tissues, isolation of polysomes (by, for example, variations on the method of Davies *et al.*, 1972) followed by extraction of RNA from the polysomes, will yield RNA free from major contaminants but representative only of expressed mRNAs and ribosomal RNAs. Many of the methods outlined here are suitable for extraction of RNA from small samples (cf. Section II.G for method of Logemann *et al.*, 1987, and modification by Cathala *et al.*, 1983; Section II.E for method of Manning, 1991) and micro methods are becoming available which yield sufficient RNA for polymerase chain reaction (PCR) amplification work.

Comparison of methods in terms of ease (or otherwise) of application is not possible in the absence of appropriate comparative experiments, but a rough indication can be obtained by comparing the relative complexities of the methods. Similarly, relative efficiencies of extraction can be determined only by comparative experiments. However, the list of methods, tissues and yields in Table 1.2 provides a rough guide.

With continuing improvements in techniques of RNA isolation and fractionation, and the development of molecular biological techniques requiring smaller and smaller amounts of RNA, it is to be hoped that many of the problems associated with plant RNA isolation will soon be things of the past.

REFERENCES

Aviv, H. and Leder, P. (1972). *Proc. Natl. Acad. Sci. USA* **69**, 1408–1412.
Bailey, J. M. and Davidson, N. (1976). *Anal. Biochem.* **70**, 75–85.
Beintema, J. J., Champagne, R. N. and Gruber, M. (1973). *Biochim. Biophys. Acta* **310** 148–160.
Bellamy, A. R. and Ralph, R. K. (1968). *Methods Enzymol.* **12B**, 156–160.
Benveniste, K., Wilczek, J. and Stern, R. (1973). *Nature* **246**, 303–305.
Berger, S. L. and Birkenmeier, C. S. (1979). *Biochemistry* **18**, 5143–5149.
Biggs, M. S., Harriman, R. W. and Handa, A. K. (1986). *Plant Physiol.* **81**, 395–403.
Bjork, G. R., Ericson, J. U., Gustafsson, C. E. D., Hagervall, T. G., Jonsson, Y. H. and Wikstrom, P. M. (1987). *Ann. Rev. Biochem.* **56**, 263–287.
Boedtker, H., Crkvenjakov, R. B., Last, J. A. and Doty, P. (1974). *Proc. Natl. Acad. Sci. USA* **71**, 4208–4212.
Boedtker, H., Frischauf, A. M. and Lehrach, H. (1976). *Biochemistry* **15**, 4765–4770.
Brown, D. D. (1967). *In* "Methods in Developmental Biology" (F. H. Wilt and N. K. Wessells, eds), pp. 685–701 Thomas Crowell Co., New York.
Callahan, A., Morgens, P. and Walton, E. (1989). *HortScience* **24**, 356–358.
Cashmore, A. R. (1982). *In* "Methods in Chloroplast Molecular Biology" (M. Edelman, R.B. Hollick, N.-H. Chua, eds), pp. 387–392. Elsevier Biomedical Press, Amsterdam.
Cathala, G., Savouret, J.-F., Mendez, B, West, B. L., Karin, M., Martial, J. A. and Baxter, J. D. (1983). *DNA* **2**, 329–335.
Chapman, R. L. and Buchheim, M. A. (1991). *Crit. Rev. Plant Sci.* **10**, 343–368.
Chirgwin, J. M., Przybyla, A. E., MacDonald, R. J. and Rutter, W. J. (1979). *Biochemistry* **18**, 5294–5299.
Christoffersen, R. E., Warm, E. and Laties, G. G. (1982). *Planta* **155**, 52–57.

Colbert, J. T., Hershey, H. P. and Quail, P. H. (1983). *Proc. Natl. Acad. Sci. USA* **80**, 2248–2252.
Cox, R. A. (1968). *Methods Enzymol.* **12B**, 120–129.
Davies, E., Larkins, B. A. and Knight, R. H. (1972). *Plant Physiol.* **50**, 581–584.
Dellaporta, S. L., Wood, J. and Hicks, J. B. (1983). *Plant Molec. Biol. Reporter* **1**, 19–21.
Dutta, S. K., Jones, A. S. and Stacey, M. (1953). *Biochim. Biophys. Acta* **1**, 613–622.
Ebeling, W., Hennrich, N., Klockow, M., Metz, H., Orth, H.O. and Lang, H. (1974). *Eur. J. Biochem.* **47**, 91–97.
Fourney, R. M., Miyakoshi, J., Day, R. S. III and Paterson, M. C. (1988). *Focus* **10**, 5–7. Published by Bethesda Research Laboratories/Life Technologies, Inc., PO Box 6009, Gaithersburg, MD 20877, USA.
Galau, G. A., Legocki, A. B., Greenway, S. C. and Dure, L. III. (1981). *J. Biol. Chem.* **256**, 2551–2560.
Geck, P. and Nasz, I. (1983). *Anal. Biochem.* **135**, 264–268.
Glisin, V., Crkvenjakov, R. and Byus, C. (1974). *Biochemistry* **13**, 2633–2637.
Grierson, D. (1982). *In* "Gel Electrophoresis of Nucleic Acids" (D. Rickwood and B.D. Hames, eds) pp. 1–38. IRL Press, Oxford.
Grinnan, E. L. and Mosher, W. A. (1951). *J. Biol. Chem.* **191**, 719–726.
Hartley, M. R. and Ellis, R. J. (1973). *Biochem. J.* **134**, 249–262.
Hughes, D. W. and Galau, G. (1988). *Plant Molec. Biol. Reporter* **6**, 253–257.
Jencks, W. P. (1969). "Catalysis in Chemistry and Enzymology", Ch. 7. McGraw-Hill, New York.
Jones, A. S. (1951). *Chem. and Ind., London*, 1067.
Jones, A. S. (1953). *Biochim. Biophys. Acta* **10**, 607–612.
Jones, A. S. (1963). *Nature* **199**, 280–282.
Kirby, K. S. (1956). *Biochem. J.* **64**, 405–408.
Kirby, K. S. (1968). *Methods Enzymol.* **12B**, 87–99.
Lee, S. Y., Mendecki, J. and Brawerman, G. (1971). *Proc. Natl. Acad. Sci. USA* **68**, 1331–1335.
Logemann, J., Schell, J. and Willmitzer, L. (1987). *Anal. Biochem.* **163**, 16–20.
Maniatis, T., Fritsch, E. F. and Sambrook, J. (1982). "Molecular Cloning : A Laboratory Manual". Cold Spring Harbor Laboratory Press, New York.
Manning, K. (1991). *Anal. Biochem.* **195**, 45–50.
Marmur, J. (1961). *J. Mol. Biol.* **3**, 208–218.
McMaster, G. K. and Carmichael, G. G. (1977). *Proc. Natl. Acad. Sci. USA* **74**, 4835–4838.
Morgens, P. H., Grabau, E. A. and Gesteland, R. F. (1985). *Nucl. Acids Res.* **12**, 5665–5684.
Morgens, P. H., Pyle, J. B. and Callahan, A. M. (1987). *In* "Molecular Biology of Plant Growth Controls" (J. E. Fox and M. Jacobs, eds), UCLA Symposium on Molecular and Cellular Biology, New series, Vol. 44, pp. 157–166. Alan Liss, New York.
Mozer, T. J. (1980). *Plant Physiol.* **65**, 834–837.
Murray, M. G. and Thompson, W. F. (1980). *Nucl. Acids Res.* **8**, 4321–4325.
Newbury, H. J. and Possingham, J. V. (1979). *Plant Physiol.* **60**, 543–547.
Noll, H. and Stutz, E. (1968). *Methods Enzymol.* **12B**, 129–155.
Noller, H. F. (1984). *Ann. Rev. Biochem.* **53**, 119–162.
Orth, H. D. (1976). *Kontakte* **3**, 35.
Palmiter, R. D. (1973). *J. Biol. Chem.* **248**, 2095–2106.
Palmiter, R. D. (1974). *Biochemistry* **13**, 3606–3615.
Parish, J. H. (1972). "Principles and Practice of Experiments with Nucleic Acids". Longman Group Ltd., London.
Penman, S. (1966). *J. Mol. Biol.* **17**, 117–130.
Perry, R. P., La Torre, D. E., Kelley, D. E. and Greenberg, J. R. (1972). *Biochim. Biophys. Acta* **262**, 220–226.
Rezaian, M. A. and Krake, L. R. (1987). *J. Virol. Methods* **17**, 277–285.
Sambrook, J., Fritsch, T. and Maniatis, T. (1989). "Molecular Cloning: A Laboratory Manual," 2nd edn. Cold Spring Harbor Laboratory Press, New York.
Scheele, G. and Blackburn, P. (1979). *Proc. Natl. Acad. Sci. USA* **76**, 4898–4902.
Sela, M., Anfinsen, C. B. and Harrington, W. F. (1957). *Biochim. Biophys. Acta* **26**, 502–512.

Slater, R. J. (1983). *In* "Techniques in Molecular Biology" (J. M. Walker and W. Gaastra, eds), pp. 113–133, Croom Helm, London and Canberra.
Slater, R. J. (1991). *Methods Plant Biochem.* **5**, 121–146.
Smith, D. A., Martinez, A. M. and Ratliff, R. L. (1970). *Anal. Biochem.* **38**, 85–89.
Southern, E. M. (1975). *J. Mol. Biol.* **98**, 503–517.
Speirs, J. and Brady, C. J. (1981). *Aust. J. Plant Physiol.* **8**, 603–618.
Speirs, J. and Grierson, D. (1978). *Biochim. Biophys. Acta* **521**, 619–633.
Speirs, J., Brady, C. J., Grierson, D. and Lee, E. (1984). *Aust. J. Plant Physiol.* **11**, 225–233.
Sugiura, M. and Wakasugi, T. (1989). *Crit. Rev. Plant Sci.* **8**, 89–101.
Taylor, B. and Powell, A. (1982). *Focus* **4**, 4–6. Life Technologies, Inc., PO Box 9418, Gaithersburg, MD 20898, USA.
Tesniere, C. and Vayda, M. E. (1991). *Plant Molec. Biol. Reporter* **9**, 242–251.
Ullrich, A., Shine, J., Chirgwin, J., Pictet, R., Tischer, E., Rutter, W. J. and Goodman H. M. (1977). *Science* **196**, 1313–1319.
Volkin, E. and Carter, C. F. (1951). *J. Am. Chem. Soc.* **73**, 1516–1519.
Wiegers, U. and Hilz, H. (1971). *Biochem. Biophys. Res. Commun.* **44**, 513–519.
Wiegers, U and Hilz, H. (1972). *FEBS Lett.* **23**, 77–82.
Wilcockson, J. (1973). *Biochem. J.* **135**, 559–561.
Wilcockson, J. and Hull, R. (1974). *J. Gen. Virol.* **23**, 107–111.
Yeh, K.-W., Juang, R.-H. and Su, J.-C. (1991). *Focus* **13**, 102–103. Life Technologies, Inc., PO Box 9418, Gaithersburg, MD 20898, USA.

2 *In Vitro* Translation of Plant Messenger RNA

JIM SPEIRS

CSIRO Division of Horticulture, PO Box 52, North Ryde, Sydney, NSW 2113, Australia

I.	Introduction………………………………………………………………………	33
II.	Selecting an appropriate translation system……………………………………	35
III.	Preparation and use of *in vitro* translation systems…………………………	36
	A. The *E. coli* system……………………………………………………………	36
	B. The wheat germ system………………………………………………………	39
	C. The rabbit reticulocyte lysate system………………………………………	41
	D. The *Xenopus* oocyte system………………………………………………	42
	E. Translation of polysomes and the S-100 system…………………………	42
	F. RNase and proteinase inhibitors……………………………………………	44
IV.	Quantitation and analysis of *in vitro* translation products…………………	45
	A. Labelled amino acid incorporation…………………………………………	45
	B. Gel fractionation and autoradiography……………………………………	45
	C. Antibody precipitation………………………………………………………	48
	D. Hybrid arrest and hybrid release……………………………………………	49
	E. *In vitro* processing…………………………………………………………	52
	References……………………………………………………………………………	53

I. INTRODUCTION

Although being replaced, to some extent, by the synthesis and analysis of cDNAs, *in vitro* protein synthesis is still a powerful technique used for studying the mechanisms and components of protein synthesis and for the identification, characterisation and quantitation of mRNAs from all sources. A number of these aspects have been reviewed in depth (Stewart and Letham, 1977; Adams *et al.*, 1981; Grierson and Speirs,

1983). This chapter aims to look at *in vitro* protein synthesis as applied specifically to studies on plant mRNAs.

In vitro translation of mRNAs requires two groups of components, a purified cell extract (or whole cell, in the case of *Xenopus* oocytes) containing the ribosomes and associated components, and the mRNA supplied together with a source of energy, salts and a buffer to adjust the ion concentration and pH of the mixture, and additional amino acids, including a radioactive tracer. Isolation and purification of plant mRNA has been detailed in the previous chapter. The purity of the RNA is of considerable importance for efficient translation, and care should be taken to ensure that the RNA is sufficiently pure.

Plant cells contain both cytoplasmic and organellar mRNAs which translate differently under different conditions. Also, the RNA can be presented to the *in vitro* translation system either simply as total RNA, or fractionated into poly(A)$^+$ and poly(A)$^-$, or as polysomes (see Section III.E). The various types of RNA are listed in Fig. 2.1, together with the types of systems most suited to their translation.

There are numerous *in vitro* translation systems, some of which are referenced below. The most commonly used systems are those derived from *Escherichia coli*, wheat germ and rabbit reticulocytes. *Xenopus* oocytes, also used to translate exogenous mRNA, are of interest and will be discussed briefly, although they are not strictly an *in vitro* system.

In recent years, the *E. coli*, wheat germ and rabbit reticulocyte translation systems have become readily available from a number of biochemical companies. Commercial systems are well defined and highly active and are by far the most convenient way to approach *in vitro* translation experiments. However, the *E. coli* and wheat germ systems are relatively simple to make, and laboratories undertaking large numbers of translation experiments may prefer to make their own. Methods for making these two systems are included here.

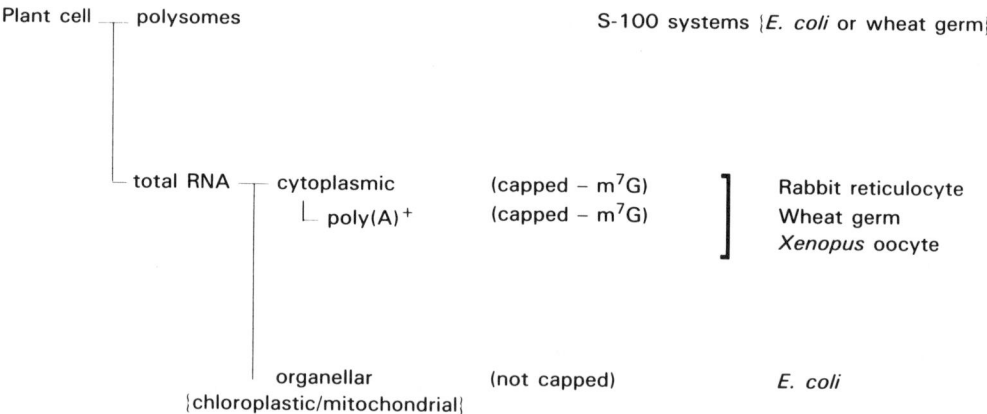

FIG. 2.1. Plant RNA classes and appropriate *in vitro* translation systems.

II. SELECTING AN APPROPRIATE TRANSLATION SYSTEM

In vitro translation systems are derived from extracts of prokaryotic or eukaryotic cells, and provide all the factors required for protein synthesis (see Fig. 2.2). They are fractionated to varying degrees to have low levels of endogenous mRNA and low concentrations of amino acids (particularly the amino acid to be used as a radioactive tracer), while retaining the ability to translate added mRNAs. Translation conditions are optimised by addition of an energy source, salts and amino acids (Fig. 2.2).

The *E. coli in vitro* translation system, which is the most common bacterial system, provides 70S (prokaryotic) ribosomes together with associated prokaryotic cofactors, and translates mRNAs of a prokaryotic nature more efficiently than those of a eukaryotic nature. In terms of plant mRNAs therefore, the *E. coli* system would efficiently translate organellar (chloroplastic or mitochondrial) RNAs, but not cytoplasmic RNAs. Conversely, the wheat germ and rabbit reticulocyte systems provide 80S (eukaryotic) ribosomes and cofactors, and translate typical eukaryotic (cytoplasmic) mRNAs more efficiently than organellar mRNAs.

Selection between the wheat germ and rabbit reticulocyte systems is less critical, although the reticulocyte system does appear to translate larger mRNAs more efficiently than the wheat germ system.

Translation of polysomes can be undertaken using standard *in vitro* systems, or can be carried out in systems depleted of ribosomes, so that only mRNAs already attached to polysomes will be translated to give *in vitro* products. Depleted systems, called S-100 systems, are described in Section III.E.

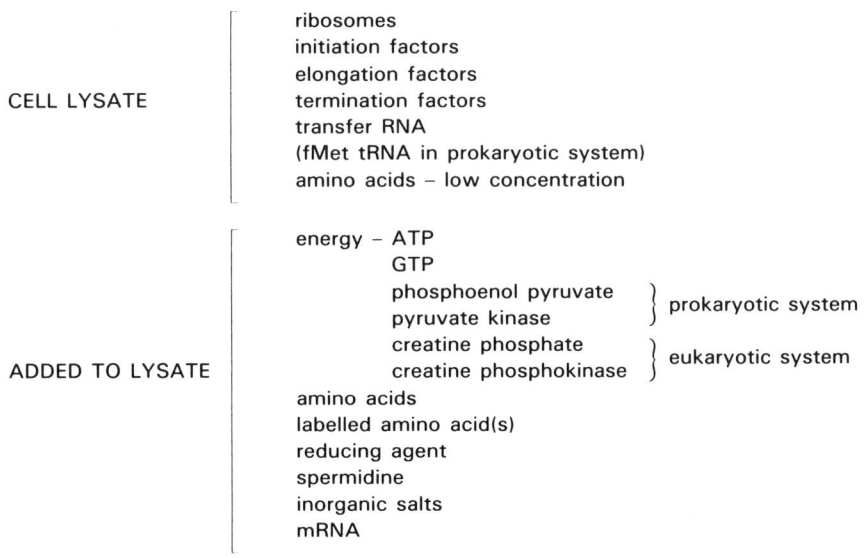

FIG. 2.2. Components of *in vitro* translation systems.

III. PREPARATION AND USE OF *IN VITRO* TRANSLATION SYSTEMS

A. The *E. coli* System

In 1961, Nirenberg and Mathaei directed the synthesis of polyphenylalanine in a cell-free lysate of *E. coli,* by the addition of the synthetic polynucleotide poly(U). Since then messenger-dependent, cell-free translation systems have been made from a number of prokaryotes including, for example, *Bacillus subtilis* (Doi, 1971) and *Halobacter* (Bayley, 1971). The *E. coli* system has been the most widely used over the years and is the system of choice for most prokaryotic applications. The method given below for the preparation and use of an *E. coli* lysate is based on methods described by Zubay *et al.* (1970), Modolell (1971) and Bottomley and Whitfeld (1979).

The term S-30 fraction (or extract), often applied to this preparation and many other cell-free translation systems, derives from a 30 000 × g centrifugation step from which the clarified supernatant (S-30 fraction) is obtained.

Bacteria. A strain of *E. coli* deficient in ribonuclease activity should be used. Strains such as MRE600 and PR7 are recommended.

Nutrient broth

Bacto tryptone	10 g
yeast extract	5 g
NaCl	5 g

Made to pH 7.0 with NaOH and autoclaved. Sterile 20% glucose added to 0.1% before use.

Extraction buffer

Tris/acetate	10 mM, pH 8.2
Mg acetate	14 mM
KCl	60 mM
dithiothreitol	0.1 mM

20 × stock amino acids. Solution containing 4 mM of all 20 essential amino acids.

20 × depleted amino acids. Same as 20 × stock, but lacking the tracer (normally methionine).

20 × energy

ATP	20 mM
GTP	10 mM
phosphoenol pyruvate	100 mM

Neutralised to *c.* pH 7.0 with NaOH.

20 × salts

Tris/acetate	1 M, pH 7.8
ammonium chloride	1.5 M
Mg acetate	224 mM

Method

Culture of cells

- Grow a 200 ml overnight culture of *E. coli* (MRE600 or PR7) in nutrient broth.
- Innoculate 5 × 1 litre of nutrient broth with 10–20 ml of the overnight culture.
- Incubate cultures with aeration and vigorous stirring until mid-log phase (OD_{650} of *c.* 2.0).
- Harvest cells by centrifugation at 4000 *g* for 10 min.
- Wash cell pellets by suspension in extraction buffer and recentrifugation.
- Drain washed pellets well and weigh.

Lysis

- Resuspend cells in an equal volume of extraction buffer at 4°C (i.e. 8 g cells + 8 ml buffer).
- Lyse cells by passing through a cold French Pressure Cell twice at a pressure of 10 000 p.s.i. (67 MPa).
- Immediately after second passage, add (for 8 g cells, starting) 10 μl 1 M dithiothreitol.
- Centrifuge at 30 000 × *g*, 30 min, 4°C.
- Collect the top 4/5ths of the supernatant and transfer to a fresh centrifuge tube.
- Recentrifuge at 30 000 × *g*, 30 min, 4°C.
- Carefully decant the supernatant fraction (S 30 fraction) and discard the pellet.

Preincubation

- To the S-30 fraction add, per ml:

100 μl	1 M Tris/acetate, pH 8.2
20 μl	140 mM Mg acetate
1 μl	1 M dithiothreitol
20 μl	20 × stock amino acids
1 μl	pyruvate kinase (*c.* 4 mg ml^{-1})
50 μl	20 × energy mix
110 μl	H_2O

- Incubate for 60 min, in the dark, at 37°C.
- Dialyse for 18 h, 4°C against extraction buffer, changing the dialysate 3 or 4 times.
- A convenient method of storing the lysate is to drip it slowly from a Pasteur pipette into liquid N_2, in such a way as to freeze it in small balls (each *c.* 40μl) which are then stored over liquid N_2 and gently thawed individually when required.

Use of the E. coli *S-30*

- For four reactions, make up the following reaction mixes, in ice:

	Preincubation[d] mix	Incubation mix
E. coli S-30	20.0 μl	–
20 × energy[a]	2.5 μl	2.5 μl
pyruvate kinase	0.25 μl	0.25 μl
100 mM dithiothreitol	1 μl	1 μl
20 × depleted amino acids	5 μl	–
20 × salts[b]	1.9 μl	3.1 μl
labelled amino acid[c]	–	2.0 μl (100–1000 kBq)
mRNA	–	–
H$_2$O	19.4 μl	1.15 μl
	50.0 μl	10.0 μl

- Incubate the preincubation mix for 5 min at 37°C then add to the incubation mix giving 60 μl.
- Divide into 15 μl aliquots.
- To each aliquot, add mRNA (10–30 μg) and water to make the volume up to 25 μl.
- Incubate 30 min, 37°C.
- Analyse incorporation (see Sections IV.A and B).

Notes

(*a*) *Energy.* ATP inhibits protein synthesis in high concentrations (it sequesters Mg^{2+}). ATP is therefore provided to the system in low concentration and is recycled via phosphoenol pyruvate/pyruvate kinase. Generally, the *E. coli* lysate contains its own pyruvate kinase and it is often beneficial not to add additional, as some preparations of pyruvate kinase are contaminated with RNase.

(*b*) *Salt optima.* Salt requirements for the *E. coli* system are not critical but should be optimised for each new preparation of lysate. Generally a Mg^{2+} concentration of 14 mM, and a total monovalent cation concentration (ammonium + potassium) of *c.* 100 mM are optimal.

(*c*) *Labelled amino acid.* ^{3}H- and ^{14}C-labelled amino acids are readily available and can be used for *in vitro* translation studies. However, the energy levels and specific activities of ^{35}S-labelled methionine and cysteine are far higher, making them the labelled amino acids of choice, particularly for analysis of the *in vitro* products by gel electrophoresis and autoradiography. ^{35}S-methionine is most commonly used, but in some cases, where methionine is rare in the proteins of interest, ^{35}S-cysteine can be used to replace it, although extra care must be taken during analysis of the products to prevent spurious binding of the ^{35}S-cysteine to proteins in the translation mixture (see Section IV).

(*d*) *Preincubation.* The preincubation steps, during preparation of the lysate and immediately prior to its use, are included for the purpose of removing ribosomes from endogenous mRNA in the lysate, and encouraging the degradation of the endogenous mRNA. If the translation background remains high after preincubation, the endogenous mRNAs can be inactivated by incubation with micrococcal nuclease in the presence of Ca^{2+}, following the procedure of Pelham and Jackson (1976) for the rabbit reticulocyte system. After incubation and before the addition of exogenous mRNA, the calcium-dependent micrococcal nuclease is inactivated by chelation of the calcium with an excess of EGTA.

B. The Wheat Germ System

A number of *in vitro* systems have been developed for translation of eukaryotic mRNAs. The most commonly used are those from wheat germ (Roberts and Paterson, 1973; Marcu and Dudock, 1974) and from rabbit reticulocytes (Schimke *et al.*, 1974; Hunt and Jackson, 1974; Pelham and Jackson, 1976). Systems have also been derived from Krebs ascites cells (Matthews and Korner, 1970), wheat embryo (Zagorski, 1978), yeast (Tuite *et al.*, 1980), rye embryo (Carlier and Peumans, 1976), mung bean primary axes (Carlier *et al.*, 1978a), pea primary axes (Peumans *et al.*, 1980a), *Saccharomyces cerevisiae* (Szczesna and Filipowicz, 1980; Tuite *et al.*, 1980), Chinese hamster ovary cells (Fischer and Moldave, 1981), and other sources.

The wheat germ system developed by Marcu and Dudock (1974) which is described in this section is simple to make and efficient to use.

1. Preparation of wheat germ S-30

Wheat germ. Germ from a hard winter wheat is best for this preparation. Ideally, a local grain mill should be approached for the germ and arrangements should be made to have it 'cold rolled' (i.e. rolled at the beginning of the day before the rollers have heated up). Some health food shops carry a suitable wheat germ or can be helpful in obtaining it. Hard winter wheat is not generally available in Britain or Europe and supplies may have to be obtained from overseas laboratories. Endosperm (which is rich in ribonuclease), should not be present in the wheat germ. If present, it can be removed by flotation (Marcus *et al.*, 1974).

Extraction buffer

	HEPES*/KOH	20 mM, pH 7.6
	KCl	100 mM
	Mg acetate	1 mM
	$CaCl_2$	2 mM
	dithiothreitol	1 mM

Elution buffer

	HEPES*/KOH	20 mM, pH 7.6
	KCl	120 mM
	Mg acetate	5 mM
	dithiothreitol	1 mM

* HEPES is 4-(2-hydroxyethyl)-l-piperazineethanesulphonic acid.

Method

- Vigorously grind together 2 g of wheat germ and 2 g of powdered glass in a chilled mortar and pestle, for about 60 s.
- Add 4 ml of extraction buffer at 4°C and mix into a paste with a spatula.
- Scrape the paste into a chilled centrifuge tube and centrifuge at 30 000 × g, 4°C, for 12 min.
- Carefully remove the supernatant with a Pasteur pipette, avoiding the fatty skin. Dilute an aliquot 1:1000 with H_2O and measure A_{260} and A_{280}. A_{260} should be c. 400 (undiluted), while A_{260}/A_{280} should be c. 1.33.
- Prepare a 16 cm × 1 cm column of Sephadex G25 coarse, equilibrated with elution buffer, 4°C.
- Apply 1 ml of wheat germ extract to the column and elute at 4°C with a flow rate of about 3 ml min^{-1}.
- Collect 0.5 ml fractions over the first peak of turbidity that elutes.
- Pool fractions with an A_{260} >90 OD ml^{-1}. The A_{260}/A_{280} ratio of this fraction should be c. 1.5.
- Centrifuge the pooled fractions at 30 000 × g, 4°C, for 20 min.
- Decant the supernatant (S-30 fraction). For this fraction, the A_{260} should be c. 75-90 ODs ml^{-1} with an A_{260}/A_{280} ratio of c. 1.6-1.65.
- Slowly drip the wheat germ S-30 fraction from a Pasteur pipette into liquid N_2 so that it freezes into small balls (approximate volume 40 µl).
- The frozen aliquots can be stored for long periods of time over liquid N_2, or for shorter periods at −80°C. Balls are gently thawed individually when required.

2. *Use of the wheat germ S-30 fraction*

- Prepare the following solutions, which can be kept frozen as stocks:

20 × energy:	ATP	20 mM
	GTP	2 mM
	Creatine phosphate	160 mM
	dithiothreitol	40 mM
	(Adjust to c. pH 7.0 with KOH)	
spermine (or spermidine)		5 mM
creatine kinase		5 mg ml^{-1} (in 50% glycerol)
HEPES/KOH		1 M, pH 7.8
50 × amino acids (depleted)		1 mM of each
Mg acetate		100 mM
KCl		1 M

- Make up a standard 50 µl reaction mix, in ice, as follows:

20 × energy[a]	2.5 µl
spermine[b]	2.5 µl
creatine kinase	0.5 µl
HEPES/KOH	0.6 µl
50 × depleted amino acids	1.0 µl

Mg acetate (100 mM)^c	0.5 μl
K acetate (1 M)	3.2 μl
labelled amino acid^d	0.75 μl (100–500 kBq)
wheat germ S-30	15.0 μl
RNA^e	
H$_2$O	
	50.0 μl

- Incubate reaction mix at 30°C for 60 min.
- Analyse translation products (see Section IV).

Notes

(a) Energy. As with the *E. coli* translation system, inhibitory concentrations of ATP are avoided. A low concentration of ATP is therefore provided and recycled, in this case via creatine phosphate and creatine kinase.

(b) Spermine. Addition of the polyamine spermine, or its related compound spermidine, tends to enhance wheat germ *in vitro* translations, in our hands. Marcu and Dudock (1974) found spermine tetrahydrochloride to stimulate translation of tobacco mosaic virus RNA in wheat germ S-30 by some 60%, but found it had no effect on the translation of oviduct or moth folicular cell mRNAs. Spermine is known to be involved in stabilising the secondary structure of transfer RNAs (Hyde and Reid, 1985) and ribosomes (Lehninger, 1975), and is an inhibitor of RNase (Payne and Loening, 1970). Its precise role in translation systems is not known.

(c) Salt optima. Mg^{2+} and K^+ optima are slightly more critical for wheat germ systems than for *E. coli* systems, but are generally around 2.0–2.5 mM for Mg^{2+} and 100–120 mM for K^+. The counter-ion is also important, acetate being preferable to chloride. The effects of varying concentrations of different ions on *in vitro* translation have been published for the wheat germ system (Wyn Jones *et al.*, 1979; Gibson *et al.*, 1984) and for a number of other eukaryotic systems (Weber *et al.*, 1977).

(d) Labelled amino acid. See notes for *E. coli* system.

(e) RNA concentration. In our hands, c. 10 μg of total RNA or 600–1000 ng of poly(A)$^+$ RNA are optimal for 25 μl assays (20 μg and 1.2–2.0 μg, respectively, per 50 μl assays). As with salts, optima should be determined for different RNA preps and different S-30 extracts.

C. The Rabbit Reticulocyte Lysate System

The rabbit reticulocyte system is probably the most efficient, and most widely used, eukaryotic *in vitro* translation system. It translates eukaryotic mRNAs, and in particular the larger mRNAs, more efficiently than does the wheat germ system. Detailed descriptions of its preparation and use are to be found in Schimke *et al.* (1974), Hunt and Jackson (1974) and Pelham and Jackson (1976). For laboratories undertaking large numbers of rabbit reticulocyte translation, there are merits in preparing one's own lysates. However, preparation requires access to live rabbits and the lysate produced contains significant amounts of endogenous mRNA, which must be removed by an additional purification step, using Ca^{2+}-dependent micrococcal nuclease (Pelham and Jackson, 1976). It is generally preferable to use one of the many commercial preparations, which are of high quality and efficiency, and are available pretreated with nuclease and with correspondingly low endogenous mRNA activity.

Commercial reticulocyte lysates are supplied frozen in 200 μl or 500 μl aliquots. With the larger volume, we find it convenient to gently thaw the lysate on receipt, and to re-aliquot it into more convenient volumes by dropping into liquid N_2, as for the *E. coli* and wheat germ preparations above, before storing at −80°C or, preferably, over liquid N_2.

D. The *Xenopus* Oocyte System

Micro-injection into *Xenopus* oocytes or eggs can be used to achieve translation of heterologous mRNAs including plant mRNAs. Not strictly an *in vitro* system, the method is mentioned here because, in addition to translating the mRNA, it has some capacity to undertake post-translational processing, packaging and, in some cases, excretion of appropriate peptide products (cf. Lane *et al.*, 1970; Gurdon *et al.*, 1971; Larkins *et al.*, 1979; Colman, 1982).

The *Xenopus* system has limited applicability. It requires the establishment and maintenance of a colony of *Xenopus* African clawed toads for a regular supply of oocytes or eggs, and micro-injection requires specialised apparatus and some skill. Oocytes will accept only small amounts of RNA and will produce correspondingly small amounts of product, the detection of which is complicated by the coincident synthesis of labelled oocyte peptides. Additionally, as with other systems capable of processing newly synthesised peptides (see Section IV.E — *In vitro* Processing), signals directing transport of preproteins across specific membranes, and subsequent removal of signal or transit sequences, will differ between animals and plants, as will the various steps in addition of sugars to glycoproteins. Hence, plant proteins synthesised, processed and packaged in (or with the use of) animal systems will differ, albeit subtly, from those synthesised and processed *in vivo*.

Despite the difficulties inherent in using the *Xenopus* oocyte system, it remains an important method for the analysis of precursor processing, glycosylation and packaging. Detailed methods for the use of *Xenopus* oocytes and eggs are to be found in Lane *et al.* (1971) and Stephens *et al.* (1981).

E. Translation of Polysomes and the S-100 System

Polysomes, like purified mRNAs, can be used to prime protein synthesis in cell-free translation systems. Generally they will synthesise a similar spectrum of peptides to those primed by mRNA from the same tissues, with a tendency to synthesise the larger peptides more efficiently (see, for example, Fig. 2.3). Translation of polysomes is sometimes the *in vitro* method of choice if difficulties in purifying the mRNA are experienced, or because of a specific interest in high molecular weight products.

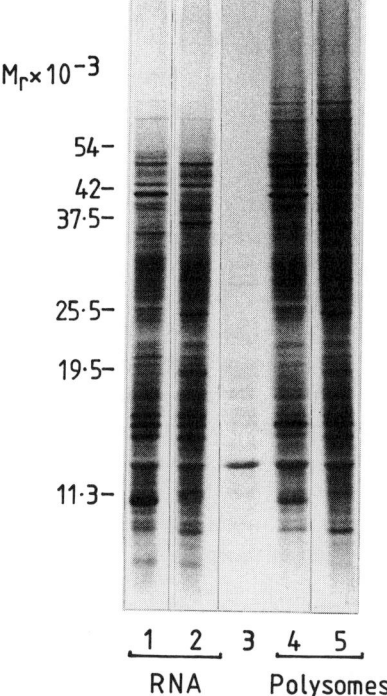

FIG. 2.3. Comparison of products from wheat germ *in vitro* translations primed with RNA or with polysomes. RNA and polysomes were obtained from pericarp tissue of tomato plants (*Lycopersicon esculentum* Mill, cv. de Ruiter 83G38), harvested green at about 80% maturity (27 days post anthesis). Wounding was carried out by cutting tissue into 2 mm slices and incubating for 3 h at 20°C in high humidity. Translations were in a standard wheat germ cell-free system. Products were labelled during synthesis with ^{35}S-methionine and were analysed by electrophoresis on an SDS-polyacrylamide gel and fluorography at -70°C. Full experimental details can be found in Lee *et al.* (1987).

The tracks contain products of translations primed with: track 1, total cell RNA; track 2, total cell RNA from wounded tissue; track 3, no added RNA or polysomes; track 4, polysomes; track 5, polysomes from wounded tissue.

In vitro translation of polysomes can also be used to learn more about mRNA utilisation *in vivo*. For this purpose, polysomes are introduced into translation systems either reduced or totally inhibited in their capacity to initiate translation. The object is to allow 'run off' of ribosomes already attached to mRNAs, but no re-initiation. The small but detectable production of peptides synthesised in this way is representative, in identity and relative quantity, of those actually being synthesised in the tissue at the time of polysome isolation.

There are two ways of reducing or inhibiting re-initiation in cell-free translation systems. The first is to use a standard S-30 translation system to which is added an antibiotic such as aurintricarboxylic acid (ATA), which, at concentrations below 100 μM (Nierhaus and Wittmann, 1980) specifically blocks initiation (Leblau *et al.*, 1970; Pestka, 1977; Beachy *et al.*, 1978). ATA concentrations between 20 and 50 μM have been shown to be most effective in *in vitro* systems primed with exogenous mRNAs (Carlier and Peumans, 1976; Higgins and Spencer, 1977; Peumans *et al.*, 1980b). Also effective inhibitors of initiation are m^7G at 200 μM (Darzynkiewicz et al. 1987) and edeine at 10 μM (Kozak and Shatkin, 1978).

An alternative method for initiation-inhibited translation of polysomes is to introduce them into a ribosome depleted cell-free system or S-100 system (also known as S-137, S-150, etc. systems). These are prokaryotic or eukaryotic cell lysates which have been centrifuged at 100 000 $\times$ g, or higher, and contain all the factors required for protein synthesis with the exception of ribosomes. In such systems, initiation events are almost completely inhibited due to the lack of available ribosomes, and newly synthesised (labelled) products represent elongation and termination of nascent peptides associated with the added polysomes.

S-100 extracts are prepared by centrifuging S-30 extracts at 100 000 $\times$ g, 4°C, for 60 min, carefully decanting the supernatant, and storing the ribosome-depleted extract as frozen beads in liquid N_2 as for the S-30 extracts.

Polysomes are added to S-30 or S-100 systems at concentrations of c. 10 μg per 25 μl assay (for polysomes, an $OD_{260} = 1$ is a concentration of c. 91 $\mu g\ ml^{-1}$; Hsiao, 1964). Translation is exactly as for mRNA translation in S-30 systems, but care must be taken to adjust the Mg^{2+} and K^+ concentrations in the assay mixes to compensate for Mg^{2+} and K^+ added with the polysomes. Examples of the use of S-100 translation systems are to be found in Fikuski et al. (1977), Rhyanen et al. (1978) and Musk et al. (1979).

F. RNase and Proteinase Inhibitors

With some translation systems, and some applications of translation systems, it may be necessary to add inhibitors of RNases or proteinases to optimise the performance of the system.

Scheele and Blackburn (1979) found evidence of the presence of RNase in cell-free translation systems derived from wheat germ, rabbit reticulocyte and Krebs ascites. In the reticulocyte and ascites systems the RNase activity was latent, due to the presence of an excess of endogenous RNase inhibitor complexed with, and inhibiting the enzyme(s). In their wheat germ system however, RNase activity was evident and was significantly reduced by the addition of 10 $\mu g\ ml^{-1}$ human placental RNase inhibitor prior to the addition of RNA. The presence of RNase in wheat germ extracts has been noted by other authors (Pelham and Jackson, 1976; Hunter et al., 1977; Hickey et al., 1978; Carlier et al., 1978b). The effect it has on the efficiency of translation will be a function of its activity in the cell-free extract, which in turn may vary from preparation to preparation, possibly due to varying amounts of RNase-rich endosperm contaminating the wheat germ. RNase inhibitor is not detrimental to translation in cell-free systems (cf. Scheele and Blackburn, 1979) and can be added as a standard component of such systems.

In vitro translation of mRNAs from tissues synthesising large amounts of RNase or protein hydrolysing enzymes, has obvious inherent problems. The *in vitro* synthesis of RNase can be neutralised by the inclusion of RNase inhibitors, as above. In cases where protein hydrolysing enzymes are synthesised in significant quantities during the *in vitro* translation, steps must be taken to inhibit their activities. Palmiter et al. (1977), for example, found it necessary to include lysozyme inhibitor in cell-free translations of enriched chicken lysozyme mRNA, in order to obtain good yields of pre-lysozyme in the translation products. Similarly Paull and Chen (1990) only obtained discrete peptide products from translations of polysomes from Papaya mesocarp in wheat germ extracts, when they included a mixture of protease inhibitors (leupeptin 50 μg ml^{-1}, antipain 20 μg ml^{-1} and chymostatin 25 μg ml^{-1} — Sigma) in the translations.

IV. QUANTITATION AND ANALYSIS OF *IN VITRO* TRANSLATION PRODUCTS

A. Labelled Amino Acid Incorporation

Progress and extent of *in vitro* translation reactions are determined by analysing the incorporation of labelled amino acid into trichloroacetic acid (TCA) insoluble peptide. This can be done by precipitation of aliquots of the incorporation assays either in solution or dried on filter paper. A simple method is as follows:

- Remove 2 μl aliquots from the translation mixture at the beginning and the end of the incubation and dry onto small squares or disks of filter paper.
- Wash filters for 10 min at 0°C, in 10% trichloroacetic acid (TCA) containing 50 mM unlabelled amino acid (corresponding to the labelled amino acid in the translation).
- Transfer filters to boiling 5% TCA for 5 min followed by 5 min in 5% TCA at 0°C.
- Transfer filters to 100% ethanol at 0°C for 5 min.
- Air dry filters and count in a scintillation cocktail containing 3 g 2,5-diphenyl-oxazole (PPO) and 0.5 g 1,4-bis-[4-methyl-5-phenyl-2-oxazolyl]benzene (POPOP) per litre of toluene.

Analysis of the incorporation of ^{35}S-cysteine requires a modified washing procedure, as the amino acid reacts with free SH groups in proteins in the incubation mixture. A method which works quite well (Grierson and Speirs, 1983) is to dilute the translation reaction, after completion, with an equal volume of 8 M urea, 200 mM unlabelled cysteine, 200 mM dithiothreitol, then spot aliquots onto filter paper and wash with TCA and ethanol as above.

B. Gel Fractionation and Autoradiography

Further analysis of translation products is generally undertaken by fractionation on denaturing polyacrylamide gels, followed by autoradiography. Detailed procedures for electrophoresis on polyacrylamide gels are available (Laemmli, 1970; Maizel, 1971) and variations in acrylamide concentrations are described in Sambrook *et al.* (1989). The method outlined below is a modification of the discontinuous buffer system of Laemmli (1970) as described by Speirs and Brady (1981).

1. Gel solutions

Stock acrylamide

	acrylamide	50.0 g
	bisacrylamide	1.0 g

Make to 100 ml with deionised distilled H$_2$O and filter.

Gel buffer

Tris	1.5 M
HCl	0.5 M
SDS	0.4% (w/v)
urea (electrophoresis grade)	2.0 M

Make to 200 ml, check pH 8.6 (solids will not dissolve if pH above 9.0).

Stacking gel buffer

Tris/HCl	0.5 M, pH 6.8
SDS	0.4% (w/v)
urea	2.7 M

Make to 100 ml.

Running buffer

Tris	50 mM
glycine	380 mM
SDS	0.1% (w/v)

Make to 1 litre, check pH 8.5.

Ammonium persulphate

For the gel	0.4% (w/v) (freshly made)
For the stacking gel	0.14% (freshly made)

Sucrose. For making gradient gels, 50% (w/v) (freshly made, or autoclaved).

2. Making the gel

Good resolution of peptides can be achieved on gels of 5% to 20% acrylamide, depending on the sizes of the peptides of interest. Gels containing gradients of acrylamide have improved resolution and are more flexible in the sizes of peptides resolved. We normally run proteins on gradient gels, cast 220 mm × 180 mm × 0.6 mm, and made with the use of a simple, two chamber, open gradient maker and pump.

3. Typical 15%–20% gradient acrylamide gel

Separating gel

	15%	20%
stock acrylamide (50%)	4.2 ml	5.6 ml
gel buffer	3.4 ml	3.4 ml
sucrose	1.4 ml	3.5 ml
H$_2$O	4.0 ml	0.5 ml
TEMED	8 µl	8 µl
ammonium persulphate (0.4%)	1.0 ml	1.0 ml
	14 ml	14 ml

Gel pumped under an overlay of 7 ml H_2O containing 0.01% SDS, and allowed to set for 45 min.

Overlay then removed, surface of gel washed thoroughly with H_2O containing no SDS, and stacking gel poured on top.

Stacking gel

stock acrylamide (50%)	1.2 ml
stacking gel buffer	2.5 ml
H_2O	1.3 ml
TEMED	5 μl
ammonium persulphate (0.14%)	5.0 ml
	10 ml

The stacking gel is set for 30 min, either flat, with a water overlay, or with a comb inserted.

As it is a discontinuous buffer system, the gel is not prerun prior to loading.

Sample loading

Wheat germ and E. coli. Per 20 μl reaction add 30 μl stop solution A. Boil for 5 min. Load up to 30 μl per gel track.

Stop solution A:	SDS	3% (w/v)
	dithiothreitol	100 mM
	sucrose	10% (w/v)

Reticulocyte lysate. Instructions for analysing reticulocyte translations are supplied with commercial kits and should be followed. We find that the abundant globin peptide in reticulocyte extracts distorts fractionation on gels when tracks are loaded with more than 3 μl reaction mix. We therefore stop reactions by adding stop solution B to the mix in a ratio of 2:1, boil for 5 min, and load a maximum of 9 μl per gel track.

Stop solution B:	SDS	10% (w/v)
	β-mercaptoethanol	5% (v/v)
	sucrose	15% (w/v)

The gel is run at 20 mA constant current, for 6–8 h, at 22°C.

4. Staining and autoradiography

Gels are stained in 30% acetic acid, 40% methanol, 0.25% Coomassie brilliant blue R and are destained in 7% acetic acid, 30% methanol (all in a fume hood).

Peptides labelled with ^{75}Se-selenomethionine, which is a γ-emitter, can be detected by autoradiography of the stained gel, either directly or after drying. The β-radiation of ^{3}H-, ^{14}C- or ^{35}S-labelled peptides is less easily detected by autoradiography and is normally detected by impregnating the gels with a fluor as described by Bonner and Laskey (1974) and Laskey and Mills (1975), or with a commercial preparation such as Enhance (New England Nuclear), and exposing the gels to X-ray film at −80°C, after drying.

C. Antibody Precipitation

With few exceptions, translation of mRNA or polysomes in an *in vitro* system results in the synthesis of a spectrum of peptides in insufficient quantity to be identified in terms of enzyme activity, and not necessarily related by size with their counterparts synthesised *in vivo*. In some cases, for example the subunits of ribulosebisphosphate carboxylase or the chlorophyll *a/b* binding proteins, tentative identification can be made on the basis of relative sizes and abundances. However, for identification of less abundant translation products, and for firm identification of even the abundant products, more stringent criteria are required.

Immunoprecipitation of *in vitro* synthesised peptides is a simple and usually stringent method for identifying specific products. Direct immunoprecipitation of peptides from *in vitro* translation mixes is inefficient and, in most cases, ineffectual. However, an effective and sensitive method is produced when the specificity of the antibody is combined with the use of a second antibody, or the immunoabsorbant protein A (Goding, 1978), to co-precipitate antigen/antibody complexes from the translation mixtures (see review by Anderson and Blobel, 1983). The method outlined

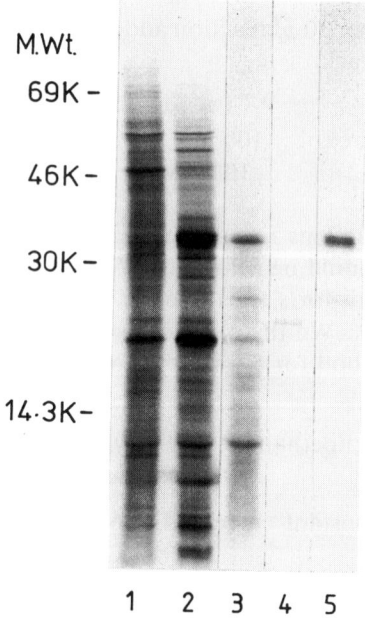

FIG. 2.4. *In vitro* synthesis of precursors to the chlorophyll *a/b* binding proteins of wheat. Antibody precipitation and hybrid select analysis. Tracks, from left to right: track 1, translation of total RNA from 5-day-old wheat leaf, grown and harvested in darkness; track 2, translation of total RNA from 5-day-old wheat leaf grown in a 14 h light/10 h dark cycle; track 3, translation of poly(A)$^+$ RNA selected by hybridisation to a cDNA encoding wheat chlorophyll *a/b* binding protein (unpublished data, Olive and Speirs; selection method as described in Section IV.D); track 4, translation system primed with 240 µg *E. coli* tRNA; track 5, labelled protein precipitated from translation as shown in track 2, by treatment with antibody to maize chlorophyll *a/b* binding protein followed by protein A–Sepharose (Pharmacia) according to the method outlined in Section IV.C.

Details of materials and methods can be found in Speirs and Brady (1981), Smith *et al.* (1983).

below has been used with good success in our laboratory for the identification, from wheat germ translation systems, of several products including the precursor small subunit of wheat Rubisco (Speirs and Brady, 1981) and wheat chlorophyll *a/b* preproteins (Fig. 2.4, track 5).

Method

- Translate RNA in 200 µl wheat germ translation system.
- Centrifuge at 160 000 × g for 30 min, 4°C.
- Carefully remove the supernatant and add the following:

saturated Tris/HCl	0.1 vol, pH 8.6
Triton X-100	to 1.0%
SDS	to 0.2%
dithiothreitol	to 10 mM
phenylmethylsulphonyl fluoride	to 1 mM

- To 50 µl aliquots add: (a) nothing.
  (b) 6 µl control IgG serum.
  (c) 6 µl specific IgG serum.
- Incubate samples for 12 h, 22°C.
- To each sample add 6 mg Protein A–Sepharose 4B (Pharmacia) in 0.1 M phosphate buffer, pH 7.5.
- Incubate for 1 h, 22°C, with gentle inversion.
- Centrifuge at 700 × g for 5 min.
- Recover the supernatant and make to 2% SDS, 100 mM dithiothreitol, 5% sucrose.
- Wash pellet 3× by suspension in 10 mM Tris/HCl, pH 8.6, 0.3 M NaCl, 0.1% SDS.
- Resuspend pellet with 30 µl of 3% SDS, 10% sucrose, 100 mM dithiothreitol.
- Boil supernatant and pellet fractions for 2 min, prior to loading on polyacrylamide gel.

D. Hybrid Arrest and Hybrid Release

Hybrid arrest and hybrid release are methods of relating cDNAs to specific *in vitro* translation products which, in turn, can be identified by antibody precipitation from the translation reaction. Hence a cDNA can be identified in terms of the antigenicity of the peptide it encodes. The methods have been mostly superseded by the use of expression vectors and colony immunoscreening (cf. Sambrook *et al.*, 1989), but are still worthy of mention as they can be used to relate *in vitro* products to cDNAs characterised, for example, on the basis of their nucleotide sequence.

In hybrid arrest, translation of a specific mRNA in an *in vitro* translation system is inhibited by annealing the mRNA with its complementary cDNA. Analysis of the translation products, compared with products from a non-arrested translation, will reveal a missing peptide (or group of missing peptides), whose mRNA is closely related to the cDNA being examined. Details of the hybrid arrest method are to be found in Paterson *et al.* (1977).

In hybrid release, or hybrid selection, a cDNA is denatured and bound irreversibly to a matrix (nitrocellulose, DBM paper, etc.). A preparation of mRNA is then hybridised with the bound cDNA, the filter washed thoroughly, and the annealed mRNA eluted and translated in an *in vitro* system. The mRNA selected in this way, will be closely related to the cDNA and will translate *in vitro* to give a single peptide (or small group of related peptides). Hybrid selection has been used to identify a number of plant and animal cDNAs, usually those encoding abundant proteins and complementary to abundant mRNAs (cf. Smith *et al.*, 1979; Goldberg *et al.*, 1979; Bedbrook *et al.*, 1980; Smith *et al.*, 1983). However, it has the capacity to identify cDNAs derived from low abundance mRNAs, as demonstrated by Parnes *et al.* (1981) who isolated and identified cDNAs complementary to a message constituting some 0.03% of the total mRNA in mouse liver cells.

Hybrid release (selection) is labour intensive, but relatively straightforward. The method outlined below has been used in our laboratory to identify wheat cDNAs encoding the small subunit of Rubisco (Smith *et al.*, 1983), and one of the chlorophyll *a/b* binding proteins (Fig. 2.4, track 3; unpublished data of Olive and Speirs).

Materials

DBM paper. Diazobenzyloxymethyl paper was used for covalently binding the cDNA prior to hybridisation. DBM paper is available as Transa-Bind from Schleicher and Schuell Inc., or can be prepared as described by Alwine *et al.* (1979). Nitrocellulose filters (Millipore, 2.5 cm diameter, 0.45 μm pore) can also be used for binding the DNA, as described by Parnes *et al.* (1981).

cDNA. Plasmids containing cDNA should be linearised by digestion with a restriction enzyme, and cleaned by extraction with phenol/chloroform (1:1; v/v), and precipitation from ethanol. 40 μg of cDNA are required.

RNA. 40 μg of poly(A)$^+$ RNA are required.

Hybridisation buffer

PIPES*/NaOH	20 mM, pH 6.4
EDTA	1 mM
SDS	0.2% (w/v)
NaCl	0.9 M
formamide (deionised)	50% (v/v)

* 1,4-Piperazinediethanesulphonic acid.

Washing buffer

Na citrate	16 mM
NaCl	40 mM
EDTA	2 mM
SDS	0.4 % (w/v)
formamide (deionised)	50% (v/v)

Elution buffer

PIPES/NaOH	10 mM
EDTA	1 mM
SDS	0.5% (w/v)
formamide (deionised)	90% (v/v)

Method

(a) Binding cDNA to the filter

- Resuspend 40 µg of linearised cDNA in 8 µl of 25 mM NaPO$_4$, pH 6.3.
- Add 32 µl of dimethylsulphoxide and denature the DNA by heating at 80°C for 10 min.
- Cool rapidly in ice and immediately add to a 2.5 cm disk of DBM paper (freshly made) in the bottom of a sealable glass tube, with a diameter just larger than the DBM paper disk.
- Seal the tube and incubate at room temperature overnight.
- After incubation, remove the disk and wash 3× with H$_2$O.
- Incubate in 0.4 M NaOH for 30 min at 37°C.
- Wash a further 3× with H$_2$O.
- Store in 50% formamide, 20 mM Tris/HCl, pH 6.3, at 4°C.

(b) Hybridisation

- Pre-equilibrate the filter in 100 µl of hybridisation buffer for 60 min.
- Precipitate 40 µg of poly(A)$^+$ RNA under ethanol.
- Wash the RNA pellet with 80% ethanol and dry.
- Dissolve the pellet in:

		final concentrations
20 µl	100 mM PIPES/NaOH, pH 6.4	(20 mM)
10 µl	2% SDS	(0.2%)
1 µl	100 mM EDTA	(1 mM)

- Add:

50 µl	formamide (deionised)	(50%)
20 µl	4.5 M NaCl	(0.9 M)

- Remove buffer from prehybridised DNA filter, and replace with hybridisation solution containing RNA.
- Incubate, with shaking, at 37°C, for 5 h.
- Remove filter from hybridisation mix and wash by shaking in 5 ml wash buffer, 37°C, for 15 min.
- Repeat wash 3×.
- Drain filter thoroughly and transfer to 100 µl elution buffer, in the bottom of a sealable glass tube (narrow diameter).
- Seal tube and incubate at 37°C for 30 min.

- Withdraw eluate to an Eppendorf tube and rinse filter with 250 μl H_2O.
- Pool rinse water with eluate and add:

50 μl	E. coli tRNA (120 μg ml^{-1})
20 μl	4 M NaCl
1.0 ml	ethanol

- Precipitate RNA by storage overnight at $-20°C$.
- Pellet RNA by centrifugation and resuspend in 200 μl, 200 mM HEPES/KOH, pH 7.6.
- Reprecipitate RNA by adding 20 μl 4 M NaCl and 450 μl ethanol, and storing overnight at $-20°C$.
- Pellet RNA by centrifugation and wash with 80% ethanol.
- Dry pellet, resuspend in small volume of H_2O and translate in wheat germ or reticulocyte cell-free translation system, or in *Xenopus* oocytes or eggs.
- Analyse by polyacrylamide gel electrophoresis, as described above.

E. *In vitro* Processing

Many peptides are synthesised *in vivo* as precursor forms destined for cotranslational or post-translational (cf. Verner and Schatz, 1988) transport across membranes and subsequent modification before taking on their final form and role. Modification can involve a number of processing steps including the removal of the N- and/or C-terminal regions, glycosylation, insertion into membranes, packaging, etc. (cf. Blobel and Dobberstein, 1975a,b; Verner and Schatz, 1988; Wickner, 1989; Pelham, 1990; Abeijon and Hirschberg, 1992).

In vitro synthesised precursor peptides, which would be destined *in vivo* for transport into plant cell organelles (cf. Schmidt and Mischkind, 1986), can be transported into and accurately processed by intact, isolated chloroplasts and mitochondria. In many cases, the directed processing of the protein precursors, involving cleavage of the *N*-terminal 'leader' sequence, can also be accomplished in the absence of intact organelles by the addition of a soluble enzyme extract from within the organelle (Highfield and Ellis, 1978; Smith and Ellis, 1979; Pichersky *et al.*, 1987; Chaumont *et al.*, 1990; Kohorn, 1986; Clark and Lamppa, 1992).

Peptides destined to pass through the rough endoplasmic reticulum (RER), a process generally resulting in a reduction in size and often in the addition of sugar moieties, appear to undergo obligatory cotranslational transport across the membrane followed by processing in the luminal spaces enclosed by the RER, although this is still under debate in some cases (Wickner, 1989). The coordinated synthesis, transport and processing requires membranes, specific endopeptidases and glycosylating enzymes, which are either in insufficient quantities or inactive in normal cell-free translation systems. Translation of exogenous mRNA in *Xenopus* oocytes or eggs does promote processing and, in some cases, excretion of mature protein products (cf. Colman, 1982), but the *Xenopus* system is demanding to maintain and to use (see Section III.D). An efficient and more convenient method is to supplement cell-free translation systems with microsomal membranes, which are closed vesicles derived from RER, and are most usually obtained from dog pancreas.

Methods for the preparation and use of dog pancreatic microsomal membranes have

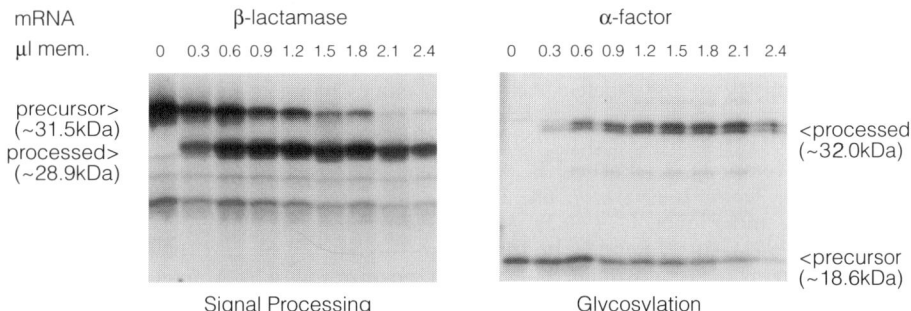

FIG. 2.5. Processing and glycosylation activity of canine pancreatic microsomal membranes. Figure kindly provided by Promega Corporation. The positive control mRNAs (0.5 μg each of *E. coli* β-lactamase and *S. cerevisiae* α-factor) were translated using Promega's rabbit reticulocyte lysate in a 25 μl reaction for 60 min in the presence of the indicated amounts of microsomal membranes. Aliquots (3 μl) were then analysed by gel electrophoresis and autoradiography of the ^{35}S-labelled proteins.

A detailed protocol is available in Promega's Protocols and Applications Guide.

been described (Blobel and Dobberstein, 1975a; Higgins and Spencer, 1981; Walter and Blobel, 1983; Scheele, 1983), as have the preparation and use of microsomal membranes from pea cotyledon (Higgins and Spencer, 1981), and from yeast (Rothblatt and Meyer, 1986). These methods are not included here, as commercial preparations of dog microsomal membranes are readily available and are the most convenient and commonly used method of promoting cotranslational transport and processing *in vitro*. Examples of *in vitro* processing of precursor peptides by dog pancreas microsomes are shown in Fig. 2.5, kindly supplied by Promega Corporation. In the figure, the presence of microsomes during *in vitro* synthesis of β-lactamase precursor is seen to result in the cleavage of its signal sequence, with the production of a smaller processed peptide. Conversely, the inclusion of microsomes during *in vitro* synthesis of *Saccharomyces cerevisiae* α-factor precursor results in an increase in the apparent size of the peptide, caused by the addition of glycosyl groups.

The use of dog pancreas microsomes *in vitro* to induce cotranslational translocation and leader sequence cleavage appears to mimic essentially the *in vivo* process, and can provide much useful information. Glycosylation by the microsomal fraction, however, appears to be core glycosylation only. That is, core oligosaccharides [composed of *N*-acetylglucosamine (GlcNac), mannose (Man) and glucose (Glc); $GlcNAc_2Man_9Glc_3$] are attached by *N*-linkage to specific asparagine residues in the target peptide. These are not further modified, as would often occur *in vivo,* nor do other possible forms of glycosylation occur. *In vitro* glycosylation must therefore be considered informative but limited.

Examples of *in vitro* processing of plant precursor peptides can be found in Higgins and Spencer (1981) and DellaPenna and Bennett (1988).

REFERENCES

Adams, R. L. P., Burdon, R. H., Campbell, A. M., Leader, D. P. and Smellie, R. S. (1981). "The Biochemistry of the Nucleic Acids", 9th edn. Chapman & Hall, London and New York.
Abeijon, C. and Hirschberg, C. B. (1992). *TIBS* **17**, 32-36.
Alwine, J. C., Kemp, D. J., Parker, B. A., Reiser, J., Renart, J., Stark, G. R. and Wahl, G. M. (1979). *Methods Enzymol.* **68**, 220-242.
Anderson, D. J. and Blobel, G. (1983). *Methods Enzymol.* **96**, 111-120.
Bayley, S. T. (1971). In "Methods in Molecular Biology", Vol. 1, Protein Biosynthesis in Bacterial Systems (J. A. Last and A. I. Laskin, eds), pp. 89-110, Marcel Dekker Inc., New York.
Beachy, R. N., Thompson, J. F. and Madison, J. T. (1978). *Plant Physiol.* **61**, 139-144.
Bedbrook, J. R., Smith, S. M. and Ellis, R. J. (1980). *Nature* **287**, 692-697.
Blobel, G. and Dobberstein, B. (1975a). *J. Cell Biol.* **67**, 835-851.
Blobel, G. and Dobberstein, B. (1975b). *J. Cell Biol.* **67**, 852-862.
Bonner, W. M. and Laskey, R. A. (1974). *Eur. J. Biochem.* **46**, 83-88.
Bottomley, W. and Whitfeld, P. R. (1979). *Eur. J. Biochem.* **93**, 31-39.
Carlier, A. R. and Peumans, W. J. (1976). *Biochim. Biophys. Acta* **447**, 436-444.
Carlier, A. R., Peumans, W. J. and Manickam, A. (1978a) *Plant Sci. Lett.* **11**, 207-216.
Carlier, A. R., Caers, L. I. and Peumans, W. J. (1978b). Société Belge de Biochemie, Reunion de Coutrai, Octobre. Abst., pp. 6-7.
Chaumont, T., O'Riordan, V. and Boutry, M. (1990). *J. Biol. Chem.* **265**, 16 856-16 862.
Clark, S. E. and Lamppa, G. K. (1992). *Plant Physiol.* **98**, 595-601.
Colman, A. (1982). *TIBS* **7**, 435-437.
Darzynkiewicz, E., Ekiel, I., Lassota, P. and Tahara, S. M. (1987). *Biochemistry* **26**, 4372-4380.
DellaPenna, D. and Bennett, A. B. (1988). *Plant Physiol.* **86**, 1057-1063.
Doi, R. H. (1971). In "Methods in Molecular Biology", Vol. 1, Protein Biosynthesis in Bacterial Systems (J. A. Last and A. I. Laskin, eds), pp. 67-88. Marcel Dekker Inc., New York.
Fikuski, S., Ishikawa, K. and Sasaki, K. (1977). *Plant Cell Physiol.* **18**, 969-977.
Fischer, I. and Moldave, K. (1981). *Anal. Biochem.* **113**, 13-26.
Gibson, T. S., Speirs, J. and Brady, C. J. (1984). *Plant, Cell Environ.* **7**, 579-587.
Goding, J. W. (1978). *J. Immunol. Methods* **20**, 241-253.
Goldberg, M. L., Lifton, R. P., Stark, G. R. and Williams, J. G. (1979). *Methods Enzymol.* **68**, 206-220.
Grierson, D. and Speirs, J. (1983). In "Techniques in Molecular Biology" (J. M. Walker and W. Gastra, eds), pp. 135-158. Croom Helm, London and Canberra.
Gurdon, J. B., Lane, C. D., Woodland, H. R. and Marbaix, G. (1971). *Nature* **233**, 177-182.
Hickey, E. D., Weber, L. A. and Baglioni, C. (1978). *Biochem. Biophys. Res. Commun.* **80**, 373-383.
Higgins, T. J. V. and Spencer, D. (1977). *Plant Physiol.* **60**, 655-661.
Higgins, T. J. V. and Spencer, D. (1981). *Plant Physiol.* **67**, 205-211.
Highfield, P. E. and Ellis, R. J. (1978). *Nature* **271**, 420-424.
Hsiao, T. C. (1964). *Biochim. Biophys. Acta* **91**, 598-605.
Hunt, T. and Jackson, R. J. (1974). In "Modern Trends in Human Leukaemia" (R. Neth, R. C. Gallo, S. Spiegelman and F. Stohlman, eds), pp. 300-307. J. F. Lehmans Verlag, Munich.
Hunter, A. R., Farrell, P. J., Jackson, R. J. and Hunt, T. (1977). *Eur. J. Biochem.* **75**, 149-157.
Hyde, E. I. and Reid, B. R. (1985). *Biochemistry* **24**, 4315-4325.
Kohorn, B. D., Harel, E., Chitnis, P. R., Thornber, J. P. and Tobin, E. M. (1986). *J. Cell Biol.* **102**, 972-981.
Kozak, M. and Shatkin, A. J. (1978). *J. Biol. Chem.* **253**, 6568-6577.
Laemmli, U. K. (1970). *Nature (London)* **227**, 680-685.
Lane, C. D., Coleman, A., Mohun, T., Morser, J., Champion, J., Koundes, I., Craig, R., Higgins, S., James, T. C., Applebaum, S. W., Ohlsson, R. I., Paucha, E., Houghton, M., Mathews, J., Leblau, B., Marbaix, G., Werenne, J., Burney, A. and Huez, G. (1970). *Biochem. Biophys. Res. Commun.* **40**, 731-739.

Lane, C. D., Marbaix, G. and Gurdon, J. B. (1971). *J. Mol. Biol.* **61**, 73-91.
Larkins, B. A., Pedersen, K., Handa, A. K., Hurkman, W. J. and Smith, L. D. (1979). *Proc. Natl. Acad. Sci. USA* **76**, 6448-6452.
Laskey, R. A. and Mills, A. D. (1975). *Eur. J. Biochem.* **56**, 335-341.
Leblau, B., Marbaix, G., Werenne, J., Burney, A. and Huez, G. (1970). *Biochem. Biophys. Res. Commun.* **40**, 731-739.
Lee, E., Speirs, J., McGlasson, W. B. and Brady, C. J. (1987). *J. Plant Physiol.* **129**, 287-299.
Lehninger, A. L. (1975). "Biochemistry". Worth Publishers Inc., New York.
Maizel, J. V. Jr (1971). *Methods Virol.* **5**, 179-246.
Marcu, K. and Dudock, B. (1974). *Nucl. Acids Res.* **1**, No. 11, 1385-1397.
Marcus, A., Effron, D. and Weeks, D. P. (1974). *Methods Enzymol.* **30**, 749-754.
Matthews, M. B. and Korner, A. (1970). *Eur. J. Biochem.* **17**, 328-338.
Modolell, J. (1971). In "Methods in Molecular Biology", Vol. 1, Protein Biosynthesis in Bacterial Systems (J. A. Last and A. I. Laskin, eds), pp. 1-65. Marcel Dekker Inc., New York.
Musk, P., Chakravorty, A. K. and Scott, K. J. (1979). *Plant Cell Physiol.* **20**, 1359-1369.
Nierhaus, K. H. and Whittman, H. G. (1980). *Naturwissenschaften* **67**, 234-250.
Nirenberg, M. W. and Matthaei, J. H. (1961). *Proc. Natl. Acad. Sci. USA* **47**, 1588-1602.
Palmiter, R. D., Gagnon, J., Ericsson, L. H. and Walsh, K. A. (1977). *J. Biol. Chem.* **252**, 6386-6393.
Parnes, J. R., Velan, B., Felsenfeld, A., Ramanathan, L., Ferrini, U., Appella, E. and Seidman, J. G. (1981). *Proc. Natl. Acad. Sci. USA* **78**, 2253-2257.
Paterson, B. M., Roberts, B. E. and Kuff, E. L. (1977). *Proc. Natl. Acad. Sci. USA* **74**, 4370-4374.
Paull, R. E. and Chen, N. J. (1990). *J. Am. Soc. Hort. Sci.* **115**, 623-631.
Payne, P. I. and Loening, U. E. (1970). *Biochem. Biophys. Acta* **224**, 128-135.
Pelham, H. R. B. (1990). *TIBS* **15**, 483-486.
Pelham, H. R. B. and Jackson, R. J. (1976). *Eur. J. Biochem.* **67**, 247-256.
Pestka, S. (1977). In "Molecular Mechanisms of Protein Biosynthesis" (H. Weissbach and S. Pestka, eds), pp. 467-553. Academic Press, New York.
Peumans, W. J., Carlier, A. R. and Delaey, B. M. (1980a). *Plant Physiol.* **66**, 584-587.
Peumans, W. J., Carlier, A. R. and Schreurs, J. (1980b). *Planta* **147**, 302-306.
Pichersky, E., Hoffmann, N. E., Malik, V. S., Bernatzky, R., Tanksley, S. D., Szabo, L. and Cashmore, A. R. (1987). *Plant Molec. Biol.* **9**, 109-120.
Rhyanen, L., Graves, P. N., Bressan, G. M. and Prockop, D. J. (1978). *Arch. Biochem. Biophys.* **185**, 344-351.
Roberts, B. E. and Paterson, B. M. (1973). *Proc. Natl. Acad. Sci. USA* **70**, 2330-2334.
Rothblatt, J. A. and Meyer, D. I. (1986). *Cell* **44**, 619-628.
Sambrook, J., Fritsch, E. F. and Maniatis, T. (1989). "Molecular Cloning. A Laboratory Manual", 2nd edn. Cold Spring Harbor Laboratories, New York.
Scheele, G. (1983). *Methods Enzymol.* **96**, 94-111.
Scheele, G. and Blackburn, P. (1979) *Proc. Natl. Acad. Sci. USA* **76**, 4898-4902.
Schimke, R. T., Rhoads, R. E. and McKnight, S. (1974). *Methods Enzymol.* **30**, 694-701.
Schmidt, G. W. and Mishkind, M. L. (1986). *Ann. Rev. Biochem.* **55**, 879-912.
Smith, D. F., Searle, P. F. and Williams, J. G. (1979). *Nucl. Acids Res.* **6**, 487-506.
Smith, S. M. and Ellis, R. J. (1979). *Nature* **278**, 662-664.
Smith, S. M., Bedbrook, J. and Speirs, J. (1983). *Nucl. Acids Res.* **11**, 8719-8734.
Speirs, J. and Brady, C. J. (1981). *Aust. J. Plant Physiol.* **8**, 603-618.
Stephens, D. L., Miller, T. J., Silver, L., Zipser, D. and Mertz, J. E. (1981). *Anal. Biochem.* **114**, 299-309.
Stewart, P. R. and Letham, D. S. (eds) (1977). "The Ribonucleic Acids", 2nd edn. Springer-Verlag, New York.
Szczesna, E. and Filipowicz, W. (1980). *Biochem. Biophys. Res. Commun.* **92**, 563-569.
Tuite, M. H., Plesset, J., Moldave, K. and McLaughlin, C. S. (1980). *J. Biol. Chem.* **255**, 8761-8766.
Verner, K. and Schatz, G. (1988). *Science* **241**, 1307-1313.

Walter, P. and Blobel, G. (1983). *Methods Enzymol.* **96**, 84–93.
Weber, L. A., Hickey, E. D., Moroney, P. A. and Baglioni, C. (1977). *J. Biol. Chem.* **252**, 4007–4010.
Wickner, W. (1989). *TIBS* **14**, 280–283.
Wyn Jones, R. G., Brady, C. J. and Speirs, J. (1979). *In* "Recent Advances in the Biochemistry of Crops" (D. L. Laidman and R. G. Wyn Jones, eds), pp. 63–103. Academic Press, London, New York and San Francisco.
Zagorski, W. (1978). *Anal. Biochem.,* **87**, 313–333.
Zubay, G., Chambers, D. A. and Cheong, L. C. (1970). *In* "The Lactose Operon" (J. Beckwith and D. Zipser, eds), pp. 375–391. Cold Spring Harbor Laboratories, New York.

3 cDNA Cloning and Screening

RACHEL P. HODGE and ROD J. SCOTT

Department of Botany, University of Leicester, Leicester LEI 7RH, UK

I.	Introduction..	57
II.	cDNA synthesis..	62
	A. RNA quality...	62
	B. Primers for first strand synthesis...................................	62
	C. First strand synthesis..	63
	D Second strand synthesis...	63
III.	Cloning of cDNA..	65
	A. Preparation of cDNA for cloning..................................	65
	B. Choice of vector...	66
IV.	Principles of using lambda vectors...................................	68
	A. Preparation and infection of bacterial host cells............	68
	B. Making the master plate...	69
	C. Transfer to supporting membranes..............................	70
	D. Preparing replicate lifts...	70
	E. Orientation marks..	71
	F. Identifying and purifying selected plaques...................	71
V.	Screening techniques...	72
	A. Differential screening..	72
	B. Cold-plaque screening..	74
	C. Screening with heterologous DNA probes and oligos...	76
	D. Screening for cDNAs encoding specific peptides..........	76
	References..	78

I. INTRODUCTION

The techniques of cDNA cloning and library screening represent invaluable tools for the molecular biologist. This is equally true for all those engaged in research with eukaryotes. In this laboratory, we decided to use these approaches to further our understanding of anther development and function. The production of male gametophytes is a complex developmental process involving a tightly controlled series of cytological and biochemical processes, and the coordinated expression of a large number of both sporophytic and gametophytic genes (Scott *et al.*, 1991a). For example, Kamalay and Goldberg (1980, 1984) estimated that about 11 000 different anther-specific transcripts were present in *Nicotiana tabacum* anthers immediately after microspore mitosis. Similarly, mature pollen grains were found to contain about 20 000 different transcripts (Willing and Mascarenhas, 1984) of which, according to one estimate 36%, or 7200, are pollen-specific (Willing *et al.*, 1988). Although it is not known to what extent these two mRNA pools overlap, and there are no estimates for the period prior to microspore mitosis, these studies suggest that a very large number of genes are expressed during andro- and gameto-genesis.

Clearly, therefore, the construction and differential screening of cDNA libraries derived from anther mRNA offered a straightforward means of cloning anther-specific genes. Numerous laboratories initiated such programmes employing a range of species and anthers or pollen grains at various stages of development. In this laboratory, we exploited the strong relationship that exists between anther length and the stage of gametogenesis in *Brassica napus*, to construct defined cDNA libraries, one enriched for sporogenesis (S) and the other for microspore development (MD) associated transcripts (Scott *et al.*, 1991b). The S library was constructed from poly(A)$^+$ RNA extracted from anthers dissected from 1.2–1.8 mm long buds and consisted of over 100 000 recombinants. The 1.2 mm bud length boundary of the S library was chosen for practical reasons since these are the smallest buds from which it is possible to dissect anthers in the necessary quantities.

Differential hybridisation was used to identify cDNA clones for genes with elevated levels of expression in developing anthers. One filter from each replicate plaque lift was hybridised with ^{32}P-labelled single-stranded cDNA probe synthesised from the mRNA used in library construction. The second, control, filter was hybridised with ^{32}P labelled probe synthesised from mRNA isolated from 3-week-old seedlings of *B. napus*. Seedlings at this stage have two partially expanded leaves, a shoot apex and a well-developed root system, and therefore probably contain a large proportion of all mRNAs expressed from so-called 'house-keeping' genes.

A total of 8000 recombinant plaques were plated, at a density of 1200–1500 plaques per 14 cm plate. The initial screen yielded 36 plaques that showed hybridisation to the anther probe stock alone. Figure 3.1 shows part of a typical pair of autoradiographs of the same area of replicate filters hybridised to seedling (A) and anther (B) probe stocks. The boxed spots shown in Fig. 3.1(B) do not appear on the autoradiograph depicted in Fig. 3.1(A), and therefore represent potential anther-specific cDNAs. Each of these initial isolates was cored from the master plate and the phage replated on 9 cm dishes and rescreened against both seedling (C) and anther (D) probes. Figure 3.1(D) shows the autoradiograph of the anther-probed filter for the boxed spot at the bottom of Fig. 3.1(B), which was labelled A9. The boxed spots in Fig. 3.1(D) are not detectable on the replicate seedling-probed filter, Fig. 3.1(C), and therefore represent A9 recombinant phage. One of these plaques, which was well isolated from all the others, was chosen to establish a plaque pure phage stock. Of the original 36 initial isolates, 12 survived this second round of differential screening. Further evaluation of

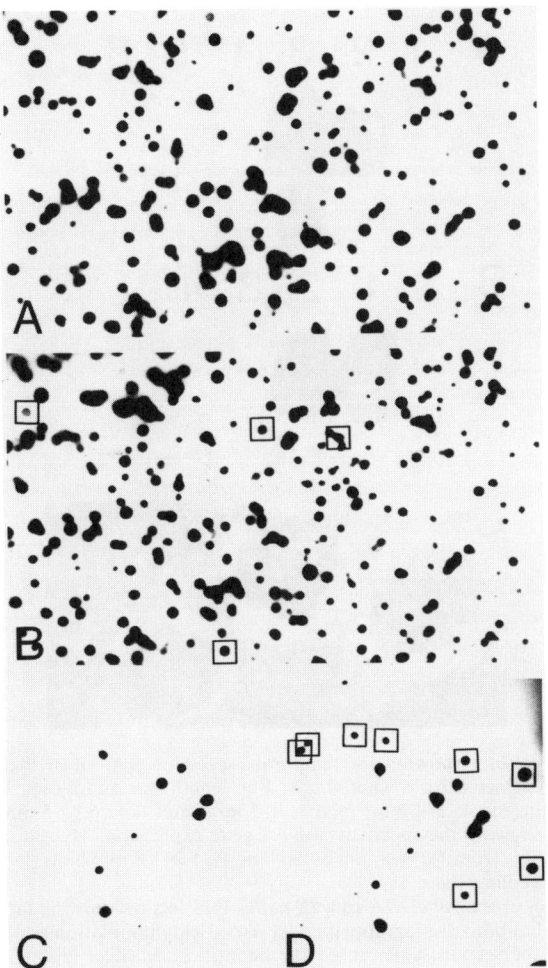

FIG. 3.1. (A) and (B) Autoradiographs of the same area of replicate filters hybridised to seedling (A) and anther (B) probe stocks. The boxed spots shown in (B) represent potential anther-specific cDNAs.
(C) and (D) Autoradiographs of replicate filters from a secondary screen of the plaque represented by the boxed spot at the bottom of (B) which was labelled A9. Filters were hybridised to seedling (C) and anther (D) probe stocks. The boxed spots on (D) are not detectable on the replicate seedling-probed filter (C), and therefore represent A9 recombinant phage.

the putative anther-specific cDNAs, including A9, was carried out with recombinant plasmids derived by *in vivo* excision of pBluescript from the lambda ZAP-II phage stock.

The temporal expression pattern of the A9 gene was assessed on an RNA gel blot with RNA isolated from buds sorted arbitrarily into six size groups: <1 mm, 1-2 mm, 2-3 mm, 3-4 mm, 4-5 mm and >5 mm in length (Fig. 3.2(A)). The organ specificity of the A9 cDNA clones was confirmed by performing a gel blot with RNA isolated from anthers, buds and pistils, each of the same developmental stage as the anthers used in library construction, and also from seedlings (Fig. 3.2(B)).

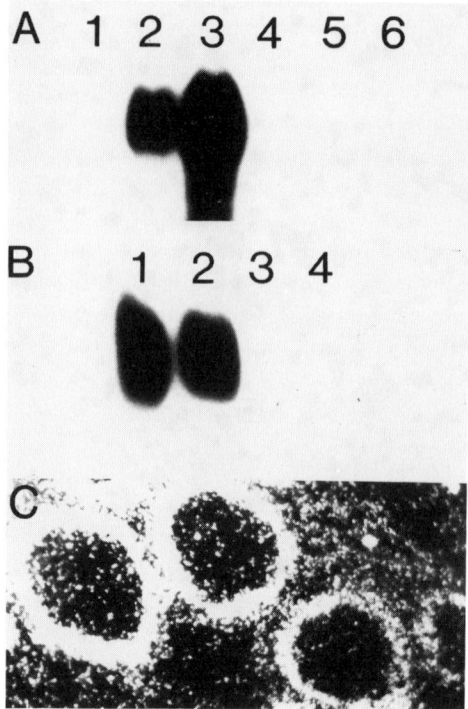

FIG. 3.2. (A) Northern gel blot showing the temporal expression pattern of the A9 gene. Hybridisation was to 10 μg total RNA extracted from whole buds. Bud length was as follows: lane 1, < 1 mm; lane 2, 1-2 mm; lane 3, 2-3 mm; lane 4, 3-4 mm; lane 5, 4-5 mm; and lane 6, >5 mm.
(B) Northern gel blot showing the specificity of A9 gene expression. Hybridisation was to 10 μg total RNA extracted from: anthers from 1.2-1.8 mm buds (lane 1); 1.2-1.8 mm buds (lane 2); pistils from 1.2-1.8 mm buds (lane 3); and seedling (lane 4).
(C) *In situ* hybridisation of the A9 cDNA to a *B. napus* bud section showing tapetum-specific expression of the A9 gene. The slide has been photographed under dark-field illumination and therefore the areas where the probe has bound to the section, and caused the deposition of silver grains, appear white.

Although the developmental programme executed within the anther is very complex, the organ consists of relatively few specialised cell types. However, a striking feature of anther development is that the number of different cell types, and in some cases the appearance of these cells, change throughout the lifetime of the organ. For example, the sporangia of very immature anthers contain, in addition to the vascular bundle, only three cell types: archesporial cells and those of the parietal and epidermal layers. Later, the archesporial cells differentiate to produce two further types, tapetal and microsporogenous cells. The microsporogenous cells differentiate further to form the tricellular gametophyte, during which time the tapetum degenerates. In order to determine in which of these cell types the transcripts corresponding to the anther-specific cDNA are expressed, selected clones, including A9, were used for *in situ* hybridisation to thin sections of *B. napus* buds. The A9 probe clearly hybridises to transcripts located within the tapetal cells surrounding the microsporangia (Fig. 3.2(C)). Since no hybridisation to the microsporangenous cell mass was detected, the *in situ* hybridisation results strongly suggest that the expression of this gene is tapetum-specific. The process described above illustrates how differential screening coupled with various other hybridisation techniques can be used to identify cDNAs derived from transcripts expressed in very specific cell types. By isolating the promoter of the *Arabidopsis thaliana* homologue of the *B. napus* A9 gene, and linking this to the potent cytotoxin Barnase, we have subsequently confirmed the extreme temporal and spatial specificity of A9 gene expression (Paul *et al.*, 1992).

Of course a primary route to gene cloning for such applications, is provided by cDNAs. Once a cDNA is available to use as a probe, isolating the corresponding gene becomes very simple. Another application of cDNAs is in identifying the possible function of the originating gene. This is achieved by obtaining the nucleotide sequence of the cDNA and searching databases for significant matches with entries of known identity and function. Alternatively, where no clues are forthcoming, other computer programs can be used to reveal physical features of the encoded polypeptide such as signal sequences, isoelectric point, etc.

Whilst this example illustrates the potency of differential screening, the approach has at least one shortcoming. This is exemplified by the observation that a high proportion of the differentially expressed transcripts identified from immature anther libraries are tapetum-specific. For example, Koltunow *et al.* (1990) examined the spatial expression patterns of a number of anther-specific transcripts using cDNAs isolated from an *N. tabacum* immature anther cDNA library and found that three (TA26, 29 and 32) out of five were localised to the tapetum. The situation is even more striking in libraries prepared from younger anthers. Smith *et al.* (1990) and Scott *et al.* (1991b) performed spatial analysis on a small number of anther-specific cDNAs by *in situ* hybridisation and found that the corresponding transcript was localised within the tapetum in every case. Why should tapetal messages dominate the mRNA population of the young anther when several other cell types, including the sporocytes and derivatives, are also present? During early microsporogenesis the tapetum is extremely active metabolically and is undoubtedly the transcriptionally dominant cell type in the anther. This dominance may be in part due to the very high expression of a relatively small number of genes. The composition of cDNA probe stocks used for differential screening will obviously reflect the mRNA population from which they are derived. This fact, coupled with the sensitivity threshold of differential screening, results in the preferential identification of high abundance tapetum-expressed transcripts. Therefore, differentially expressed transcripts of lower abundance, presumably including many of those expressed in other anther cell types such as the microsporocytes, are missed during screening.

However, a modification of differential screening, termed cold-plaque screening (Section V.B) provides a partial remedy. Cold-plaques are the product of those recombinant phage clones that fail to give a signal after hybridisation with an appropriate probe stock, which in our experiments was derived from anther mRNA. We found that a high proportion of randomly selected 'cold-cDNAs' are derived from differentially expressed mRNAs, in this case from anther-specific mRNAs. As expected, the peak concentrations of the majority of these mRNAs are below 0.02%, and they therefore fall within the medium–low abundance class of messages.

Where specific target cDNAs are sought within a library, and molecular probes are available, a variety of suitable screening techniques have been developed. These include the use of antibodies prepared to the target polypeptide (Section V.D.2), heterologous DNA probes (Section V.C) and more recently, tandem arrays of DNA binding motifs to isolate transcription factors (Section V.D.3).

II. cDNA SYNTHESIS

There are now a number of excellent kits available commercially which take the researcher from mRNA to cDNA library in a matter of days. It is of course possible to buy all the necessary enzymes, DNAs and reagents and successfully produce cDNA libraries independently, but in most cases it will be cheaper and much more convenient to use a kit. The following sections on cDNA synthesis and cloning are therefore intended primarily as background reading, but also as a discussion of the various options available when planning library construction.

A. RNA Quality

Successful cDNA synthesis is dependent on the isolation of high quality poly(A)$^+$ RNA (see Chapter 2 for poly(A)$^+$ isolation protocols). A variety of procedures can be used to check RNA quality before library construction is undertaken. If sufficient poly(A)$^+$ RNA is available the RNA can be run on a gel (Draper et al., 1988). A smear of RNA should be seen from c. 300 bp up to around 10 kb. In vitro translation can also be used as an indicator of RNA quality (see Chapter 2). High-quality poly(A)$^+$RNA should translate to give a complex mixture of polypeptides ranging from 5 kD to over 100 kD in size. If antibodies are available to specific peptides of interest, immunoprecipitation of the in vitro translation products can be carried out. Before bulk cDNA synthesis for library construction is undertaken it is advisable to perform a radioactively labelled (usually with [^{32}P]2'-deoxy-cytidine-5'-triphosphate (dCTP)) first strand synthesis test reaction. This enables calculation of cDNA yield and allows the experimenter to check the size of cDNA produced by running on an agarose gel. The first strand synthesis reaction should be oligo(dT) dependent and incorporation of [^{32}P]dCTP should be reduced by at least 95% in the absence of primer.

In some cases it may be possible to enrich the poly(A)$^+$ RNA population for sequences of interest. For instance, if the size of transcript is known, RNA can be size fractionated prior to cDNA synthesis.

B. Primers for First Strand Synthesis

The primer oligo(dT)$_{12-18}$, which binds to the 3' poly(A) tail of eukaryotic mRNAs, is generally used to prime first strand cDNA synthesis. However, other primers are now available which allow the directional cloning of cDNA by combining an oligo(dT) region with an adaptor-like sequence, usually one or two restriction sites unique to the polylinker of the cloning vector. This can be important when constructing expression libraries as it enables all cDNAs to be cloned in the correct orientation for expression of fusion proteins, thus reducing by half the number of clones which have to be screened. Other primer–adaptor combinations are available which are designed to allow polymerase chain reaction (PCR) amplification of the cDNA following synthesis. Random hexadeoxyribonucleotides are also used to prime cDNA synthesis, either in combination with oligo(dT) or alone, and can be used where secondary structure in the RNA is likely to inhibit synthesis of the 5′ regions of transcripts. Annealing of RNA and primer is carried out prior to first strand synthesis by heating to 70°C for 5 min and cooling slowly to room temperature.

C. First Strand Synthesis

Synthesis of first strand cDNA from RNA depends on the activity of reverse transcriptase. There are two enzymes available to the molecular biologist: avian reverse transcriptase (AMV-RT) and murine reverse transcriptase (M-MLV RT). The AMV-RT has two polypeptide subunits with single-stranded RNA- or DNA-dependent DNA polymerase activity and RNase H activity. The murine enzyme is a single subunit with the same DNA polymerase activities as the avian enzyme but a much lower level of RNase H activity. The lower RNase H activity of M-MLV RT makes it the enzyme of choice when synthesis of full length first strand cDNA is required. The two enzymes have different temperature and buffer optima and the manufacturers' instructions should be followed.

D. Second Strand Synthesis

AMV-RT generates a 3′ hairpin loop in first strand cDNA which can be used to prime second strand synthesis (Efstratiadis et al., 1976). Up until the replacement synthesis protocol of Okayama and Berg (1982), this was the only method available for generating second strand cDNA. Following first strand cDNA synthesis the cDNA–RNA hybrid is denatured by boiling or alkaline hydrolysis of the RNA. This allows the single-stranded cDNA to form a 3′ hairpin loop which acts as a primer for second strand synthesis by the Klenow fragment of DNA polymerase 1. The cDNA loop is then digested using S1 nuclease to allow cloning of the double-stranded cDNA. However, S1 nuclease digestion of the loop is difficult to control and often many bases corresponding to the 5′ region of the mRNA are lost. In addition cDNA yield may be reduced causing problems in construction of representative libraries.

The second strand cDNA synthesis procedure first described by Okayama and Berg (1982) and subsequently modified by Gubler and Hoffman (1983) is in most cases the method of choice (see Fig. 3.3). Also known as replacement synthesis, this method involves the action of RNase H to generate gaps in the RNA strand of the RNA–DNA hybrid. The resulting RNA fragments then act as primers for DNA polymerase 1 to synthesise the second strand of cDNA. Gaps between the DNA fragments are closed by T4 DNA ligase and the ends of the cDNA made blunt by the Klenow fragment of DNA polymerase 1. The major disadvantage of this system is that it is not possible to synthesise second strand cDNA corresponding to the extreme 5′ region of the transcript and so a few nucleotides will be lost. This is because DNA synthesis can only proceed from the 3′ end of RNA primers generated by RNase H activity. The RNA fragment 5′ to the 5′-most nick in the RNA–DNA hybrid is unlikely to remain annealed to the first strand cDNA and subsequently the 3′–5′ exonuclease activity of DNA polymerase 1 will remove the single-stranded overhang. In practice, as most

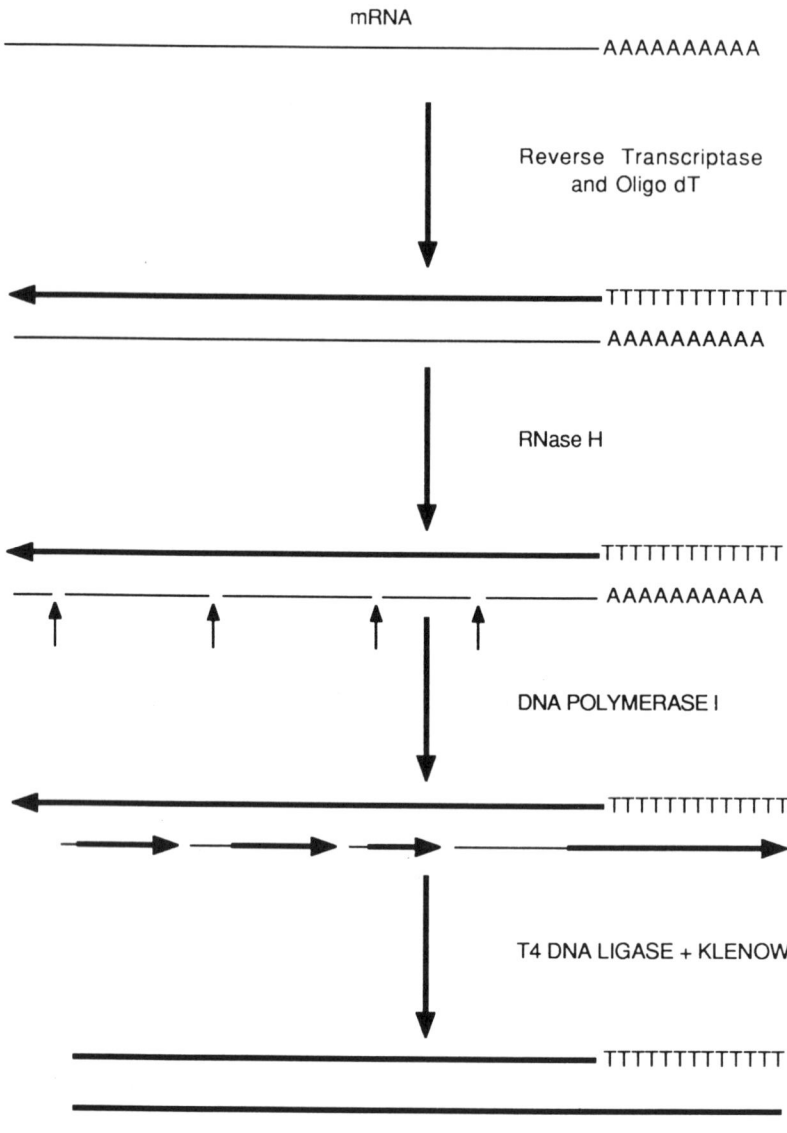

FIG. 3.3. Replacement synthesis of double-stranded cDNA.

eukaryotic mRNAs have quite long untranslated leaders, the loss of a few nucleotides at the 5' end is not likely to be of concern. However, where problems are encountered, other strategies can be employed. For example, terminal transferase can be used to add tails of dC residues to the 3' end of the first strand of cDNA, after which second strand synthesis can be primed using an oligo(dG) primer. However, there is some evidence to suggest that cDNAs vary in the efficiency with which they can be tailed by terminal transferase. It is also possible that the dG/dC tails may interfere with subsequent procedures such as expression of fusion proteins.

III. CLONING OF cDNA

A. Preparation of cDNA for Cloning

1. Size selection

Since small cDNA fragments will be cloned more efficiently than larger ones it is generally necessary to size-select cDNA prior to cloning. Usually cDNA is purified on a column which simply removes small cDNA fragments, i.e. those less than 200 bp. However, if cDNA corresponding to a particular transcript of known size is required, more specific size selection can be undertaken.

2. Subtraction

Subtraction of cDNA populations prior to cloning to remove common sequences, such as those corresponding to 'housekeeping' transcripts, is becoming routine, particularly since PCR amplification of cDNA now allows the production of large quantities of cDNA from very small amounts of starting material. A number of subtraction protocols are available, many based on the ability of streptavidin to remove biotin-labelled DNA from solution (Sive and St. John, 1988). Briefly, the minus cDNA stock is labelled with biotin, following hybridisation with the unlabelled plus cDNA in which homologous sequences, i.e. those shared between the two cDNA populations, will anneal. The biotinylated hybrids are removed from solution by the addition of streptavidin. Repeated rounds of subtraction leave only those sequences unique to the plus cDNA in solution and allow cloning of a cDNA population enriched for specific sequences. Other subtraction procedures are available which use hydroxyapatite to selectively remove RNA–DNA hybrids (Davis *et al.*, 1984).

3. Tailing of cDNA and addition of linkers/adaptors

In order to facilitate cloning of cDNA into plasmid or lambda vectors, homopolymer tails or restriction sites must be added to the 5′ and 3′ termini of the double-stranded cDNA. There are essentially two methods for achieving this. The first is homopolymer tailing of the cDNA by terminal transferase, and the second is ligation of linkers or adaptors containing restriction sites, to the ends of the cDNA.

Homopolymer tailing was commonly used in cDNA library construction prior to the availability of synthetic linkers and adaptors. Complementary tails, usually dG and dC, are added to cDNA and linearised plasmid using terminal transferase. Hydrogen bonding between the tails forms open circular plasmid molecules which are then transformed into *Escherichia coli* to produce a library of clones. By linearising the plasmid at a unique *Pst*I site and tailing the plasmid with dG residues, the cDNA fragment can be released from the vector by digestion with *Pst*I. The main disadvantage of this system is that it effectively restricts library construction to plasmid vectors, which in turn limits the size and ease of handling of the library.

Synthetic linkers, which are small blunt-ended double-stranded DNA molecules usually containing one or two restriction sites, can be ligated to blunt-ended cDNA. Digestion with a restriction site in the linker and purification away from excess linker, usually by size selection on a column, then enables cloning into a vector which has been digested with a restriction enzyme generating compatible termini with that of the linker. Cloning efficiency is greatly increased when the restricted vector is dephosphorylated to prevent self-ligation. Obviously, in some cases there will be a restriction enzyme site within the sequence of the cDNA. In these instances it will be difficult, if not impossible, to clone full-length cDNAs unless the extra step of methylating the cDNA prior to linker addition is undertaken. Alternatively, the use in the linker of very rare restriction enzyme sites, such as *Not*I, should minimise these problems.

Adaptor molecules are short double-stranded DNA molecules which have one blunt end, for ligation to the cDNA, and one cohesive end to allow cloning into the vector. As the cohesive end is already present in the adaptor, no digestion of the cDNA is necessary following ligation of the adaptor. To prevent formation of multimers of the adaptor molecule, only the shorter of the two oligos forming the adaptor is phosphorylated. This prevents ligation of the cohesive ends and means that multimer formation is limited to dimers created by ligation between the blunt ends of the molecules. To allow ligation into dephosphorylated vector the cDNA is purified away from excess adaptor and the cohesive termini are then phosphorylated by T4 polynucleotide kinase.

B. Choice of Vector

Originally almost all cDNA libraries were constructed in plasmid vectors which had the advantages of relative simplicity of cloning, usually by homopolymer tailing, and ease of manipulation of the clones. However, in recent years, the availability, in particular of synthetic linkers and adaptors and efficient packaging extracts for bacteriophage DNA, has allowed the development of bacteriophage vectors so that they now incorporate all of the features associated with plasmid systems. In addition, lambda vectors offer greater cloning efficiencies, and libraries constructed in phage vectors can be screened at much higher densities and stored and amplified with greater ease.

1. Lambda vectors

Numerous commercially available lambda phage-based vectors are suitable for cDNA cloning. Some are designed for particular applications such as polypeptide expression, e.g. lambda gt11. Others combine many of the features of specialised vectors into a single multipurpose DNA. A good example of this type of vector is lambda ZAP-II developed by Stratagene. Whilst not wishing to endorse one particular product over any other, this vector has been used very successfully in this laboratory over a number of years. One of the principle advantages of lambda ZAP-II over other lambda vectors is that the previously irksome task of subcloning selected cDNA inserts into a plasmid has been made a trivial step. In general, reduction to a plasmid form is necessary because the large size and complexity of the lambda vectors makes further characterisation of cDNA inserts extremely difficult. This has been achieved by incorporating pBluescript, a 2.9 kb high copy-number multipurpose plasmid, into the lambda vector. This so-called phagemid contains the multicloning site into which cDNA is inserted and is capable, in the presence of protein products from a helper phage, of *in vivo* excision from the viral DNA and circularisation to form a recombinant plasmid. The procedure takes 24 h and avoids the *in vitro* manipulation of DNA completely. In common with some other vectors, lambda ZAP-II incorporates blue/white selection. This allows the identification of both non-recombinant phage (blue plaques) when assessing the quality of a new library and of recombinant (white plaques) during cold-plaque screening. Allied to this, lambda ZAP-II can be induced to synthesise fusion proteins and is therefore suitable for constructing expression libraries.

pBluescript contains all of the features normally required in the characterisation and modification of cDNA inserts. These include a polylinker containing 21 unique restriction sites which is flanked by T3 and T7 promoters. These promoters enable the synthesis of sense or antisense RNA transcripts from the cDNA insert for such applications as *in situ* hybridisation and *in vitro* translation. DNA sequencing is made simple and efficient because single-stranded DNA, the best substrate for sequencing, is easily generated from pBluescript. Single-stranded DNA is also useful in site directed mutagenesis. Finally, since the plasmid contains the the *lacZ* promoter, fusion proteins can be generated for Western blot analysis or protein purification. However, other plasmids that have been designed specifically for this application, e.g. pEX (Clontech) and pGEMEX (Promega) may be more suitable in some cases.

Lambda ZAP-II DNA is supplied by the manufacturer in two forms: either ready to receive insert cDNA at the *Eco*RI site, or as the complete viral genome which must be pretreated before use. This latter form is used where either double-cut vector is required, e.g. in unidirectional cloning, or insert DNA is not prepared with *Eco*RI sticky-ends. The ready-to-use form is prepared by digesting the viral DNA at the unique *Eco*RI site to produce two vector arms and then dephosphorylating these to prevent self-ligation during the cloning step. The vector is capable of accepting cDNAs of up to 10 kb that have phosphorylated *Eco*RI sticky-ends.

2. Preparation of vector and ligation

Ligation conditions need to be carefully arranged so as to maximise the number of recombinant molecules produced while minimising the possibility of a single vector molecule containing more than one cDNA insert. If sufficient cDNA is available it is advisable to do a number of test ligations and packaging reactions in order to optimise the procedure. When using insertion vectors, such as lambda ZAP-II digested with a single restriction enzyme, which can religate to produce non-recombinant phage, dephosphorylation following restriction is almost essential. If the vector is to be double-cut to produce non-cohesive sticky-ends it is obviously less important, providing of course that the DNA is fully digested. In other vectors such as lambda gt10, following ligation and packaging, non-recombinant phage are selected against by plating on a strain of *E. coli* which carries the hfl^- mutation.

3. Packaging of bacteriophage lambda DNA in vitro

Following ligation of cDNA into a lambda vector, it is necessary to package the DNA in order to produce complete bacteriophage particles capable of infecting phage competent cells. Highly efficient packaging extracts are available commercially and purchasing a small amount for library construction is advisable. However, if large amounts are required it is possible for researchers to prepare their own extracts following quite simple protocols (Scalenghe et al., 1981; Rosenberg, 1987).

4. Measuring library titre

Before screening a new library, it is important to determine its titre in order to plate the correct number of plaque forming units (pfu) per Petri dish. This is done by making serial dilutions of the phage stock in SM buffer. Typically, the titre of a good library is in the region of 500–1000 pfu μl^{-1}. Aliquots (10 μl) of the dilutions are plated (see Section IV.B), incubated overnight, and the resultant plaques counted to determine the library titre. The same procedure should be repeated following library amplification (Section III.B.5.) and after prolonged storage.

5. Amplification of cDNA libraries

cDNA libraries constructed in lambda vectors may be amplified and stored in order to maintain a permanent stock. Amplification procedures vary slightly depending on the type of vector used and instructions supplied by the manufacturer should be adhered to. Although libraries are stable for several months if stored at 4°C, longer term storage is also possible by adding dimethyl sulphoxide (DMSO) to a final concentration of 7% and freezing at −70°C.

IV. PRINCIPLES OF USING LAMBDA VECTORS

Lambda-based cDNA cloning vectors are now the choice of most researchers. The advantage of viral over plasmid vectors is most evident during library screening. The objective here is to maximise both sensitivity, by giving good access of probe molecules (DNA, antibodies) to individual recombinant clones, and throughput, by enabling large numbers of clones to be screened at reasonable cost. Regardless of the type of vector, the first step in the screening procedure is to plate out a suitable number of recombinants (the precise number of clones screened and the optimum density of plaques on the plates depends on the application (see Section V). In the case of viral vectors, an appropriate host bacterial strain (usually supplied as part of the cloning kit) is infected with an appropriate number of viral particles. The initial or primary screen is usually large (10 000–100 000 plaques) and is therefore normally carried out in 14 cm Petri dishes.

A. Preparation and Infection of Bacterial Host Cells

Escherichia coli cells are made competent for infection with bacteriophage lambda in the following way. A single colony of an appropriate strain is grown overnight with vigorous shaking at 30°C in media supplemented with 0.2% maltose and 10 mM $MgSO_4$. Cells are pelleted (1000 × g, 10 min) and gently resuspended in 0.5 vols of cold (4°C) 10 mM $MgSO_4$. Prior to use, the cells are diluted to an OD_{600} of 0.5 with 10 mM $MgSO_4$. Although cells prepared this way can be used for up to a week if stored at 4°C, the highest plating efficiencies are obtained from freshly prepared cells. Competent cells for plating are infected by mixing aliquots of the phage with 600 μl (14 cm Petri dish) or 200 μl (9 cm Petri dish) of host cells and incubating at 37°C for 15 min.

B. Making the Master Plate

For successful screening, it is important to start with good master plates, i.e. an even distribution of equal-sized plaques in a flat 'lawn'. To achieve this, the, first important step is to ensure that the thick bottom agar layer (70–80 ml for a 14 cm dish, and 20–25 ml for a 9 cm dish) is level. If it is not, when the thin agarose layer, which contains the virally infected bacteria, is poured on top, a layer of uneven thickness will form. Since plaques in shallow areas grow much larger than those in deeper areas, plaques of different sizes form. The problem here is that the subsequent hybridisation signal is proportional to the size of the plaque and therefore small plaques give a reduced signal, which makes the detection of any differences more difficult. Flat lawns that cover the whole surface of the bottom agar are achieved by ensuring that the top agarose remains molten long enough to complete the spreading procedure. This is simple provided that the aliquot of top agar and bacteria is at the highest permissible temperature (48°C) and the dishes containing the bottom agar are heated to 37–40°C. It is also important that the bottom agar is dry. If it is not, during the incubation period excessive moisture is released from the agar and accumulates on the surface of the top agarose layer. This may then run across the surface and disturb the pattern of developing plaques. Plates should be dried in an oven or incubator (30 min at 37°C) or prepared at least 24 h in advance.

Agarose is used for the top layer because it forms a more rigid gel than agar. High gelling temperature agarose, at a concentration of about 1%, should be used for this application. Appropriate volumes of top agar are 6.5 ml for a 14 cm plate and 2.5 ml for a 9 cm plate. If blue/white colour selection is being used IPTG and X-GAL should be added to the top agar/bacteria mixture immediately before pouring. Once poured, the top agar hardens within 2–3 min, during which time the plate should not be disturbed. Plates should then be inverted and incubated at 37°C until clear plaques form.

The length of the incubation period depends on several factors, such as the bacterial strain used to create the lawn and, most importantly, the final plaque density. At low densities, plaques can be allowed to develop fully without becoming confluent (i.e. all the plaques merging). In this case, plates can be incubated overnight (15–16 h) without any problems. However, at plating densities above approximately 2000 pfu per 14 cm dish, complete plaque development would lead to significant merging between individual plaques. At very high densities this results in the complete clearing of the lawn. This is more serious than it appears since the signal from individual plaques is almost completely dissipated under these conditions and the screen will consequently fail. Therefore, at high densities the plaque development is monitored continuously after approximately the first 3–4 h of incubation. Once small, but clear plaques have formed the plates are transferred to 4–8°C to stop further growth.

C. Transfer to Supporting Membranes

The next stage is to transfer the target molecules to a supporting membrane which has the capacity to preserve faithfully the pattern of plaques throughout whatever subsequent screening procedure is employed. Depending on whether the vector is plasmid or viral, the procedure is termed a colony or plaque lift, respectively. Examples of suitable membranes include Hybond-N (Amersham) and Duralon (Stratagene). In the case of plasmid vectors, the recombinant cells of bacterial colonies must, of course, be lysed prior to transfer in order to release the target molecules. However, this is not required for viral vectors which naturally lyse the host bacterial cells during plaque formation. This represents a significant advantage over plasmid vectors since access of probe molecules to target DNAs or polypeptides is achieved without any of the harsh treatments necessary to lyse bacteria. Aside from ensuring that target molecules remain unaltered, the absence of a chemical lysis step greatly increases the resolution of the processed lift. This enables the practical screening of recombinant clones at high densities (up to 50 000 plaques per 14 cm plate).

Several companies supply circular membranes suitable for use with 14 or 9 cm diameter round Petri dishes. Alternatively, circles can be cut from sheets, but it is worth noting that these should be slightly smaller than the diameter of the dish, since advertised diameters usually refer to the lid and not to the base. However, perhaps a more economical method of making home-made filters is to use rectangular Petri dishes. Identification marks, in pencil, should be made on each membrane before taking the lift. Again, before making the lift, the Petri dishes should be placed at 4°C for a minimum of 1 h. This ensures that the top agarose layer is rigid enough to resist peeling away with the membrane during the lifting procedure. If small areas of top agarose do adhere to the membrane, remove these by rinsing in 3 × SSC (SSC; 3 M sodium chloride, 0.3 M tri-sodium citrate adjusted to pH 7.0 with sodium hydroxide) following the denaturation and neutralisation steps.

Since primary screening is usually carried out in 14 cm Petri dishes, positioning filters of this size centrally within the dish is initially quite a challenge. The simplest method is to hold the edge of the membrane at two points (diametrically opposed) between the thumb and forefinger (of gloved hands) and to form a well-rounded U-shape by bringing the hands 4–5 cm apart. Contact between the membrane and agar surface is then first made at the bottom of the U, which is placed across the middle of the plate, and each raised side is gradually lowered as the membrane wets from the middle outward. This both locates the membrane correctly and avoids any bubbles of air becoming trapped beneath the membrane. Once in contact with the agar the membrane cannot be repositioned. If an error is made, for example in positioning, the membrane must be discarded. Seal master plates with clingfilm and store at 4°C until screening is complete.

D. Preparing Replicate Lifts

For differential screening, each plated clone population is screened with two or more probes (Section V.A). The most straightforward way to do this is to make replicate plaque lifts from each plate, and to screen these with the different probes simultaneously. By progressively doubling the time for which the membrane is in contact with the plaques for each additional filter, up to four replicate lifts can be taken without any loss of signal. An alternative strategy, but one which takes more time, is to make a single lift, screen with the first probe and then strip the filter before screening with the second probe.

E. Orientation Marks

An important component of making plaque lifts is to ensure that following completion of the screening procedure it is possible to align precisely the pattern of spots on the finished autoradiograph or Western blot with the original plaques on the bacterial lawn. Reference marks therefore need to be made whilst the filter is in contact with the bacterial lawn. There are numerous techniques for doing this. In this laboratory, we simply push a hypodermic needle (1.1 × 40 mm) through the filter and the agar, close to the edge of the plate, at three asymmetric positions. An alternative strategy, which avoids the problem of occasionally making three symmetrical marks, is to make a single hole at the first point, two at the second and three at the third. When the filter is removed from the plate, the holes remain in the agar. Unfortunately, the marks within the agar are not very durable and since the results of the screen may not become available for anything up to a week, it is advisable to mark the position of the holes on the base of the Petri dish with a marker pen. In practice, this is done immediately after making the reference holes and whilst the filter is still in contact with the agar. This method avoids any frustrating mistakes. Where autoradiography forms part of the screening procedure, a second set of orientation marks must be made to allow the transfer of the first set of marks from the filter to the developed photographic film. Remember that it is the autoradiograph and not the filter that is used to locate interesting clones. This second set of marks is made in exactly the same way as for a standard Southern blot. For example, the filter is wrapped in clingfilm, taped to the photographic film, and then sets of two parallel lines, again at asymmetric positions, are drawn from the filter and onto the film using a marker pen which is resistant to the photographic development process. Once the film is developed these lines are used to relocate the filter onto the film, and the first set of marks are transferred from the filter to the film. Of course, this second orientation stage is not necessary where the location of probe binding is developed on the membrane itself, such as in many immuno-detection systems and some non-radioactive nucleic acid hybridisation techniques.

F. Identifying and Purifying Selected Plaques

Once the autoradiograph or membrane is developed, and any interesting signals have been identified, the original plate with its pattern of plaques is overlain in the correct orientation. A plug of agar containing the plaque is then cored from the plate and dropped into a 1.5 ml Eppendorf containing SM buffer (100 mM sodium chloride; 15 mM $MgSO_4$, 50 mM Tris-HCl, pH 7.5; 0.01% gelatin). A drop of chloroform (10–20 μl) is then added to kill the *E. coli*. The size of the plug, and therefore the number of plaques it contains, is related to the certainty of identifying the positive plaque. In a differential screen, where a large proportion of the plaques give a signal, it is often possible to line up the pattern of plaques precisely with the corresponding pattern of spots on the autoradiograph or membrane. The plaques of interest are then easily identifiable, especially in a low density screen. In this case, a small plug that contains a single plaque may be taken. However, screening procedures employing antibodies or heterologous probes give perhaps only one or two positive signals per plate. Consequently, it is very much more difficult to line up spots with the corresponding plaques with any degree of certainty. Under such circumstances, much larger plugs are taken that contain all the plaques in the vicinity of the spot. Positive plaques are then identified by performing a second round of screening at a relatively low density (aim for 100–200 plaques per 9 cm plate). However, care should be taken to screen sufficient plaques to guarantee recovering the positive plaque. Of course, during the second round a higher proportion of the plaques will be positive, which aids alignment.

Regardless of the core size, a minimum of two rounds of screening should be performed. This ensures that the correct plaque was taken and that the new phage stock is pure. Impure stocks are easily identified since the screen will consist of both negative and positive plaques. Plaque purity is essential before embarking on any further work with the cDNA.

V. SCREENING TECHNIQUES

A. Differential Screening

Plating of the library for differential screening should be done at reasonably low density to enable the clear identification of individual plaques, between 1500 and 2000 plaques per 14 cm Petri dish is ideal.

1. Preparation of lifts for differential screening

Replicate lifts will be required for differential screening. For the first lift the filter should remain on the agar for 30 s; each subsequent filter should be left on for an additional 30 s. Plaque lifts for nucleic acid hybridisation are placed 'DNA side up' on filter paper soaked in 0.5 M sodium hydroxide, 1.5 M sodium chloride for 7 min to denature the double-stranded DNA and then neutralised for 7 min on filter paper soaked in 0.5 M Tris-HCl, pH 7.4, 3 M sodium chloride. The filters are then briefly rinsed in 2 × SSC (0.3 M sodium chloride, 0.03 M sodium citrate, pH 7.0), air-dried and then treated as per the manufacturers' instructions to fix the DNA to the support. In the case of nylon membranes this involves UV irradiation for a predetermined period. The filters are then ready for hybridisation, which is carried out exactly as for gel-blotted equivalents.

2. Preparation of first strand cDNA probe stocks

Poly(A)$^+$ RNA is prepared as described in Chapters 1 and 2 and diluted to a concentration of 0.2 mg ml^{-1}. In a sterile Eppendorf tube 1.2 µg poly(A)$^+$ RNA is diluted to 10 µl with RNase-free sterile water and 1 µl of oligo(dT) (10 mg ml^{-1}) added. The mixture is heated to 70°C for 5 min and cooled slowly to room temperature to allow annealing of primer to mRNA. The tubes are then spun for a few seconds to bring the contents to the bottom and placed on ice. The reaction conditions vary according to the type of reverse transcriptase used and the manufacturers' instructions should be adhered to. Between 1 and 2 µl of [^{32}P]dCTP (>3000 Ci mmol^{-1}, Amersham) should be added to a 20–25 µl reaction. After incubation at the recommended temperature (37°C for M-MLV RT, 42°C for AMV-RT) for 20–30 min, 1 µl chase mixture (10 mM dATP, 10 mM dCTP, 10 mM dGTP, 10 mM dTTP) is added and incubation continued for a further 15 min. The reactions are then spun for a few seconds in a microfuge and placed on ice. The reaction volume is made up to 50 µl with sterile water then 50 µl of 0.6 M sodium hydroxide are added and the tubes placed at 65°C for 15 min to hydrolyse the RNA and so denature the RNA–cDNA hybrids. The reactions are again spun down and placed on ice while the columns for probe purification are prepared. Prior to column purification it is important to check the incorporation of ^{32}P into cDNA by simple trichloroacetic acid (TCA) precipitation of a small aliquot of the reaction. Incorporations of 60% plus should be achieved.

3. Preparation and running of Sephadex G50 columns to purify first stand cDNA probes

Sephadex G50 columns (supplied by Pharmacia as 'Nick Columns') are used to purify first strand cDNA probes. They size-select nucleic acids and hence allow separation of cDNA from unincorporated nucleotides and small fragments of DNA which, if not removed, result in high background hybridisation. The columns come prepacked and simply need washing through with STE Buffer (10 mM Tris-HCl, pH 8.0, 1 mM ethylenediamine tetraacetic acid (EDTA), 100 mM sodium chloride) before use. A column will be needed for each probe stock generated. Probe stocks are made up to 200 µl by the addition of 100 µl of STE. The columns are washed through and allowed to run dry. The probe stock is loaded onto the top of the column and the eluate (which should be approximately 200 µl) collected. Five 200 µl aliquots of STE are then added to the column and the eluate from each collected in a separate tube to give six samples for each probe stock. The samples are Cherenkov counted in a scintillation counter. The first peak of counts, which is usually in samples 3 and 4, represents first strand cDNA and the second peak, usually sample 6, unincorporated nucleotides. The samples containing first strand cDNA should be pooled and used in subsequent hybridisation.

4. Nucleic acid hybridisation

This is carried out using standard methods. In this laboratory, prehybridisation and hybridisation are performed in appropriate sized Petri dishes with very good results. Preheated prehybridization solution (hybridisation lacking probe) is poured into the base of the dish and each membrane is then individually immersed in the liquid and any trapped air bubbles gently coaxed to the edges and expelled. The volume of prehybridisation solution required depends on the number of membranes being screened. A rough guide, for 13 cm membranes in a 14 cm Petri dish, is 20 ml for the first filter + 1–2 ml for each subsequent one. Each dish can accommodate 10–12 membranes. The lid is then placed on the dish and the unit sealed with tape capable of withstanding the incubation temperature. Following prehybridisation, the probe is introduced into the dish through a small hole previously melted in the lid (there is no necessity to change the hybridisation solution).

B. Cold-plaque Screening

Cold-plaque screening is a simple technique enabling the identification of clones corresponding to low abundance organ- or tissue-specific transcripts from cDNA libraries (Hodge *et al.*, 1992). Clones which correspond to specifically expressed high

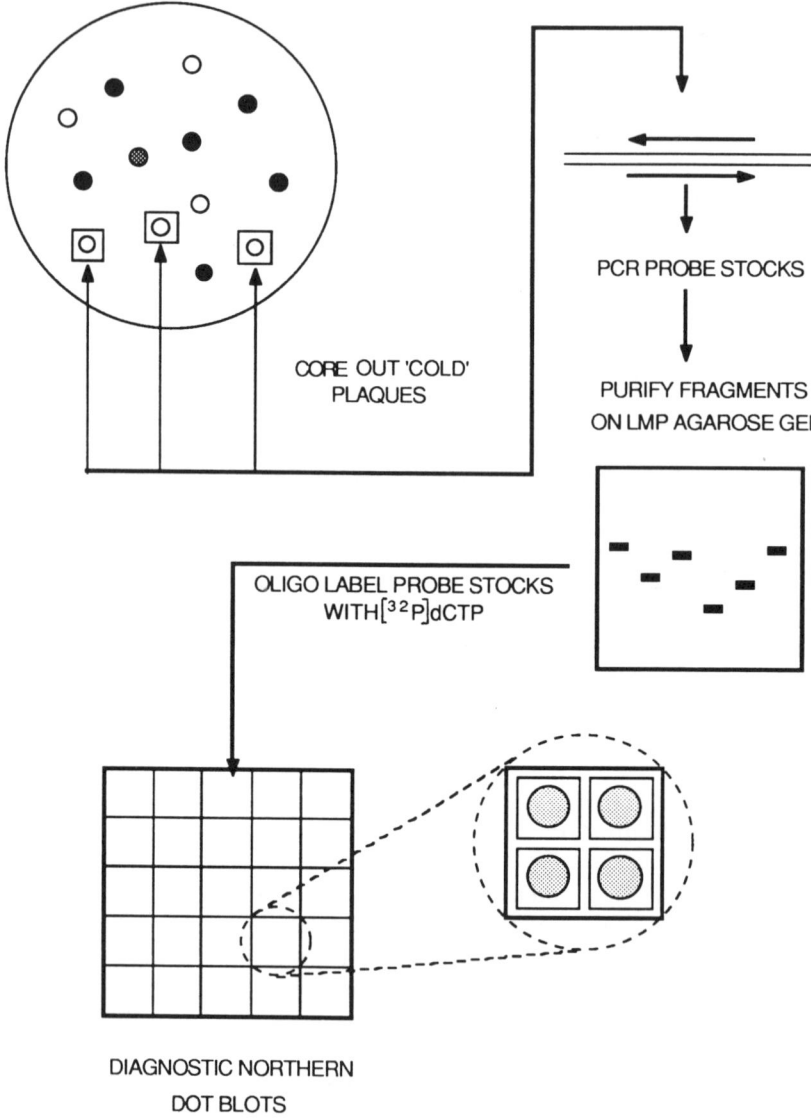

FIG. 3.4. Cold-plaque screening.

abundance transcripts can be identified by conventional differential screening methods; however, a large proportion of differentially expressed transcripts cannot be identified by this approach because they are of low or medium abundance in the mRNA pool. This technique can be used alongside differential screening in order to isolate cDNAs derived from organ-specific transcripts with a range of expression levels. Kamalay and Goldberg (1980, 1984) demonstrated that in specialised organs such as the anther a large proportion of the unique transcripts present are specific to that organ, for example they estimate that 11 000 (40%) of the 25 000 anther transcripts (present at stage 3 — immediately after the first pollen grain mitosis) are organ-specific. However, the majority of these organ-specific messages are present at medium or low abundance in the mRNA pool and hence are not identified by differential screening. We have used cold-plaque screening of a *B. napus* cDNA library to test and confirm the prediction that 40% of unique transcripts in the anther are organ-specific and to identify low abundance anther-specific mRNAs.

Initially the library is screened with a positive probe, usually first strand cDNA derived from the mRNA with which the library was constructed. Following autoradiography plaques are selected which fail to give a positive hybridisation signal ('cold-plaques'). Since all clones should be represented in the probe stock 'cold-plaques' are assumed to be derived from transcripts of low abundance in the mRNA pool. The cDNA inserts of these clones are then amplified by PCR and used to probe diagnostic Northern dot blots (see Fig. 3.4).

1. Preparation of probe stocks

Initial screening of the lifts with positive probe is exactly as described for differential screening. It is useful to plate on X-gal and IPTG where blue/white colour selection is available in order to distinguish between recombinants and non-recombinants. Recombinant plaques which fail to give a hybridisation signal are cored into 100 μl SM buffer. After incubation at room temperature for 4 h to allow elution of the phage particles from the agar plug, 10 μl of phage stock is Proteinase K treated by addition of an equal volume (10 μl) of 2 × PK Buffer (100 mM Tris-HCl, pH 8.5, 2 mM EDTA, 1% Triton-X100, 400 μg ml^{-1} Proteinase K) and incubation at 37°C for 1 h. The Proteinase K is then inactivated by incubation at 95°C for 10 min and the resulting phage DNA stock diluted five-fold. PCR amplification of the cDNA using 4 μl of the phage DNA stock and primers based on the polylinker region of the cloning vector is carried out and the PCR products separated on a low melting point agarose gel (0.8%). The amplified cDNAs are then excised and stored at -20°C to be used as probe stocks. Any phage stocks which give a mixed product or poor amplification should be discarded.

1. Diagnostic Northern dot blots

RNA dot blots should be prepared using 5 μg total RNA dotted onto Hybond N (Amersham International) or an equivalent membrane according to the manufacturers' instructions. Probe stocks are oligo-labelled with [^{32}P]dCTP using standard techniques and hybridisation carried out at 42°C for 16 h in a volume of 1 ml using

C. Screening with Heterologous DNA Probes and Oligos

One of the most straightforward means of identifying a cDNA of interest in a library is to use a DNA probe derived from a heterologous source. However, care should be taken to assess the feasibility of the approach before attempting to screen the library. The principal problem is that since the nucleotide sequence of the probe and target cDNA sequences are merely related, and the degree of this relatedness is unknown at the outset, appropriate hybridisation conditions must be determined empirically. Unfortunately, the degree of nucleotide conservation within and between species for particular genes does not follow any useful predictive rules. For example, in some instances probes derived from human genes have been used successfully to identify plant homologues via heterologous screening (Shorrosh and Dixon, 1992). Conversely, in this laboratory, we have found that within *B. napus*, sporophytically (seed) expressed oleosins differ in nucleotide sequence so markedly from their gametophytic (pollen) counterparts, that one could not have been used as a probe to identify the other (Roberts *et al.*, 1993). Thus, in certain instances, probes derived from one organ system may not prove useful in identifying related genes expressed in a different organ, even within the same species. In fact, it appears that sporophytically expressed oleosins of widely diverged species share a more recent ancestor than that shared by the sporophytic and gametophytic gene families. Consequently, possibly the best approach is to test prospective probes and hybridisation conditions using Northern gel blots of the same RNA from which the library was constructed. If a positive signal is obtained the cDNA is, at least potentially, represented in the library. The stringency of hybridisation can be varied by lowering the hybridisation temperature and/or increasing the ionic strength of the hybridisation solution.

D. Screening for cDNAs Encoding Specific Peptides

The availability of vectors, such as lambda ZAPII and lambda gt11, which allow expression of cDNAs generally producing β-galactosidase fusion proteins, enables the researcher to screen directly for cDNAs encoding proteins of interest. With the availability of synthetic primers it is now straightforward to construct libraries in which the cDNAs are cloned unidirectionally. This is obviously preferable as it effectively reduces by half the number of plaques which need to be screened.

1. Preparation of filters

For antibody screening, the optimal plaque density is approximately two- to three-fold higher than that required to produce a very dense field of individual plaques without actually reaching confluence. Therefore, it is important to titrate the phage stock carefully (see Section III.B.4). Master plates are made in the normal way (see Section IV.B) and incubated at 42°C until plaques are just visible (approx. 3.5 h). At this stage the plaque lawn is overlain with blotting membrane soaked in isopropyl-β-D-thiogalactopyranoside (IPTG). This induces infected, but, as yet, unlysed bacteria, to produce fusion proteins from the cDNA inserts. IPTG-treated membranes are prepared about 30 min before they are needed by submersing them into IPTG solution (10 mM in distilled water) until completely wet. Membranes that fail to wet properly should be discarded. Filters are air-dried on blotting paper and then given identification marks in pencil. The membranes are applied to the corresponding plate (see Section IV.C) and incubated at 37°C for 3.5 h. During this period fusion proteins are produced and released from the bacteria as lysis progresses. Orientation marks should be made prior to the removal of the membrane. To prevent the top agar sticking to the membrane during its removal, place the plates at 4°C for 4 h. To guard against false positives it is advisable to make a duplicate lift for each plate. This is done by applying a second IPTG-treated filter and a further 4 h incubation at 37°C. The filters are then handled according to the subsequent screening protocols being employed.

2. Antibody screening

Successful antibody screening is dependent on the ability of the primary antibody to detect very small amounts (nanograms) of native antigen immobilised on blotting membrane in the presence of bacterial and phage proteins. Therefore, it is advisable to check that the primary antibody possesses the required specificity and avidity characteristics before embarking on screening. The optimum concentration is usually the greatest dilution that gives a strong signal. Adopting this approach reduces background signal and false positives. Suitable tests can be done with simple dot blots. Serial dilutions of both antigen (1 μl aliquots containing between 1 μl and 10 pg antigen) and primary antibody (1:100–1:1000 for a polyclonal serum in 1 μl aliquots) are spotted onto a suitable membrane, together with a single spot of *E. coli*/phage lysate. Replicates are made to enable the evaluation of various primary antibody dilutions. Following air-drying and blocking, each membrane is incubated in a different dilution of primary antibody (1:100–1:1000 for a polyclonal serum) for 1–2 h. The strips are then incubated in secondary antibody diluted according to the manufacturers' instructions. Following signal development, the results are evaluated. Generally, the working dilution of primary antibody chosen is that which produces a clear signal with the 10 ng antigen spot. If signal develops on the primary antibody spots but not the antigen spots, this indicates that the concentration of the primary antibody or the period of its incubation with the membrane should be increased. A strong signal against the *E. coli*/phage spot indicates that the primary antibody contains antibody that cross-reacts with *E. coli* or phage proteins. In this case, the serum should be preabsorbed against *E. coli*/phage lysate. Where no signal develops against the primary antibody spots, the dot blot should be repeated with increased concentrations of secondary antibody. There are many examples of plant cDNAs which have been cloned by screening expression libraries with an appropriate antibody. A recent example is the cloning of a tomato cDNA encoding polyphenoloxidase reported by Shahar *et al.* (1992).

3. Screening for cDNAs encoding DNA binding proteins

Analysis of the upstream regulatory sequences of genes may result in the identification of short sequences or boxes, which confer specificity or inducibility on the promoter.

Gilmartin et al. (1992) used a tetramer of the pea rbcS-3A box II, which had been shown previously to confer light inducibility on the promoter, to screen an expression library. Using a protocol slightly modified from Singh et al. (1989), these authors identified cDNAs encoding a DNA binding protein with specificity for the box II element. In order to eliminate cDNAs encoding non-specific DNA binding proteins it is important to use a mutated box as a negative screen when using this technique.

REFERENCES

Davis, M. M., Cohen, D. I., Neilsen, E. A., Steinmetz, M., Paul, W. E. and Hood, L. (1984). *Proc. Natl. Acad. Sci. USA* **81**, 2194-2198.
Draper, J., Scott, R., Armitage, P. and Walden, R. (1988). "Plant Genetic Transformation and Gene Expression: A Laboratory Manual." Oxford: Blackwell Scientific Publications.
Efstratiadis, A., Kafatos, F. C., Maxam A. M. and Maniatis, T. (1976). *Cell* **7**, 279-288.
Gilmartin, P. M., Memelink, J., Hiratsuka, K., Kay, S. A. and Chua, N.-H. (1992). *Plant Cell* **4**, 839-849.
Gubler, U. and Hoffman, B. J. (1983). *Gene* **25**, 263-269.
Hodge, R., Paul, W., Draper, J. and Scott, R. (1992). *Plant J.* **2**, 257-260.
Kamalay, J. C. and Goldberg, R. B. (1980). *Cell* **19**, 935-946.
Kamalay, J. C. and Goldberg, R. B. (1984). *Proc. Natl. Acad. Sci. USA* **81**, 2801-2805.
Koltunow, A. M., Truettner, J., Cox, K. H., Wallroth, M. and Goldberg, R. B. (1990). *Plant Cell* **2**, 1201-1224.
Okayama, H. and Berg, P. (1982). *Mol. Cell. Biol.* **2** 161-170.
Paul, W., Hodge, R., Smartt, S., Draper, J. and Scott, R. (1992). *Plant Mol. Biol.* **19**, 611-622.
Roberts, M. R., Hodge R., Ross J. H. E., Sorensen, A., Murphy, D. J., Draper J. and Scott, R. (1993). *Plant J.* (in press).
Rosenberg, S. M. (1987). *Methods Enzymol.* **153**, 95-103.
Scalenghe, F., Turco, E., Edstrom, J. E., Pirrotta, V. and Melli, M. (1981). *Chromosoma*, **82**, 205-216.
Scott, R. J., Hodge, R., Paul, W. and Draper, J. (1991a). *Plant Sci.* **80**, 167-191.
Scott, R., Dagless, E., Hodge, R., Paul, W., Soufleri, I. and Draper, J. (1991b). *Plant Mol. Biol.* **17**, 195-207.
Shahar, T., Hennig, N., Gutfinger, T., Hareven, D. and Lifschitz, E. (1992). *Plant Cell* **4**, 135-147.
Shorrosh, B. S. and Dixon, R. A. (1992). *Plant Mol. Biol.* **19**, 319-321.
Singh, H., Clere, R. G and Lebowitz, J. H. (1989). *Biotechniques* **7**, 252-261.
Sive, H. L. and St. John, T. (1988). *Nucl. Acids Res.* **16**, 10 937.
Smith, A. G., Gasser, C. S., Budelier, K. A. and Fraley, R. T. (1990). *Mol. Gen. Genet.* **222**, 9-16.
Willing, R. P. and Mascarenhas, J. P. (1984). *Plant Physiol.* **75**, 865-868.
Willing, R. P., Bashe, D. and Mascarenhas, J. P. (1988). *Theor. Appl. Genet.* **75**, 751-753.

4 Nucleic Acid Blotting and Hybridisation

PAOLO A. SABELLI and PETER R. SHEWRY

Department of Agricultural Sciences, University of Bristol, AFRC Institute of Arable Crops Research, Long Ashton Research Station, Bristol BS18 9AF, UK

I.	Introduction	80
II.	Method	81
	A. DNA blotting	81
	B. Probe labelling	82
	C. Prehybridisation and hybridisation	84
	D. Washing and exposure	84
	E. Probe removal	85
III.	Comments on the method	85
	A. DNA transfer and immobilisation	85
	B. Probe labelling	87
	C. Hybridisation	87
	D. Posthybridisation	88
	E. Modification for northern analysis	88
	F. T_m and stringency	89
IV.	Applications	91
	A. Gene copy number	91
	B. Gene localisation	93
	C. Identification of members of multigene families	94
	D. Analysis of mutants	97
V.	Summary	98
	Acknowledgements	99
	References	99

I. INTRODUCTION

Since its introduction by Southern (1975), the annealing (or hybridisation) of nucleic acids, immobilised onto a solid support, to complementary labelled nucleic acid probes, has become a routine procedure in any molecular biology laboratory. The principle is based on the fact that, under suitable conditions, complementary nucleic acid strands will anneal (hybridise) to form a duplex molecule. The extent to which molecular hybridisation occurs is a function of the sequence similarity (complementarity) between the two single-stranded molecules. Generally, in filter hybridisation experiments, single-stranded nucleic acids are transferred and immobilised onto a membrane (or filter) and are available for hybridisation with labelled nucleic acid probes. This technique is usually utilised to detect specific sequences and the intensity of the hybridisation signal is related to the abundance of the target sequences within the sample and to the degree of similarity between target and probe sequences. Additional information, mostly regarding the size of the target sequence, is also obtained when filter hybridisation is coupled to nucleic acid fractionation by gel electrophoresis.

Although an impressive number of improved methods and variations has been published, unsatisfactory results are often obtained, especially when high sensitivity is required. It is not our intention to review the many different protocols which have been described and evaluated elsewhere (Hames and Higgins, 1985; Sambrook et al., 1989; Dyson, 1991), nor do we intend to duplicate available information regarding theoretical aspects (Meinkoth and Wahl, 1984; Hames and Higgins, 1985; Wetmur, 1991). Instead, we will describe the method which is currently used in our laboratory and which has proved to be remarkably versatile and suitable for many applications in plant molecular biology including Southern, northern and colony/plaque hybridisations. The method will be described in detail to ensure that maximum reproducibility will be obtained, but it is worth emphasising that, as with any long and laborious procedure, the best results will be achieved when care and good laboratory practice are observed throughout the entire procedure rather than only in steps thought to be critical. This protocol has been optimised for Southern analysis of sequences within large plant genomes, such as that of bread wheat (*Triticum aestivum* L.) (which is approximately five-fold larger than the human genome; Bennett and Smith, 1976) and therefore emphasis has been put on sensitivity. For this reason we prefer to use methods based on isotopic probes and detection procedures. Although highly sensitive non-radioactive protocols and kits are becoming available, there are as yet very few published reports of their use. However, such non-radioactive methods have several advantages including safety for the operator, long-term stability of probes and fast detection of the hybridisation signal, and are certainly worth considering, especially when sensitivity is not the main requirement (i.e. when the abundances of neither of the two nucleic acid strands involved in the formation of the duplex could be limiting). They have been recently reviewed by Mundy et al. (1991).

In the second part of this chapter we give some examples of applications of nucleic acid filter hybridisation, based on our studies of the gene families encoding seed storage proteins in cereals. Although these systems are more complex than single gene systems, most eukaryotic genes do appear to be members of multigene families (Jeffreys, 1981; Beltz et al., 1983). Our results provide examples of the information which can be obtained when filter hybridisation is used for studies of gene characterisation

4. NUCLEIC ACID BLOTTING AND HYBRIDISATION

and organisation, for the estimation of gene copy numbers, for the determination of the chromosomal locations of structural genes, and for the analysis of mutant lines.

II. METHOD

In this section we describe the method for analysis of DNA separated by gel electrophoresis (Southern blotting). Quantities given are for large gels (20 cm × 20 cm) and can be reduced accordingly for smaller gels. Modifications for analysis of RNA will be given subsequently. The hybridisation method can also be used to screen gene libraries transferred to membranes (Fig. 4.1).

A. DNA Blotting

(1) After electrophoresis the gel is photographed with a Polaroid MP4 camera using Polaroid 667 film and a long-wave UV transilluminator. A ruler placed at one edge of the gel (with zero set relative to the wells) indicates the migration distance, while a cut corner serves as a useful reference point. The size-marker lane is cut off, and also the bottom of the gel if the length is more than 20 cm.

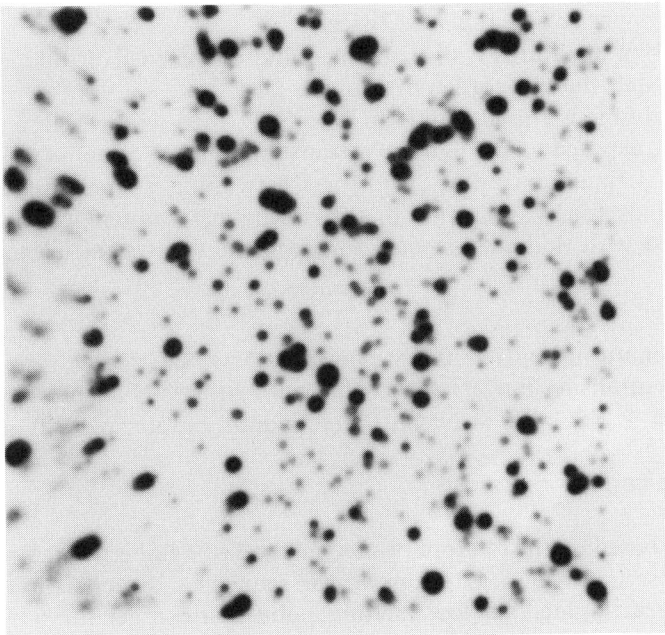

FIG. 4.1. cDNA library screening by plaque hybridisation. A polymerase chain reaction (PCR)-amplified cDNA library derived from maize root cap RNA is hybridised with the corresponding ^{32}P-labelled PCR-amplified cDNA pool from root caps. Hybridising plaques are revealed by autoradiography. The intensities of the signals vary according to the abundance of the individual cDNAs within the fraction used as a probe. The membrane was washed in 0.05 × SSC, 0.1% (w/v) SDS at 65°C for 1 h and exposed for 1 day. The strong signal obtained in a relatively short time using an heterogeneous probe, demonstrates the sensitivity of the method (Sabelli *et al.*, 1993).

(2) The DNA is denatured by incubating the gel twice in denaturing buffer (300 ml) for 15 min each.
(3) After brief washing in distilled water, the gel is incubated twice in neutralising buffer (300 ml) for 15 min each.
(4) Hybond/N (Amersham) nylon transfer membrane is cut 5 mm wider and longer than the gel and equilibrated in 2 × SSC for 10 min. Four sheets of Whatman 3 MM filter paper are cut slightly larger than the membrane and soaked in 2 × SSC.
(5) The blotting apparatus is prepared as follows. A glass plate is placed across a large glass or plastic tray. A sheet of Whatman 3 MM paper is placed on the glass plate so that the ends hang down into the tray. 20 × SSC (500 ml) is poured into the tray and over the 3 MM paper and two of the pre-wetted 3 MM paper sheets placed on top. Any air bubbles are removed by gently rolling a glass pipette between each layer of the transfer. The gel is slid onto the paper sheets on the blotting apparatus and any exposed 3 MM paper surrounding the gel is masked with clingfilm. The nylon membrane is then placed on the gel with the top corresponding to the wells of the gel and the other two sheets of prewetted 3 MM paper placed on top. A 5-10 cm stack of absorbent paper towels is placed on top, followed by a glass plate and a 0.5 kg weight. For a diagrammatic description of the transfer set-up, see Dyson (1991) or Sambrook *et al.* (1989).
(6) The DNA is transferred to the membrane overnight by capillarity, with the wet stack of paper towels replaced twice.
(7) The blotting apparatus is disassembled and one corner of the membrane is cut off to match the cut corner of the gel.
(8) The membrane is rinsed briefly in 2 × SSC, then wrapped in a single layer of clingfilm, exposed to UV light on a transilluminator for about 2 min (DNA-side down) and finally baked at 80°C for 1 h (without a vacuum). The membrane is now ready for hybridisation or for storing at room temperature. The gel can be stained with thidium bromide to check the transfer efficiency.

Solutions

denaturing buffer: 0.5 M NaOH, 1.5 M NaCl
neutralising buffer: 1.5 M NaCl, 0.5 M Tris-HCl, pH 7.5
20 × SSC: 3.0 M NaCl, 0.3 M sodium citrate, pH 7.0

B. Probe Labelling

The protocol is based on the random primed method of Feinberg and Vogelstein (1983, 1984). DNA inserts are purified from vector sequences by restriction endonuclease digestion and fractionated on low melting point agarose gel for use as templates. 100 ng of template DNA is sufficient to hybridise to six membranes; 15 ng is required for one membrane.

(1) The template DNA is denatured by boiling for 5-10 min, then quickly cooled on ice. The labelling reaction is set up as follows:

10 × oligolabelling buffer	2 μl
10 × dNTPs mixture	2 μl

10 × primers	2 μl
bovine serum albumin (5 mg ml^{-1})	1 μl
template DNA	100 ng
[α-^{32}P]dATP (~3000 Ci mmol^{-1})	50 μCi
Klenow fragment (5 U μl^{-1})	1 μl
sterile distilled water to a total of	20 μl

(3) The reaction is incubated at room temperature for 2–6 h.
(4) The labelled insert is separated from the unincorporated nucleotides by gel filtration on Sephadex G-50 (medium) in a Pasteur pipette, eluting with TE buffer. The reaction mixture is loaded onto the dry surface of the column and run into the beads, after which 150 μl aliquots of TE buffer are added. Neutralisation of static electricity by using a zerostat antistatic instrument (Sigma), both on the pipette and the Eppendorf tubes, greatly enhances the reproducibility of the chromatography. About 15 fractions (4 drops per fraction) are collected. This separation, including the column preparation, takes about 20 min.
(5) Aliquots (2 μl) of each fraction are counted in a Beckman LS 1800 scintillation counter.
(6) The labelled DNA elutes in the first peak between fractions 5 and 10. These peak fractions are pooled (volume 300–500 μl). 2 μl of the pooled fractions are counted again and the approximate specific activity calculated (5×10^8–10^9 dpm μg^{-1}).

Example

template = 100 ng
probe after labelling ~200 ng
conversion factor for 1 μg probe = 5
activity of the 2 μl aliquot = 2×10^5 dpm (10^5 dpm μl^{-1})
volume of the pooled fractions = 5×10^2 μl
specific activity: $5 \times 10^5 \times 5 \times 10^2 = 2.5 \times 10^8$ dpm μg^{-1}

This calculation provides only approximate values, but has been shown to give rapid, reliable and reproducible results in our laboratory. More accurate values can be calculated according to Mundy *et al.* (1991).

(7) The probe is ready for denaturation and hybridisation.

Solutions

TE buffer: 10 mM Tris-HCl, 1.0 mM Na$_2$EDTA, pH 8.0
10 × oligolabelling buffer: 0.5 M Tris-HCl, pH 8.0, 0.05 M MgCl$_2$, 0.1 M z-mercaptoethanol
10 × dNTPs mix.: 0.2 mM each of dCTP, dGTP, dTTP
10 × primers: random hexadeoxyribonucleotides, resuspended at 0.4 OD$_{260}$ ml^{-1} (approx. 14 μg ml^{-1})

The concentration of the primer oligonucleotides is about 1/100 of that of the original method, resulting in longer probes (up to 2 kb) and increased hybridisation efficiency (Hodgson and Fisk, 1987; Patrick Gallois, personal communication).

C. Prehybridisation and Hybridisation

The compositions of the prehybridisation and hybridisation buffers are based on those of Boulnois (1987). The procedure is carried out in heat-sealed plastic bags (Hybaid), which is a good system for large membranes. However, where possible, the use of plastic or glass bottles in a rotating oven is preferable.

(1) The filters are incubated for 4–12 h in a water bath at 65°C with gentle shaking in 30 ml per membrane (20 cm × 20 cm) of prehybridisation buffer. Two membranes can be processed in a single plastic bag, separated by a nylon mesh. The buffer is mixed by inversion of the bag and air bubbles removed using a syringe.

(2) The prehybridisation buffer is replaced by the hybridisation buffer which has the same composition except for the denatured probe (obtained by boiling the probe for 5–10 min and cooling on ice). The probe is added at about 5×10^5 dpm ml^{-1} activity. This concentration is critical for high sensitivity and for low background hybridisation. Hybridisation is carried out for 8–24 h at 65°C with gentle shaking in a water bath.

Solutions

> hybridisation buffer: 1.5 × SSPE (from stock 20 × SSPE), 0.5% (w/v) commercial low-fat dried milk, 1% (w/v) sodium dodecyl sulphate (SDS), 6% (w/v) polyethylene glycol (PEG) 6000, 30 µg ml^{-1} denatured salmon sperm DNA.

The solution is prepared by suspending the dried milk, then adding 20 × SSPE, PEG and SDS. The pH is adjusted to 7.7 and the solution filtered through Whatman No. 1 paper. Do not autoclave. This buffer can be prepared in a large batch, divided into aliquots and stored at $-20°C$ for several months. Just before use, the solution is warmed to the required temperature and freshly denatured (by boiling) salmon sperm DNA is added.

Solutions

> 20 × SSPE: 3.6 M NaCl, 0.2 M NaH$_2$PO$_4$, 20 mM Na$_2$EDTA, pH 6.7

D. Washing and Exposure

(1) The hybridisation buffer, with the non-bound probe, is removed and the membranes are briefly rinsed (by inverting the bag) with two washes of 2 × SSC at room temp.
(2) The membranes are transferred from the bag into a plastic box and washed at the required stringency (e.g. 0.5 × SSC, 0.1% (w/v) SDS at 65°C) for 15–30 min, then monitored using a Geiger counter. If the activity exceeds 10 cps, the membranes are washed again.
(3) The membranes, between two layers of clingfilm, are placed in an autoradiography cassette with two intensifying screens and Kodak X-Omat X-ray film and

4. NUCLEIC ACID BLOTTING AND HYBRIDISATION

exposed overnight. Several rounds of washing at increasing stringency followed by exposure usually give more information than a single wash. Films may be developed manually or automatically in a film processor machine. The membranes should never be allowed to dry during the washing/exposure procedure as this can cause irreversible binding of the probe.

E. Probe Removal

(1) Up to six membranes are incubated in a large plastic box with 0.5–1 litre of alkali wash solution (0.4 M NaOH) at 45°C for 30 min with gentle shaking.

(2) The membranes are then neutralised by incubation in 0.2 M Tris-HCl, 0.1 × SSC, 0.1% (w/v) SDS, pH 7.5 at 45°C for 30 min. After a brief rinse in distilled H_2O, the membranes are ready for re-use. Up to 6–8 hybridisation cycles can be carried out with the same membrane without significant loss of signal.

III. COMMENTS ON THE METHOD

A. DNA Transfer and Immobilisation

We will limit discussion to capillary blotting methods based on that of Southern (1975), although other efficient methods have been reported including ones based on the use of alkali (Reed and Mann, 1985), positive and negative pressure (Olszewska and Jones, 1988), electroblotting (Smith *et al.*, 1984; Ishihara and Shikita, 1990), downward capillarity blotting (Lichtenstein *et al.*, 1990, Chomczynski, 1992) and tissue printing (McClure and Guilfoyle, 1989).

A photographic record is needed because, unless radiolabelled size-markers are used, it is necessary to estimate the sizes of the hybridisation products by comparison of their migration distances with those of the stained size-markers. A ruler placed at one edge of the gel indicates the distance of nucleic acid migration. Fluorescent rulers are available from many suppliers, but one can easily be made by sticking yellow adhesive tape to the bottom of a conventional drawing ruler. We prefer not to use ^{32}P- or ^{35}S-labelled size-markers as these often require different exposure times from the hybridisation products, and have relatively short half-lives which means they are of limited value in hybridisation experiments carried out several weeks or months after blotting.

We have often observed that probes hybridise to some bands of the size-marker providing a useful reference. However, some probes may bind strongly to contaminating sequences in the size-markers which are not visualised by ethidium bromide staining, resulting in misleading information regarding the sizes of the hybridisation products (Fig. 4.2). Many markers for the analysis of DNA are commercially available, covering virtually every requirement. Additional markers can easily be prepared by digestion of cloning vectors with suitable endonucleases. We normally use either bacteriophage lambda DNA digested with *Hin*dIII or a commercially available 1 kb ladder (Gibco), the former giving good size markers between 4 and 20 kb and the latter below 4 kb.

We have omitted the depurination step from the protocol. This step is usually carried out by incubating the gel in 0.25 M HCl (15–30 min) at room temperature, resulting

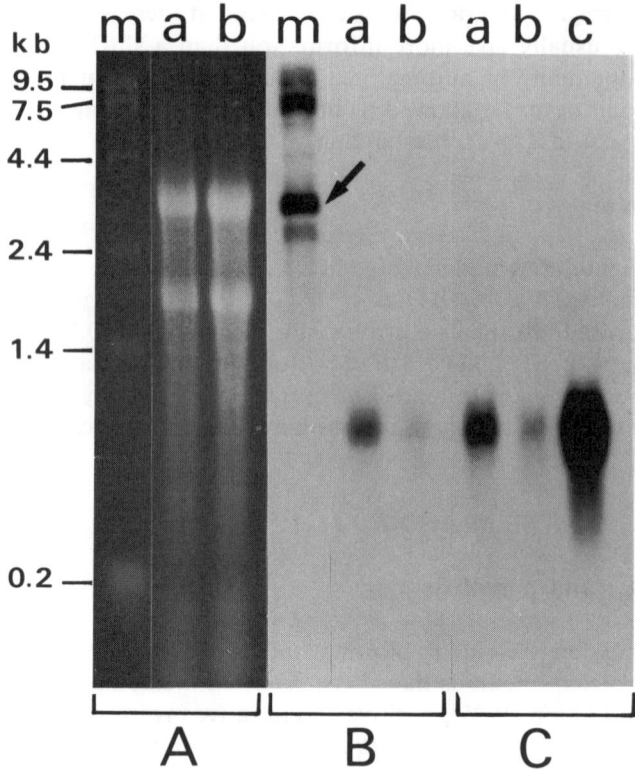

FIG. 4.2. Northern analysis of RNA extracted from barley root tips hybridised with a histone H3 cDNA probe (Chojecki, 1986). Tracks are: m, RNA size-marker; a, total RNA (10 μg); b, poly(A)⁻ RNA (15 μg), c, poly(A)⁺ RNA (1 μg)

Part (A) shows a stained gel and part (B) shows the results obtained when the same gel is blotted and probed. Part (C) shows the results obtained when a replicate gel is blotted and probed without prestaining. Comparison of (B) and (C) shows that staining results in a reduction in the hybridisation signal intensity to about half. In addition, the probe hybridises strongly to at least one contaminating RNA species in the size-marker (arrow). This is not revealed by staining with ethidium bromide, and may be misleading if used as a size reference.

Washing was in 0.5 × SSC, 0.1% (w/v) SDS at 65°C for 30 min, exposure was for 7 h.

in DNA nicking and more efficient transfer of large fragments. However, depurination for times exceeding 15 min should be avoided because excessive DNA nicking can result in degradation of short fragments and the subsequent loss of hybridisation signal. We have observed that up to 15% of the high molecular size DNA (≥15 kb) may not be blotted if depurination is omitted, which can affect the quantitative analysis of hybridisation. We therefore consider that DNA depurination is not required for the blotting of short sequences, but that nicking of DNA (either by HCl or UV irradiation) is required for transferring large DNA fragments, e.g. after pulse field gel electrophoresis (PFGE) (Sørensen, 1989; Röder et al., 1992). It has been shown by Southern (1975) that the efficient transfer of high molecular size DNA is best achieved with high salt concentration buffers (e.g. 20 × SSC), which should therefore be used for blotting genomic DNA. Lower salt concentrations (10 × SSC) can be used for short sequences (e.g. plasmids).

In assembling the blotting apparatus it is important to remove any air bubbles which could be trapped between the layers. Failure to do so may lead to localised areas of inefficient DNA transfer which can be identified on the autoradiograph as circular regions lacking any hybridisation signal (even background).

Irreversible binding of denatured DNA to nitrocellulose membranes can be performed by baking at 80°C for 2 h under vacuum. However, nylon membranes are preferable, especially when several rounds of hybridisation are to be carried out, due to their greater physical strength. In this case cross-linking of DNA by longwave UV light (e.g. 305 nm wavelength) for 2-5 min may be sufficient, but we prefer to additionally bake membranes for 1 h.

B. Probe Labelling

It is important to excise the insert probe from the vector as the latter may cross-hybridise to the target DNA. In our hands, the recovery of inserts from low melting point agarose is the most practical method. Electroelution in dialysis membranes or onto DE81 paper (Whatman) may give higher sample purity but lower yields. Many manufacturers supply prepacked resin columns for one-step purification of DNA from melted agarose. It is also possible to carry out random primed labelling in the melted agarose slice, but we have not tested this. Although several different labelling methods are available (Mundy *et al.*, 1991), random primed labelling (or oligolabelling), developed by Feinberg and Vogelstein (1983), is one of the most popular. As mentioned above, a significant improvement results from reducing the primer concentration, leading to a higher average probe length.

It is important to remove unincorporated radiolabelled nucleotides from the probe mixture, as these could be responsible for high levels of background due to the sensitivity of the hybridisation conditions. This can be accomplished by column chromatography. Although a range of push-, spin- and gravity-columns are commercially available, gel filtration on Sephadex G-50 as described above has proved to be convenient, highly reproducible and cheap, though perhaps slightly slower than other methods. Labelling kits are available from several manufacturers, the latest versions utilising T7 DNA polymerase instead of the Klenow fragment which reduces labelling times to a matter of minutes. Utilisation of such labelling kits considerably simplifies the labelling procedure, is relatively inexpensive, and is recommended.

C. Hybridisation

During the prehybridisation and hybridisation steps it is necessary to prevent the probe binding non-specifically to the membrane and to the immobilised DNA. The use of dried milk as a blocking agent has been described by Johnson *et al.* (1984); in addition we routinely use 30 μg ml^{-1} denatured salmon sperm DNA (higher concentrations up to 100 μg ml^{-1} seem to have no significant effect on the level of background). Another notable feature of the hybridisation buffer is the inclusion of PEG 6000 (Amasino, 1986). High molecular size polymers such as PEG or dextran sulphate occupy most of the buffer volume and, as a result, significantly increase the actual probe concentration. This means that a high probe concentration can be obtained even in relatively large hybridisation volumes (which simplifies the practical aspects of the procedure) and with remarkably small amounts of template DNA per membrane. As little as 1/10 of the normally recommended concentrations can be used, with considerable savings of probe and reagents.

It is essential that the hybridisation conditions are carefully controlled. The best results are obtained using probes which are prepared from purified inserts and are free of unincorporated radioactive nucleotides. The highest signal-to-noise ratios are obtained with a probe concentration corresponding to about 5×10^5 dpm ml^{-1} activity. This can represent as little as 0.5 ng ml^{-1} probe concentration. Activity above 10^6 dpm ml^{-1} may result in virtually irreversible background labelling. Localised irreversible background labelling may also be obtained if air bubbles are left in the bags. Removal of air bubbles can be difficult due to the SDS in the buffer, but the process can be assisted by adding antifoam-A (Sigma) to the hybridisation buffer.

Although the protocol describes the use of heat-sealed plastic hybridisation bags (Hybaid), these are fairly expensive and difficult to re-use. Ordinary plastic bags or specially designed bottles can also be used. Ordinary plastic bags are cheap but squeezing out bubbles and sealing are very laborious, and contamination of both the external bag surface and the bench is likely. Instead, hybridisation bottles have important advantages such as ease of adding and pouring off buffers, protection from radiation and tolerance of air bubbles which are continuously moved around during incubation in a suitable rotating oven. The main disadvantage is that it can be difficult to process large membranes because overlapping can cause severe background problems. Thin nylon meshes are available commercially, but these sometimes do not retain the membranes in the position in which they are loaded. Better results have been obtained by utilising the thicker meshes designed for hybridisation in bags, but this limits the number of membranes to three to four per bottle. This can be considered a minor disadvantage and, wherever possible, we recommend performing the hybridisation in bottles.

D. Posthybridisation

Since in most cases it is not possible to predict the degree of similarity between target and probe sequences, it is advisable to carry out several rounds of washing and autoradiography using increasing levels of stringency. The hybridisation conditions described are of low stringency (about T_m −30 for an average probe; see Section III.F), and homoduplexes can be expected to be relatively stable even at the melting temperature. High performance hybridisation experiments allow multiple washing and exposure of the hybridised membrane, which, as will be discussed later, can provide very useful information.

For routine experiments we use Fuji RX X-ray films or equivalent, whereas for higher sensitivity we prefer Kodak X-Omat AR5 films.

E. Modification for Northern Analysis

RNA can be easily degraded by RNases and solutions should be treated with the RNase inhibitor diethylpyrocarbonate (DEPC) (Kumar and Lindberg, 1972). Even the hybridisation buffer should be treated with DEPC since dried milk is an excellent source of RNases (Siegel and Bresnick, 1986). Care should be taken because DEPC is toxic and cannot be removed from the hybridisation buffer as this cannot be autoclaved. Although northern blotting follows the same principles which govern Southern blotting, minor modifications to the procedure are required (only RNA fractionated using formaldehyde gel electrophoresis will be discussed):

(1) Gel staining: ethidium bromide stains single-stranded nucleic acids less efficiently than double-stranded. In addition, the hybridisation signal from ethidium bromide-stained RNA is only about half of that from unstained RNA (Fig. 4.2). It is therefore advisable not to stain RNA gels before hybridisation, although the RNA size-marker can be cut off and stained separately. Alternatively, especially when the quality of the RNA extraction has to be assessed, replicate samples can be loaded onto the gel and stained separately. Staining can be accomplished by incubating the gel in a highly concentrated solution of ethidium bromide (5 μg ml^{-1}) for 2-3 min. Longer incubation will result in excessive staining of the whole gel. The gel can then be destained for about 1 h with at least two changes of distilled water (this also removes formaldehyde from the gel which may interfere with the hybridisation).
(2) Transfer: RNA blotting does not require any depurination, denaturation and neutralisation treatments since the RNA molecules are already relatively short and single-stranded. Therefore a lower concentration of salts can also be used (typically 10 × SSC).
(3) Probe preparation, prehybridisation and hybridisation are carried out as for hybridisation of DNA, but the probe concentration may be further reduced due to the superior stability of DNA-RNA duplexes.
(4) It is not possible to strip the probe off the membrane by incubation with alkali as this hydrolyses RNA. As an alternative procedure the membrane can be incubated in 5 mM Tris-HCl, 2 mM Na$_2$EDTA, pH 8.0 at 65°C for 1 h.

F. T_m and Stringency

The factors which affect hybridisation can be divided into those relevant to the rate of hybridisation and those which affect the stability of the duplex nucleic acid (for comprehensive reviews see Hames and Higgins, 1985 and Dyson, 1991). While for a full understanding of the hybridisation process the influence of each factor should be considered, only a few major factors play a critical role in the majority of hybridisation experiments. These include factors which affect the thermal stability of nucleic acid duplexes.

The melting temperature (T_m) is the temperature at which the hybrid is half denatured (under the given conditions). For perfectly matching DNA hybrids the T_m depends mostly on the ionic strength of the buffer, the base composition and the presence of helix-denaturing agents (e.g. formamide). It can be estimated by the following equation based on that of Schildkraut and Lifson (1965):

$$T_m(°C) = 81.5 + 16.6 \, (\log_{10} M) + 0.41 \, (\% \, G+C) - 0.72 \, (\% \, f) - 500/n$$

where M stands for the molarity of monovalent cations (e.g. Na$^+$), (% G+C) is the percentage of guanosine and cytosine nucleotides, (% f) is the percentage of formamide and n is the length of the duplex.

The above equation was derived from studies of hybridisation in solution, and therefore provides only an approximate estimate for filter hybridisation experiments. It clearly indicates that the thermal stability of nucleic acid hybrids increases with the salt concentration, the percentage of nucleotides forming triple hydrogen bonds in the hybrid, and with the length of the duplex.

In most experiments base composition has little effect on the T_m but in the case of short hybrid molecules ($n \leq 50$), such as for oligonucleotide hybridisation, it may become a major parameter and the T_m is more conveniently estimated as follows:

$$T_m(°C) = 4(G+C) + 2(A+T)$$

with a standard 0.9 M NaCl concentration.

Hybridisations can be carried out at lower temperatures with the use of helix-denaturing agents such as formamide; however, we prefer not to use this agent because it is toxic and we have not detected any improvement by using it.

The effect of mismatches has been estimated as a decrease in T_m of approximately 1°C for each 1% of mismatch (Bonner et al., 1973). However, this should be taken only as an average value as the mismatch distribution can have a significant influence.

The concept of T_m leads to the concept of stringency. In hybridisation experiments stringency is inversely related to the criterion value which is given by C (°C) = $T_m - T_i$, where C is the criterion and T_i is the experimental temperature. When T_i is low, C is high and the stringency is low; as T_i approaches T_m, C decreases and the stringency increases. As a general rule, if the experimental goal is the detection of sequences identical or closely related to the probe used, high stringency conditions should be used (e.g. $5°C \leq C \leq 10°C$), whereas stable hybridisation between more distantly related sequences can be accomplished under low stringency conditions.

Stringency can be changed most conveniently by altering the parameters of temperature and salt concentration. For most applications, and in order to achieve maximum reproducibility, we carry out hybridisation under relatively low stringency conditions and increase the stringency in the washing steps simply by reducing the salt concentration (see Section IV.C).

The following is an example of how to estimate T_m and stringency conditions for pKAPla, a γ-gliadin cDNA which will be referred to later in this paper.

pKAPla size = 1049 bp
C + G content = 49%
T_i = 65°C
Na$^+$ concentration:
   hybridisation (1.5 × SSPE, 1% SDS) = 0.32 M
   low-stringency washing (1.0 × SSC, 0.1% SDS) = 0.2 M
   high-stringency washing (0.2 × SSC, 0.1% SDS) = 0.04 M
hybridisation: $T_m = 81.5 + 16.6(-0.49) + 0.41(49) - 500/1049 = 94°C$
        $C = T_m - T_i = 94°C - 65°C = 29°C$
low-stringency washing: $T_m = 90°C$
        $C = 25$
high-stringency washing: $T_m = 79°C$
        $C = 14°C$

4. NUCLEIC ACID BLOTTING AND HYBRIDISATION

IV. APPLICATIONS

A. Gene Copy Number

Once a fragment of DNA (gene, cDNA clone or interspersed sequence) has been isolated, and possibly sequenced, it becomes necessary to obtain information about its position in the genome and whether it is a single-copy gene or a member of a multigene family. Although the number of copies can be estimated by hybridisation in solution (e.g. measuring the kinetics of DNA annealing) or filter hybridisation to saturation, the most common approach is based on filter hybridisation to genomic DNA. In this the intensities of the hybridisation signals of known amounts of the DNA fragment under investigation are compared with that of genomic DNA restricted with one or, preferably, several endonucleases (in independent digestions). It is important to digest the genomic DNA with enzymes which do not cut within the segment of the DNA of interest, otherwise the data obtained may be difficult to interpret.

Precise determinations of the concentrations of the genomic DNA and the reference DNA are crucial. We have found that spectrophotometric determination of UV absorbance may not be reliable, even when applied to nucleic acids purified on a CsCl gradient. We therefore confirm the results obtained by absorbance measurements by comparing the ethidium bromide staining intensities of the samples with those of known amounts of standard nucleic acids electrophoresed on an agarose gel (for comparison with genomic DNA we use bacteriophage lambda DNA on a 0.4% (w/v) agarose gel, while plasmids can be used for comparison with shorter fragments of DNA on a 1.5% (w/v) gel).

To provide a reference of 1 copy per haploid genome for copy number reconstructions, the amount of standard DNA loaded onto the gel for blotting must be equal to the amount in moles of haploid genomes loaded. Multiples of this standard reference DNA give a means of calibration. It is advisable to use dilutions of the linearised plasmid containing the sequence of interest as standards, to minimise differences in transfer efficiency between relatively large genomic sequences (1–20 kb) and cloned sequences (0.5–2 kb for cDNAs). It is necessary, however, to check that probe and vector sequences do not cross-hybridise under the chosen stringency conditions, either using filter hybridisation or dot-matrix comparison analyses (Staden, 1982).

If the hybridisation experiment shows only a single band with a range of restriction enzymes, it is likely that only one copy of the sequence is present per haploid genome. The intensity of this band should, however, be compared with the intensities of reference amounts of the cloned sequence used as probe, as multiple copies of DNA sequences with identical restriction maps may be present in the genome resulting in co-migration of fragments. Figure 4.3 shows a copy-number reconstruction experiment in which the intensity of hybridisation of hexaploid bread wheat (*Triticum aestivum* L.) cv. Chinese Spring DNA digested with three different endonucleases is compared with that of a linearised plasmid (containing the insert sequence) at different dilutions corresponding to 0.5 and 10 copies per haploid genome. Although considerable information can be obtained about the size and organisation of a given gene family,

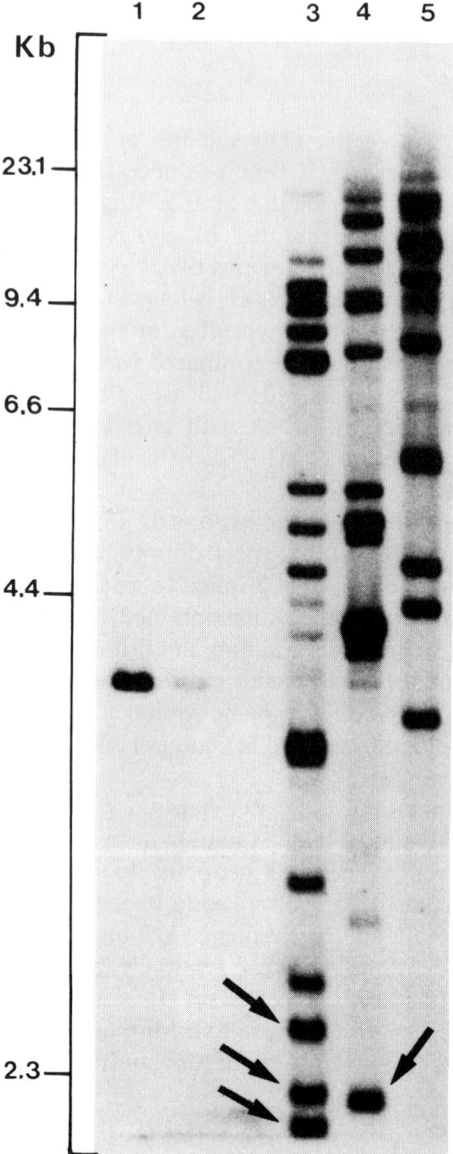

FIG. 4.3. Copy number reconstruction by Southern blotting. Genomic DNA from bread wheat cv. Chinese Spring was digested with *Hin*dIII, *Eco*RI and *Bam*HI (lanes 3, 4 and 5, respectively) and hybridised with pKAP1a, a γ-gliadin cDNA (Sabelli and Shewry, 1991). Amounts of linearised plasmid (containing the insert probe) corresponding to 10 and 0.5 copies per haploid genome (lanes 1 and 2, respectively) provide calibration standards. Arrows indicate restriction fragments which give strong hybridisation signals which contrast with their relatively small sizes (see text for discussion).

it should be emphasised that such analyses only give estimates when applied to multigene families. Variation in the degree of homology between target and probe and multiple hybridisation of probes to repetitive sequences may affect the signal intensity. In addition, these experiments do not discriminate between active and inactive genes (pseudogenes), unless inactivation is due to major mutations which also affect the hybridisation performance of the target sequences. Uncertainty may still remain. An example is given in Fig. 4.3 where the number of gene copies ($\sim$ 10), calculated for the smallest fragments (indicated by the arrows), is inconsistent with the sizes of the fragments ($\sim$ 2 kb), and the cDNA probe ($\sim$ 1 kb). A single copy of this fragment could clearly not contain 10 gene copies. This result could be due to the presence of multiple identical fragments, or to enhanced hybridisation signal due to repetitive sequences, or to both. Indeed, experiments carried out at higher stringency have shown that repetitive regions are probably responsible for the strong signals observed in Fig. 4.3 (Sabelli and Shewry, 1991).

The following is an example of how to calculate the amount of standard DNA equivalent to 1 copy per haploid genome complement (1C value) of hexaploid bread wheat (cv. Chinese Spring).

$$1C \text{ DNA content} = 17.3 \text{ pg}$$
$$8 \ \mu g \text{ DNA per lane} = 4.62 \times 10^5 \text{ moles of haploid genome}$$
$$1 \text{ bp} = 1.036 \times 10^{-9} \text{ pg mol}^{-1}$$
$$\text{plasmid size} = 3.714 \times 10^3 \text{ bp}$$
$$\text{plasmid amount corresponding to 1 copy per haploid genome}$$
$$= 4.62 \times 10^5 \times 1.036 \times 10^{-9} \times 3.714 \times 10^3$$
$$= 1.77 \text{ pg}$$

Compilations of the genome sizes of a range of wild and cultivated plant species have been reported (Bennett and Smith, 1976; Bennett et al., 1982).

B. Gene Localistion

Mapping the position(s) of cloned sequences in the genome is an important part in studies of gene characterisation. Ideally, the position in the genome should be mapped as precisely as possible.

1. Molecular location

The sequence can be mapped to a specific restriction fragment. However, in this case restriction endonucleases which cut within the cloned sequence should also be used to identify unequivocally the homologous fragment in the genome. By isolating the corresponding genomic clone the map can be extended to the level of nucleotide sequences. Such molecular maps can then be extended by means of chromosome walking. A major area of research which has been made possible only by the development of Southern blotting is the construction of genetic maps based on restriction fragment length polymorphisms (RFLPs). This procedure uses molecular probes to identify variation in the lengths of restriction fragments produced by digestion with various restriction enzymes. These are used as genetic markers for the

2. Chromosomal location

Although chromosomal location can be assigned by direct *in situ* hybridisation to metaphase chromosomes (Huang *et al.*, 1988), plants often have a large number of small chromosomes which are not readily identifiable. In addition, the location of single or low copy number sequences by *in situ* hybridisation is technically very difficult. These problems can be overcome by filter hybridisation of DNA from cytologically characterised genetic stocks, providing such stocks are available. This approach is based on comparing the hybridisation patterns of euploid and aneuploid lines as in the analysis of mammalian cell hybrids (Shows *et al.*, 1982), although the procedure used to generate the aneuploids is considerably different. Any DNA sequence can be localised to a specific chromosome (or a chromosomal region) by correlating the presence of hybridising restriction fragments with a particular chromosome or chromosomal fragment. An example the chromosomal locations of sequences related to γ-gliadin seed storage proteins, both in bread and durum wheats, is shown in Fig. 4.4. Comparison of the hybridisation pattern of the euploid (CS) line with those of nullisomic-tetrasomic lines (Sears, 1954; Harberd *et al.*, 1985), which lack one pair of homoeologous group 1 chromosomes but have four copies of another, allows the location of any *Eco*RI fragments to specific group 1 chromosomes. Analysis of ditelosomic lines ($1A^s$, $1A^l$, $1B^s$, $1B^l$) which lack specific short or long arms of chromosomes 1A or 1B allows the localisation of the same restriction fragments to specific chromosome arms. Similarly, substitution of chromosome 1D from Chinese Spring for chromosomes 1A or 1B of Langdon (a durum wheat) allows the chromosomal locations of most fragments in durum wheat to be determined. However, in some cases it is not possible to determine unequivocally the locations of Langdon fragments due to the masking effects of fragments from chromosome 1D of Chinese Spring (not shown). The left part of Fig. 4.4 shows finer band resolution, especially in the high molecular size range, following long gel separations of about 50 h (Sabelli and Shewry, 1991). This allows some apparently single bands present on normal gels (run for about 12 h) to be resolved as several bands with the same or different chromosomal locations. This simple modification, coupled with high sensitivity hybridisation experiments, allowed us to resolve approximately twice as many restriction fragments as in previously reported analyses (Bartels *et al.*, 1986) and we recommend the use of long-gel fractionation for the analysis of complex hybridisation patterns.

C. Identification of Members of Multigene Families

Multigene families often contain members which show different degrees of homology. Once one member has been cloned, or when related heterologous probes are available, it is possible to identify other related sequences within the family by Southern or northern analyses or even to identify individual clones by library screening. We have shown that it is possible to recognise subsets of sequences within such gene families by varying the stringency conditions, or by utilising specific probe fragments in hybridisation experiments.

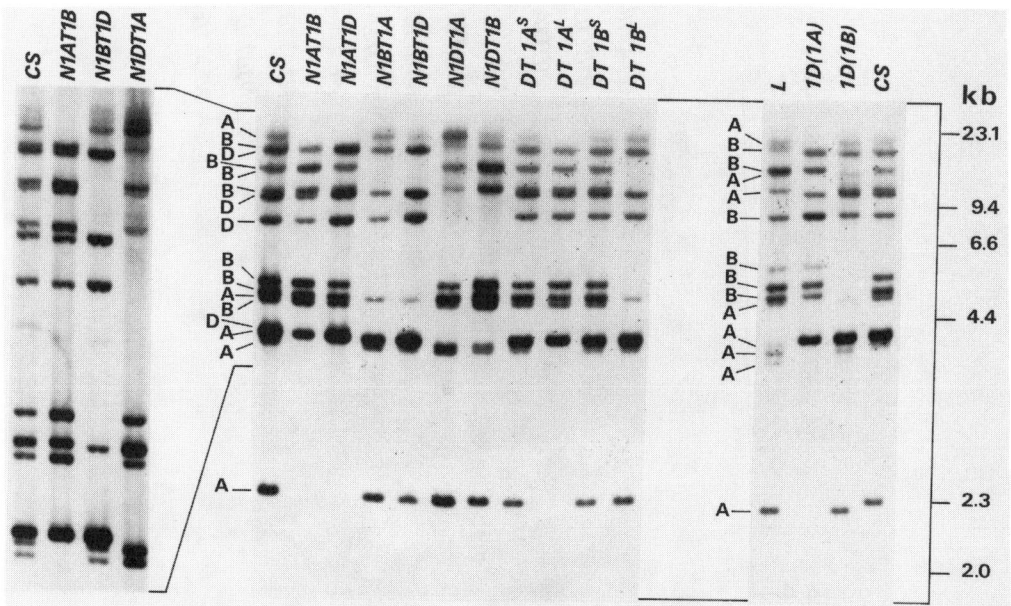

FIG. 4.4. Mapping sequences to specific chromosomes and chromosome arms by Southern blotting. The centre part compares the hybridisation patterns of *Eco*RI digests of DNA from euploid Chinese Spring, aneuploid nullisomic-tetrasomic lines and ditelosomic lines with a γ-gliadin cDNA (pKAP1a). This allows the location of hybridising sequences to the homoeologous group 1 chromosomes (A, B and D).

The left part shows the results obtained when the gel is run for a longer time (about 50 h) to increase the resolution in the high molecular size range.

The right part shows similar analyses of euploid durum wheat cv. Langdon and of lines with chromosomes 1D of Chinese Spring substituted for chromosomes 1A or 1B of Langdon.

Washing was in 0.5 × SSC, 0.1% (w/v) SDS at 65°C for 30 min, exposure was for 2 days.

1. Varying the stringency

We usually carry out the hybridisation and a first round of washing at relatively low stringency (criterion values ranging from 25–35°C depending on the probe). Under such conditions, it is possible to detect relatively distant members of gene families (which may show up to 30–40% mismatches in the duplex). It is then possible to carry out several rounds of washing at increased stringency followed by autoradiography. Although it has been suggested that hybridisation at low stringency followed by higher stringency washing may give slightly different results than when both hybridisation and washing are carried out under high stringency conditions (Beltz *et al.*, 1983), we have found the former strategy to be very convenient, especially if a large number of samples are to be processed. Although several factors affect the hybridisation stringency, we find that varying the salt concentration is simple and gives highly reproducible results. Figure 4.5 shows a Southern analysis of *Bam*HI digests of DNAs from euploid Chinese Spring and aneuploid nulli-tetrasomic lines probed with a γ-gliadin cDNA and washed at 'moderate' stringency (m) followed by 'high' stringency (h). The bands indicated by arrows hybridise weakly or not at all under high stringency conditions, indicating that they have lower sequence similarity to the probe used

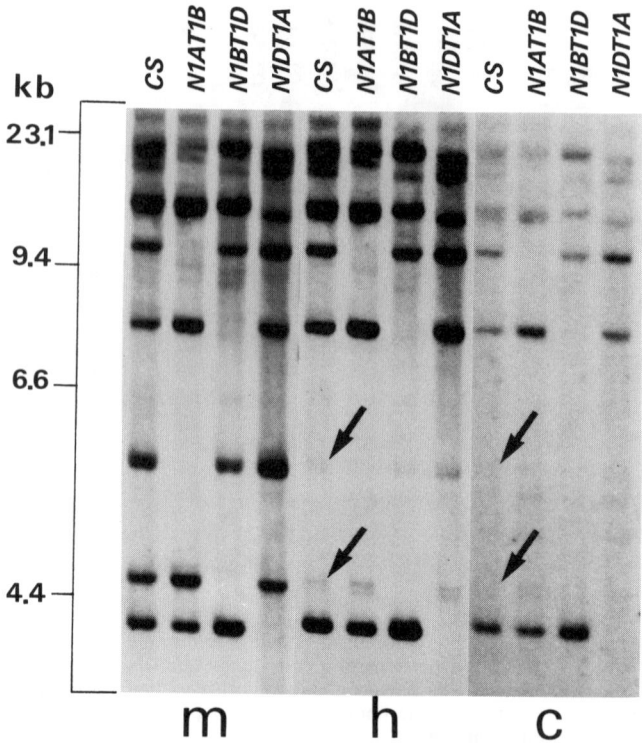

FIG. 4.5. Identification by Southern blotting of a subset of sequences within a multigene family, characterised by a low degree of homology with the probe used. *Bam*HI digests of euploid (CS) and aneuploid nullisomic-tetrasomic lines of bread wheat cv. Chinese Spring were hybridised to pKAP1a, a γ-gliadin cDNA, then washed under conditions of moderate (m) and high (h) stringency (criterion values ~20°C and ~14°C, respectively). The high stringency conditions were obtained simply by reducing the salt concentration in the washing buffer from 0.5 × to 0.2 × SSC. Arrows in (h) indicate the positions of fragments which hybridise under conditions of moderate but not high stringency. The same fragments also fail to hybridise with the non-repetitive part of the probe (pKAP1a-3′) under conditions of medium stringency (arrows in c), suggesting that they encode ω-gliadins which consist almost entirely of repetitive sequences (see Fig. 4.6).

Washing was in 0.5 × or 0.2 × SSC, 0.1% (w/v) SDS at 65°C for 30 min, exposure was for 2 days.

(Sabelli and Shewry, 1991). The same approach could also be used to isolate corresponding clones from a genomic or cDNA library.

2. Using specific fragments of the probe

Most multigene families have evolved from a single ancestral gene by mechanisms of gene duplication, inversion, translocation, etc. As a result, members of the family contain regions in which nucleotide sequences are highly conserved and regions which are more diverged or even unique. If enough sequence information is available, at the DNA and/or protein levels, it is possible to design a subset of probes which can specifically detect certain members of the gene family and not others. Figure 4.5 shows how this approach can be applied to the identification of ω-gliadin sequences within

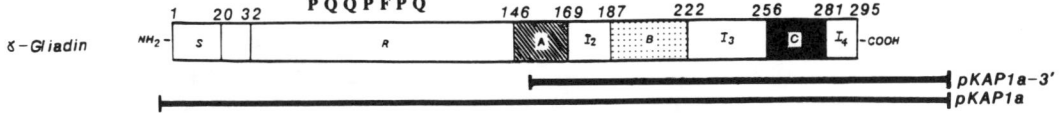

FIG. 4.6. Summary of the domain structure of the γ-gliadin encoded by pKAP1a, showing the subclone pKAP1a-3′ encoding only the non-repetitive C-terminal part of the protein. S indicates the signal peptide, and A, B, and C three regions which are conserved in related seed proteins. These are flanked and separated by three divergent regions, labelled I_2, I_3 and I_4. The repetitive domain is based on the consensus motif PQQPFPQ, where P is proline, Q is glutamine and F is phenylalanine.

the hybridisation patterns obtained with a γ-gliadin probe. The rationale lies in the fact that the γ-gliadin protein consists of a N-terminal repetitive domain (based on the heptapeptide motif PQQPFPQ) and a C-terminal non-repetitive domain, while the ω-gliadins probably consist almost entirely of repeats (consensus motif PQQPFPQQ; Fig. 4.6). The sequence similarity in the repeat motifs present in the two proteins means that the whole γ-gliadin probe cross-hybridises to ω-gliadin restriction fragments. In contrast, if a subclone encoding the C-terminal non-repetitive domain is used as a probe, only true γ-gliadin sequences should be detected (Fig. 4.6). Indeed, as shown in Fig. 4.5, several genomic fragments do fail to hybridise to this probe (pKAP1a-3′) and may, therefore, contain sequences encoding ω-gliadins. Analyses under high stringency conditions, heterologous probing and characterisation of mutants have largely confirmed this suggestion (Sabelli and Shewry, 1991; Sabelli et al., 1992).

Our studies of cereal seed proteins have been facilitated by the large amount of sequence information which has been obtained over the last decade. A more general criterion which can be used to distinguish individual members within a family or a superfamily of genes is based on utilising 5′ or 3′ untranslated sequences, as these usually show a high degree of heterogeneity.

Alternatively, oligonucleotide probes can be used (Miyada and Wallace, 1987; Henderson et al., 1991), but efficient hybridisation to specific sequences within large genomes can be very difficult.

D. Analysis of Mutants

Southern and northern analyses of mutants can provide insights into the organisation of genes and gene families, as well as information on the nature of the mutants themselves (Sabelli et al., 1992). In order to be detected, the mutation should affect the hybridisation performance of the duplex or result in size polymorphisms in the gel-fractionated nucleic acid fragments. Some mutations (such as point mutations, or the deletion, inversion, duplication or translocation of short sequences) may cause dramatic phenotypic changes, but are unlikely to be detected by standard nucleic acid hybridisation procedures. They may, however, be detected using high resolution separations on polyacrylamide gels (Préat, 1990). In other cases, even single-base mutations can be detected, for example the generation of in-frame stop codons which can result in a smaller mRNA in a northern blot.

We have studied mutant lines of bread and durum wheats which have been identified by the absence of specific ω- and γ- gliadin proteins (Lafiandra et al., 1987). These mutant phenotypes could be due to either gene deletion or repression of gene

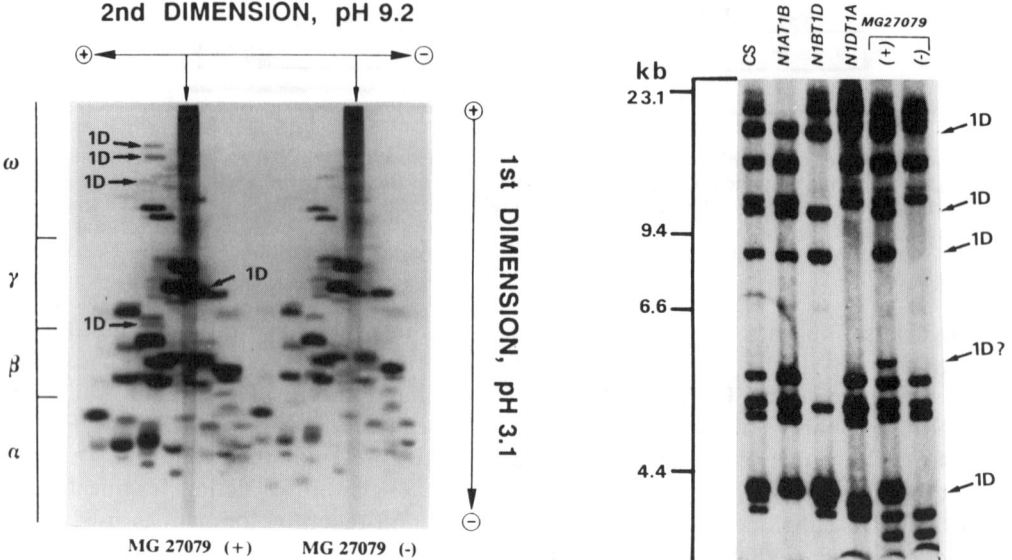

FIG. 4.7. Characterisation of a mutant form of bread wheat with a deletion at the *Gli-D1* locus on chromosome 1D.

Left: Two-dimensional (two-pH) gel electrophoresis of gliadins of bread wheat accession MG27079. 'Normal' (+) and mutant (−) forms are separated on the same gel. The 1D-encoded components absent in the mutant line are indicated.

Right: Southern blot analysis of *Eco*RI digested genomic DNAs from cv. Chinese Spring (CS), the group 1 nulli-tetrasomic aneuploids and accession MG27079 (+ and − forms). The DNAs are probed with PKAP1a, a γ-gliadin cDNA that also hybridises to ω-gliadin genes. Most fragments missing in the mutant line correspond to 1D-encoded fragments of Chinese Spring and to the missing protein components. Deletion events appear to be responsible for the mutant phenotype. Taken from IACR Annual Report for 1991, and based on Sabelli *et al.* (1992).

expression. Southern analyses have shown that partial but substantial deletions of the corresponding loci have been responsible for the lack of specific proteins, as shown in Fig. 4.7 (Sabelli *et al.*, 1992). Analyses of these mutants also allowed genes encoding specific proteins to be mapped onto specific restriction fragments.

V. SUMMARY

The wide range of applications of nucleic acid blotting and hybridisation in studies of plant, animal and microbial systems has resulted in a confusing number of different methods and protocols. In the present chapter we have tried to simplify the choice for plant molecular biologists by describing a single method which can be used, with only minor modifications, for Southern, northern and colony/plaque hybridisations. We have also tried to identify the major parameters which can affect the quality of the results, and have discussed how these can be optimised. Finally, we have discussed the applications of blotting and hybridisation to plant systems using as examples our own studies of wheat seed storage proteins. We hope that the information will be of interest to experienced plant molecular biologists, as well as acting as an encouragement to other plant scientists who wish to start using nucleic acid blotting and hybridisation in their own studies.

ACKNOWLEDGEMENTS

We thank Shirley Burgess for critically reading the manuscript.

REFERENCES

Amasino, R. M. (1986). *Anal. Biochem.* **152**, 304-307.
Bartels, D., Altosaar, I., Harberd, N. P., Baker, R. F. and Thompson, R. D. (1986). *Theor. Appl. Genet.* **72**, 845-853.
Beltz, G. A., Jacobs, K. A., Eickbush, T. H., Cherbas, P. T. and Kafatos, F. C. (1983). *Methods Enzymol.* **100**, 266-285.
Bennett, M. D. and Smith, J. B. (1976). *Proc. R. Soc. Lond. Ser. B.* **274**, 227-274.
Bennett, M. D., Smith, J. B. and Heslop-Harrison J. S. (1982). *Proc. R. Soc. Lond. Ser. B.* **216**, 179-199.
Bonner, T. I., Brenner, D. J., Neufield, B. R. and Britten, R. J. (1973). *J. Mol. Biol.* **81**, 123-135.
Boulnois, G. J. (1987). "University of Leicester Gene Cloning and Analysis: A Laboratory Guide". Blackwell Scientific, Oxford.
Chojecki, J. (1986). *Carlsberg Res. Commun.* **51**, 211-217.
Chowczynski, P. (1992). *Anal. Biochem.* **201**, 134-139.
Dyson, N. J. (1991). In "Essential Molecular Biology. A Practical Approach", Vol. II (T. A. Brown, ed.), pp. 111-156. IRL Press, Oxford.
Feinberg, A. P. and Vogelstein, B. (1983). *Anal. Biochem.* **132**, 6-13.
Feinberg, A. P. and Vogelstein, B. (1984). *Anal. Biochem.* **137**, 266-267.
Hames, B. D. and Higgins, S. J. (1985). "Nucleic Acid Hybridisation. A Practical Approach". IRL Press, Oxford.
Harberd, N. P., Bartels, D. and Thompson, R. D. (1985). *Mol. Gen. Genet.* **198**, 234-242.
Henderson, G. S., Conary, J. T., Davidson, J. M., Stewart, S. J., House, F. S. and McCurley, T. L. (1991). *Biotechniques* **10**, 190-197.
Hodgson, C. P. and Fisk, R. Z. (1987). *Nucl. Acids Res.* **15**, 6295.
Huang, P.-L., Hahlbrok, K. and Somssich, I. E. (1988). *Mol. Gen. Genet.* **211**, 143-147.
Ishihara, H. and Shikita, M. (1990). *Anal. Biochem.* **184**, 207-212.
Jeffreys, A. J. (1981). *In* "Genetic Engineering", vol. 2 (R. Williamson, ed.), pp. 1-48. Academic Press, London.
Johnson, D. A., Gautsch, J. W., Sportsman, J. R. and Elder, J. H. (1984). *Gene Anal. Techn.* **1**, 3-8.
Kumar, A. and Lindberg, U. (1972). *Proc. Natl. Acad. Sci. USA* **69**, 681-685.
Lafiandra, D., Colaprico, G., Kasarda, D. D. and Porceddu, E. (1987). *Theor. Appl. Genet.* **74**, 610-616.
Lichtenstein, A. V., Moiseev, V. L. and Zaboikin, M. M. (1990). *Anal. Biochem.* **191**, 187-191.
McClure, B. A. and Guilfoyle, T. J. (1989). *Plant Mol. Biol.* **12**, 517-524.
Meinkoth, J. and Wahl, G. (1984). *Anal. Biochem.* **138**, 267-284.
Miyada, C. G. and Wallace, R. B. (1987). *Methods Enzymol.* **154**, 94-107.
Mundy, C. R., Cunningham, M. W. and Read, C. A. (1991). In "Essential Molecular Biology. A Practical Approach", Vol. II. (T. A. Brown, ed.), pp. 57-110. IRL Press, Oxford.
Olszewska, E. and Jones, K. (1988). *Trends Genet.* **4**, 92-94.
Préat, T. (1990). *Nucl. Acids Res.* **18**, 1073.
Reed, K. C. and Mann, D. A. (1985). *Nucl. Acids Res.* **13**, 7207-7221.
Röder, M. S., Sorrells, M. E. and Tanksley, S. D. (1992). *Mol. Gen. Genet.* **232**, 215-220.

Sabelli, P. A. and Shewry, P. R. (1991). *Theor. Appl. Genet.* **83**, 209-216.
Sabelli, P. A., Lafiandra, D. and Shewry, P. R. (1992). *Theor. Appl. Genet.* **83**, 428-434.
Sabelli, P. A., Burgess, S. R., Carbajosa, J. V., Parker, J. S., Halford, N. G., Shewry, P. R. and Barlow, P. W. (1993). In "Molecular and Cell Biology of the Plant Cell Cycle" (J. C. Ormrod and D. Francis, eds), pp. 97-109. Kluwer Academic Publishers, The Netherlands.
Sambrook, J., Fritsch, E. F. and Maniatis, T. (1989). "Molecular Cloning. A Laboratory Manual". Cold Spring Harbor Laboratory Press, Cold Spring Harbor, NY.
Schildkraut, C. and Lifson, S. (1965). *Biopolymers* **3**, 195-208.
Sears, E. R. (1954). "The Aneuploids of Common Wheat", pp. 1-59. University of Missouri College of Agriculture Agricultural Experiment Station, Columbia, MO.
Shows, T. B., Sakaguchi, A. Y. and Naylor, S. L. (1982). *Adv. Hum. Genet.* **12**, 341-452.
Siegel, L. I. and Brosnick, E. (1986). *Anal. Biochem.* **159**, 82-87.
Smith, M. R., Devine, C. S., Cohn, S. M. and Lieberman, M. W. (1984). *Anal. Biochem.* **137**, 120-124.
Sørensen, M. B. (1989). *Carlsberg Res. Commun.* **54**, 109-120.
Southern, E. M. (1975). *J. Mol. Biol.* **98**, 503-517.
Staden, R. (1982). *Nucl. Acids Res.* **10**, 2951-2961.
Wetmur, J. G. (1991). *Crit. Rev. Biochem. Mol. Biol.* **26**, 227-259.

5 Applications of Protein Blotting in Plant Biochemistry and Molecular Biology

ROGER J. FIDO, ARTHUR S. TATHAM and
PETER R. SHEWRY

Department of Agricultural Sciences, University of Bristol, AFRC Institute of Arable Crops Research, Long Ashton Research Station, Long Ashton, Bristol BS18 9AF, UK

I.	Introduction	102
II.	Membrane types	102
III.	Transfer methods	102
	A. Diffusion	103
	B. Capillary action	103
	C. Vacuum blotting	103
	D. Electroblotting	104
	E. Typical protocol for electroblotting	105
IV.	Staining, blocking and detection	106
	A. Protein stains	106
	B. Membrane blocking	106
	C. Detection of protein with antibodies	107
	D. Alternative methods of detection	108
	E. Recommended protocols	108
	F. Double staining technique	109
V.	Purification of antibodies by immunoaffinity blotting	111
	A. Protocol	111
VI.	Microanalysis of blotted proteins	111
	A. Protocol	112
VII.	Detection of DNA binding proteins with 'Southwestern blotting'	113
	A. Protocol	114
VIII.	Conclusion	114
	References	114

I. INTRODUCTION

Protein blotting was initially developed for the specific detection of protein antigens using antibody probes and was named 'western blotting' in accordance with the nomenclature for nucleic acid blotting procedures (see Chapter 4) (Towbin *et al.*, 1979; Burnette, 1981). Many refinements have been introduced since to reduce the need to use radio-iodinated probes and to improve speed, resolution and sensitivity.

In addition, major applications of protein blotting have developed in two other areas: the analysis of protein–DNA interactions, and the separation of proteins for microsequencing (facilitated by the development of a new generation of automated protein sequencers sensitive at the picomole level). Protein blotting has also been used to affinity purify individual antibodies from a polyclonal antiserum by elution from blots using pH shock (Olmsted, 1981).

In the present chapter we review these applications and, where appropriate, recommend protocols for routine use.

II. MEMBRANE TYPES

Nitrocellulose (NC) is probably the most commonly used blotting membrane and is available from a large number of suppliers (Schleicher and Schuell, Amersham, Millipore, etc.). It was originally designed as a surface filter, and is available in a range of pore sizes. The most commonly used pore size is 0.45 μm but 0.22 μm is preferred for low molecular weight proteins (i.e. below $\approx$ 20 000). Nitrocellulose, which has a binding capacity of approximately 80–100 μg protein cm^{-2}, is very fragile when dry and is best cut to size with the backing paper in place. The nylon membranes are stronger than NC and have six times the binding capacity of pure NC. Zetaprobe (Bio-Rad) is highly cationic and can bind 480 μg protein cm^{-2} (Gershoni and Palade, 1982). Because of this higher binding capacity the membrane also needs increased concentrations of blocking agent to overcome high non-specific binding. However, nylon membranes are incompatible with most commonly used protein stains (Coomassie Brilliant Blue and Amido Black), but compatible with Ferridye from Janssen Life Science Products.

A more recent addition to blotting membranes is polyvinyldifluoride (PVDF) membrane, which is hydrophobic, has high mechanical strength and protein binding capacity, and can be used in essentially the same way as NC after initial prewetting in 100% methanol for a short time (seconds) or 5% Tween 20 for 15–20 min (see Immobiline Tech. Bulletin, Millipore). A comparison of different membranes for maximum protein binding, combined with efficiency for automated microsequencing analysis of the protein, showed that the PVDF-based membranes had the highest protein binding efficiency (Baker *et al.*, 1991).

III. TRANSFER METHODS

Protein blotting, developed from procedures for DNA and RNA blotting, is used extensively. Proteins which have been subjected to high resolution polyacrylamide gel electrophoresis can be transferred to a membrane to be more accessible for identification by probing using a wide variety of methods. The transferred proteins are immobilised on a membrane which is easily handled and which can be treated immediately after blotting or stored indefinitely until required.

Four general methods are applicable to the transfer of proteins from polyacrylamide gels to membranes. They are well documented and reviewed (Towbin and Gordon, 1984; Beisiegel, 1986; Hames, 1990; Harper *et al.*, 1990) and, in terms of increasing efficiency, are diffusion, capiliary action, vacuum blotting and electroblotting.

A. Diffusion

The simplest of all methods of protein transfer relies on diffusion. This can be done with two membranes, one on either side of the gel, with the membranes themselves sandwiched between filter papers and foam pads. The entire assembly is held firmly between ridged plastic sheets punched with evenly spaced holes to allow free movement of buffer, before being submerged under transfer buffer (Bowen *et al.*, 1980; Reinhart and Malamud, 1982). The buffer is changed after 24 h and the transfer allowed to continue for up to 48 h, or longer if necessary. The transfer results in the production of two identical blots.

In a recent report on cultivar identification of seed proteins (McDonald, 1991), an alternative method of simple diffusion was used to transfer non-denatured proteins (which were enzymically active) from isoelectric focusing gels to polyvinylidine difluoride (PVDF) membranes with a 5 kg weight being used to assure intimate contact between gel and membrane.

B. Capillary Action

In 1975 E.M. Southern described a method of transferring DNA fragments from agarose gels to nitrocellulose (NC) using capillary forces. The method, which became known as Southern blotting, used radioactive RNA in hybridisation experiments to detect fragments which had bound to the NC membrane and which contained homologous sequences. Subsequently, Alwine *et al.* (1977), also working with DNA and RNA, used capillary blotting for transfer, but substituted the NC with diazobenzyloxymethyl cellulose membrane (DBM). Using the same DBM membrane, Renart *et al.* (1979) then developed a method to transfer protein by capillary action but the system proved to be slow and inefficient.

The assembly for capillary blotting described by Southern (1975) has been used extensively for DNA and RNA blotting. The mass flow in the system is achieved by placing the gel on paper wicks which have been previously wetted with transfer buffer and which can be replaced at suitable intervals or may be continuously wetted by a buffer reservoir. The movement of fluid acts as the driving force to elute the proteins from the gel and to transfer them into the membrane where they are bound.

C. Vacuum Blotting

In order to improve the rather slow and inefficient transfer of proteins by capillary action, a negative pressure (low vacuum) system was developed which increased the flow of buffer through the system. This development of the capillary blotting system was made by Peferoen et al. (1982) who produced a 'cheap, easy, reproducible and fast blotting system' by adapting a slab gel drier. The system used for vacuum blotting is essentially similar to that used in capillary blotting but the system is closed and relies on the power of a suction pump to drive the separated polypeptides from the gel into the NC membrane. A carefully positioned plastic sheet is used to ensure the buffer passes through the gel and not around it. Vacuum blotting is much faster than capillary action and any commercial slab drier can be adapted for use. As with drying slab gels, the pump must be protected from possible buffer suckback. The efficiency of vacuum blotting is apparently comparable to low powered electroblotting and is certainly very inexpensive and easy to use if gel drying equipment is already available. There is also no limitation on the choice of buffers which can be used, as there is no increase in conductivity or temperature during transfer.

The method described by Peferoen et al. (1982) made use of a Bio-Rad slab gel drier and the transfer buffer as described originally by Towbin et al. (1979) was 25 mM Tris, 150 mM glycine and 20% (v/v) methanol. The transfer membrane was NC with a 0.45 μm pore size. The proteins of interest were around the 50 000–60 000 molecular weight range, but proteins of between 16 000 and 356 000 molecular weight were also shown to blot in a 45 min time period. The 0.45 μm NC membrane was reported to be less efficient in absorbing proteins with a molecular weight of 20 000 or less (Peferoen, 1988), and leakage or diffusion of blotted proteins can occur. In order to overcome this problem NC membrane with 0.22 μm pore size should be used.

D. Electroblotting

At the same time (and in the same journal) as Renart et al. (1979) described the method of transferring protein by capillary action, Towbin et al. (1979) described a transfer method using electroelution to transfer proteins from polyacrylamide gels to nitrocellulose membrane. The method gave improved transfer in a short time and has subsequently become the standard method for protein transfer. In 1981 Burnette modified the system and gave it the name 'western blotting', which now encompasses essentially all forms of protein blotting.

It must be emphasised that when transferring proteins from gels to membranes, the rate and efficiency of transfer depends upon a number of factors (Lin and Kasamatsu, 1983). Thus the molecular weight of the protein (with higher molecular weight proteins of > 80 000–90 000 transferring less efficiently), its amino acid sequence and isoelectric point, and the concentration of acrylamide, can all affect transfer.

The buffer used by Towbin et al. (1979) for transferring from SDS gels was Tris–glycine–methanol, pH 8.3. There are problems associated with the transfer of proteins to nitrocellulose in the presence of SDS, as proteins which are still complexed with SDS (to give an overall negative charge) absorb poorly to nitrocellulose, which is itself negatively charged at pH 8. Methanol was included in the transfer buffer to reduce gel swelling during transfer, and also to give improved binding to the nitrocellulose. However, methanol also contracts the pores of the gel, which reduces protein elution (Gershoni and Palade, 1982). It has subsequently been shown that the addition of low levels of SDS (0.01–0.02% w/v) can counter this effect and improve the transfer of certain proteins, especially those of high molecular weight (Neilson et al.,

1982). Alternative buffers used for transfer included Tris-boric acid, pH 8.5, without methanol, which gave improved transfer when compared to a CAPS–methanol buffer, pH 11 (Baker et al., 1991).

A wide range of apparatus is now available commercially. For example, Bio-Rad, Pharmacia-LKB and Hoefer all produce both wet tank systems (including mini-blotters) and semi-dry horizontal electroblotters. The semi-dry systems, which use a variety of materials for electrodes including graphite, platinum and stainless steel, are reported to be more efficient than wet systems. They use less buffer, have shorter blotting times and have been used to produce multiple blots (Kyhse-Anderson, 1984). However, from personal communications and experiences, equally good blots of plant proteins can be obtained using a wet tank system, especially for the transfer of high molecular weight proteins. In addition, re-use of the transfer buffer for up to five times minimises reagent costs (Gooderham, 1984).

E. Typical Protocol for Electroblotting

The method described is based upon a wet tank system with a Bio-Rad Trans-Blot cell. When preparing for electroblotting care should be taken in the preparation of the transfer membrane, which must be fully wetted to ensure complete and even transfer of protein. Disposable gloves must be worn throughout to avoid contaminating the membrane. The membrane, (typically NC with 0.45 μm pore size from Schleicher and Schuell), is wetted by carefully dipping one edge into transfer buffer (25 mM Tris, 192 mM glycine and 20% methanol, pH 8.3, with or without 0.2% SDS) and slowly lowering so as to ensure removal of air from within the matrix. For proteins of 20 000 or less a membrane with 0.2 μm pore size is preferred, to reduce losses due to poor binding. After separating the proteins on SDS-PAGE slab gels (Hames, 1990), the gel is marked by cutting one corner and equilibrated in transfer buffer for up to 30 min, depending on the gel thickness. Proteins from SDS-PAGE are eluted as anions, therefore the membrane is placed on the anodic side of the gel. Following equilibration, the gel holder of the Trans-Blot cell (Bio-Rad) is placed, cathode side down, into a shallow tray containing sufficient buffer to maintain all further steps just under liquid. The gel–membrane sandwich is made by first placing a porous Scotch-Brite pad onto the gel holder, then a sheet or sheets of filter paper Whatman (3MM) (depending on gel thickness), previously cut to the size of the gel, onto the pad.

The equilibrated gel is laid onto the filter paper, avoiding trapping air bubbles, with the reverse side down so as to generate a true replica (i.e. in the correct orientation) when the membrane is developed. The wetted membrane is carefully laid over the gel by placing from the centre outwards, and gently rollered until full contact is made. The rollering, using a rimless test tube or graduated 10 ml pipette, is again performed from the centre outwards to remove any air pockets which remain and which would seriously interfere with the transfer to give patchy areas devoid of protein.

The sandwich is completed by placing further filter paper(s) over the membrane and a second pad over the filter paper. The gel holder is held firm and closed before being placed into the transfer tank with sufficient buffer to cover the blot. For standard transfers of up to 5 h the buffer is precooled to 4°C and cold water used as coolant. Voltage is held constant at 60 V. When gels have been run during the day, with a typical run time of 3–4 h, it is convenient to transfer overnight, especially when high molecular weight proteins are present. When transfer is complete the membrane is removed from the gel and rinsed in Tris-buffered saline before proceeding. The membrane can be visualised immediately or, if necessary, stored wet at 4°C for days or stored frozen for longer periods.

IV. STAINING, BLOCKING AND DETECTION

A. Protein Stains

A number of the methods for staining total protein in gels are also used for staining protein within nitrocellulose membranes. The visualisation of membrane-bound protein can be extremely useful, either simply to check if transfer is complete by comparison with an equivalent non-blotted stained gel, or to identify molecular weight marker proteins in order to determine the molecular weights of unknown proteins.

Some of the most commonly used protein stains are Amido Black 10B (Towbin et al., 1979), Coomassie Brilliant Blue R-250 (Burnette, 1981), India ink (Hancock and Tsang, 1983), Aurodye (Moermans et al., 1985) and colloidal iron (Moermans et al., 1986). A more recent study of the quantification of proteins using three dyes, Aurodye, Ferridye and India ink, showed that Aurodye could be used to quantify proteins in the low nanogram range (100 ng), while Ferridye and India ink were both less sensitive (Li et al., 1989).

Molecular weight markers can be identified using biotinylated proteins with detection by mixing avidin-conjugated horseradish peroxidase or alkaline phosphatase with the labelled second antibody. An interesting alternative approach is to produce antisera to commercially available molecular weight marker proteins and use an immunoenzymatic method for the detection of bands (Bjerrum and Hinnerfeldt, 1987). A more simple approach is to use prestained molecular weight markers which are visible during electrophoresis and on the blotted membrane. However, we found that blue prestained markers migrated as very diffuse bands and were only useful as a general guide, whereas Rainbow Markers (Amersham) ran as very sharp bands which were also clearly visible on the NC.

For rapid visualisation of protein on NC, prior to immunodetection, the filter can be reversibly stained essentially *en passant* using either Amido Black (Harper et al., 1986) or the red dye Ponceau S (Salinovich and Montelaro, 1986). Using this stain, protein transfer can be checked and individual tracks, molecular weight marker proteins and proteins of interest identified by lightly marking the membrane with a soft grade pencil. The stain, which does not appear to interfere with the immunoreactivity of membrane-bound proteins, can be removed by washing with Tween Tris-buffered saline (TTBS) prior to detection.

Further use of India ink can be made in a double stain system where an immunoreacted stained membrane can subsequently be stained for total protein, thus revealing an immunoreactive coloured stain against the general stain of Amido Black (Ono and Tuan, 1990).

B. Membrane Blocking

Following protein transfer, all unused sites on the membrane must be blocked to prevent any non-specific binding by the immunoglobulin(s) used in the detection assay.

The most common blocking agents, especially for nitrocellulose, are protein based, often bovine serum albumin (BSA) (up to 5%), fetal calf serum (10%) or gelatin (3%). However, the cheapest of all is non-fat dried milk powder ('Marvel' type, 5% w/v), which we have found to be a very efficient blocking agent. A blocking cocktail referred to as BLOTTO (bovine lacto transfer technique optimiser), which consists of a 5% (w/v) solution of dried milk in Tris-buffered saline containing antifoam and antimicrobial agents, has also been used for blocking, as an incubation medium and as a washing agent (Johnson et al., 1984). Blocking of NC is usually done at room temperature for approximately 1 h with shaking. Alternative blocking agents are non-ionic detergents such as Tween-20, which can be used alone or in combination with other protein-based agents, and Nonidet P-40. The use of Tween-20 alone has, however, been reported to lead to false-positive reactions due to non-specific antigen-antibody complexes (Bird et al., 1988). The final choice of blocking agent depends on the membrane type and in particular the proteins of interest, so as to give improved sensitivity with the minimum of background interference. In a comparison of four commonly used blocking agents, to give efficient blocking with maximal immunoreactivity, the most powerful blocking agent was defatted milk powder. It was also shown that individual monoclonal antibodies may have specific blocking requirements to give maximal immunoreactivity (Hauri and Bucher, 1986).

The blocking must be maintained during the detection process and requires the addition of low levels of blocking agent (routinely 1% BSA, w/v) to the solutions containing primary and secondary antibodies. It has been shown that over 60% of absorbed blocking protein is lost from a membrane after the normal washing steps during immunodetection (Ono and Tuan, 1990).

C. Detection of Protein with Antibodies

The probes used for detection of absorbed protein on membranes can be antibodies, both monoclonal and polyclonal, specific binding proteins or other ligands.

The detection of proteins by antisera is very well documented and can be performed using direct or indirect methods. The direct method uses a ligand or specific antibody which must be labelled. However, the sensitivity of this system may be much reduced (Beisiegel, 1986; Bernstein et al., 1987) when compared to the more typical detection system.

The indirect method makes use of a primary antibody which binds specifically to the immobilised protein (antigen) and which is then detected by the binding of a second (or sometimes third) labelled anti-species antibody. The second antibody can be either radio-, fluorescent- or enzyme-labelled and the antibody can be specific for different immunoglobulin groups, classes or fragments of immunoglobulins.

Early workers used radioiodinated antibodies (commonly with ^{125}I) with detection by autoradiography. This is extremely sensitive and the blots can also be re-exposed for varying times. Fluorescent-labelled antibodies are also used, typically labelled with fluorescein isothiocyanate (FITC) and detected under UV illumination.

Probably the easiest and most convenient method makes use of anti-species antibodies, which are conjugated to marker enzymes of which a large range is available commercially. Commonly used enzyme labels produce insoluble end-products, and are generally based on histochemical methods, typically using horseradish peroxidase (HRP) or alkaline phosphatase (AP). The insoluble chromogen produced from HRP fades when exposed to light so the AP system is preferred. It is also reported to be more sensitive with detection limits of approximately 30 pg of protein (Blake et al., 1984) when using the substrate 5-bromo-4-chloro-3 indolyl phosphate (BCIP) and nitroblue tetrazolium (NBT). Typically the enzymic reaction is complete within minutes, although the incubation may be continued for several hours. However, the times of incubation and the dilution of primary antibody vary greatly and must usually be worked out empirically for any given antibody. A general rule of thumb is to use greater dilution of antibody combined with longer times of incubation. This has the double advantage of using minimal amounts of what may be very limited and precious antisera, as well as reducing the amount of non-specific binding.

One disadvantage of the conjugated enzyme labelled detection system is that it is essentially a 'one-off' and further blots must be prepared to optimise conditions. It is, however, possible to re-probe the membrane by stripping off both primary and secondary antibodies by washing in 62.5 mM Tris-HCl containing 100 mM 2-mercaptoethanol, 2% SDS (w/v), pH 6.7, and incubating at 50°C for 30 min. After this treatment the membrane must be thoroughly washed and re-blotted. What is not known, however, is the effect this process may have on the avidity of an antibody.

D. Alternative Methods of Detection

An alternative to an enzyme-labelled second antibody is to use enzyme-linked Protein A or Protein G, both of which bind specifically to the Fc region of the antibody. Alternative labels can be with ^{35}S or ^{125}I. However, neither Protein A nor Protein G binds to all classes or subclasses of antibody and care must be taken to ensure specificity.

Biotin can also be attached to both the second antibody and the enzyme label. Avidin or streptavidin is then used to form a link between the biotinylated peroxidase or alkaline phosphatase and antibody, with detection made by the usual colour reaction.

The enhanced chemiluminescence (ECL) western blotting system for use with NC has been developed by Amersham. ECL makes use of standard procedures and works on the principle of chemiluminescence in which primary antibody is detected using species-specific secondary antibodies or proteins conjugated with HRP. The substrate used is luminol, which is oxidised by HRP to produce a sustained emission of light which is enhanced 1000-fold. The system is reported to be 10-fold more sensitive than the BCIP/NBT system.

E. Recommended Protocols

1. Non-specific (total) protein stain

The protein stain used routinely with NC is a rapid reversible stain using Ponceau S (Salinovich and Montelaro, 1986). The stain is available from a number of suppliers, including Sigma, and need only be diluted for use. The transblotted membrane is briefly washed before adding the dilute stain. This is left in contact for several minutes before pouring off and retaining for re-use. Washing the membrane in distilled water removes background stain so the membrane can be photographed (using a green filter) or lightly marked using a soft grade pencil to define the individual tracks, molecular weight markers, etc. The stain can be removed by washing twice with TTBS before blocking the membrane.

2. *Alkaline phosphatase immunoreaction system*

The method of detecting proteins by alkaline phosphatase conjugated second antibody gives rapid results and is simple to use. The labelled antibodies commercially available have high avidity and are of consistent quality.

- Wash blotted membrane briefly in Tris buffered saline (TBS: 20 mM Tris, 500 mM NaCl, pH 7.5).
- Block membrane by immersion in a 5% (w/v) solution of Marvel in TBS with gentle shaking for 1 h at room temperature.
- Remove blocker and wash membrane 2× with TBS containing 0.05% (v/v) Tween-20 (TTBS).
- Add primary antibody in TTBS containing 1% BSA (w/v) (fraction V). Incubate with gentle shaking for 1-2 h or overnight if convenient.
- Remove primary antibody and wash membrane 2× with TTBS.
- Incubate with AP labelled second antibody (anti-species) diluted in 1% BSA (w/v) in TTBS for 1 h with shaking.
- Remove second antibody and wash 2× with TTBS and final wash with TBS.
- Add colour reagents by mixing 1 ml of a 30 mg ml^{-1} solution of Nitro blue tetrazolium (stock solution prepared in 70% (v/v) aqueous DMF) and 1 ml of 15 mg ml^{-1} 5-bromo-4-chloro-3-indolyl phosphate (stock solution prepared in 100% DMF) into 100 ml of carbonate buffer (0.1 M $NaHCO_3$, 1.0 mM $MgCl_2$ pH 9.8).
- Incubate at 37°C until colour development occurs, which can be up to several hours.
- Stop reaction by washing the membrane with several changes of distilled water.
- The membrane can be air-dried for storage.

F. Double Staining Technique

Identification of individual immunoreactive proteins on two-dimensional gels can pose major problems, as exemplified by cereal seed storage protein fractions which can consist of 50 or more components. In this and similar cases it is possible to use a double staining method, in which immunodetection using AP-labelled second antibody is coupled with a total protein stain, allowing the spots to be matched perfectly (Ono and Tuan, 1990).

The transferred membrane is immunoreacted as described above with AP to produce the insoluble coloured end-product revealing the antigenic protein. Whilst the membrane is still wet, or following re-wetting, the entire electrophoretic pattern can be stained with India ink or any suitable equivalent used at 0.1% (v/v) in TTBS (Fig. 5.1). The system is also reported to work in the reverse order, with the blot being stained first. However, immunostaining first is the preferred method as there can be no possibility of interference from loss of immunoreactivity due to interaction with

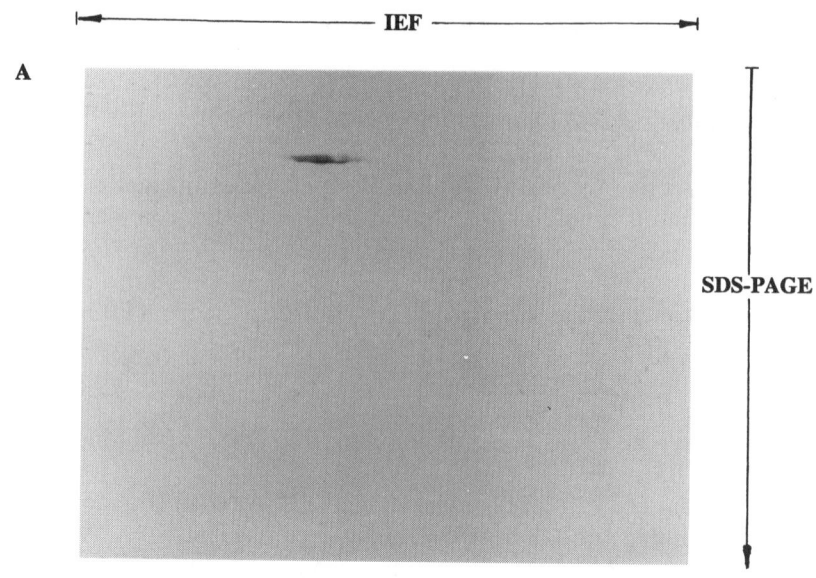

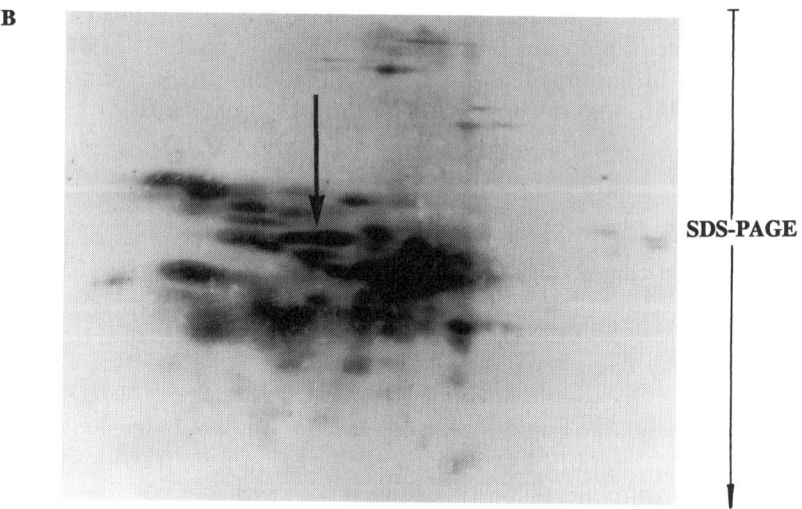

FIG. 5.1. Immunodetection (A) and double staining (B) of wheat storage proteins separated by two-dimensional IEF/SDS-PAGE. The proteins were transferred to nitrocellulose membrane and immunoreactive proteins detected using monoclonal antibody IFRN0067 followed by anti-mouse IgG labelled with alkaline phosphatase (A). Total proteins were then stained using India ink (B). The arrow indicates the immunoreactive proteins observed as purple spots, amongst the other non-reactive spots which stained black. Taken from Brett *et al.* (1992).

the dye. It also gives better results than transferring a Coomassie stained gel and immunoreacting the stained proteins (Jackson and Thompson, 1984).

V. PURIFICATION OF ANTIBODIES BY IMMUNOAFFINITY BLOTTING

Olmsted (1981) described a method for purifying antibodies by binding to proteins that have been electrophoretically transferred to diazotised paper, the antibody being released by pH shock with 0.2 M glycine HCl buffer at pH 2.8. Talian *et al.* (1983) used a similar procedure to purify antibodies to proteins bound to a nitrocellulose filter, and in addition conjugated the antibodies with fluorescein isothiocyanate (FITC) or rhodamine isothiocyanate prior to elution. Although initially developed to purify antibodies to cytoskeletal proteins from animals (tubulin and tropomyocin), these methods have since found more general applications including plant systems.

A. Protocol (Talian *et al.*, 1983)

- Wash nitrocellulose strips with several changes of 50 mM Tris-HCl, pH 7.5, 5 mM EDTA and 150 mM NaCl.
- Place strips in a 5 ml syringe and wash by forcing through two 5 ml aliquots of the same buffer.
- Add 3 ml of 0.2 M glycine HCl, pH 2.8, and leave for 2–5 min.
- Force solution through the paper twice and neutralise with 1 M NaOH.
- Dilute with buffer as appropriate.

VI. MICROANALYSIS OF BLOTTED PROTEINS

The development in the 1980s of highly sensitive gas phase and pulsed liquid phase sequencers provided an opportunity to determine the *N*-terminal sequences of proteins and peptides at levels below 20 pmol. This is of the same order as the amounts of proteins that can be separated by one-dimensional and two-dimensional electrophoresis, leading to the development of methods for direct microsequencing of blotted proteins. This has the advantages that the proteins are electrophoretically pure, and on a matrix which is easy to handle and relatively free of contaminants.

The initial studies used filters based on glass fibre with chemical activation (Aebersold *et al.*, 1986), coating with quaternary ammonium polybases (Vandekerckhove *et al.*, 1985) or siliconisation (Eckerskorn *et al.*, 1988). However, most workers now prefer to use PVDF-type membranes, either Immobilon™ (Millipore Corp.; Matsudaira, 1987) or Pro-Blott™ (Applied Biosystems Inc.; Yuen *et al.*, 1990). The latter membrane has been specifically developed for use with the Applied Biosystems Model 477A Pulsed Liquid Sequencer and a modified cycle (BLOTT 1) can be used to improve the sequencing efficiency of samples in the absence of polybrene.

Baker *et al.* (1991) compared the efficiency of six different membranes and found that PVDF-based membranes, and especially Pro-Blott, outperformed membranes based on glass fibre or polypropylene. Similarly Yuen *et al.* (1990) reported that Pro-Blott offered several improvements over other available membranes (including Immo-

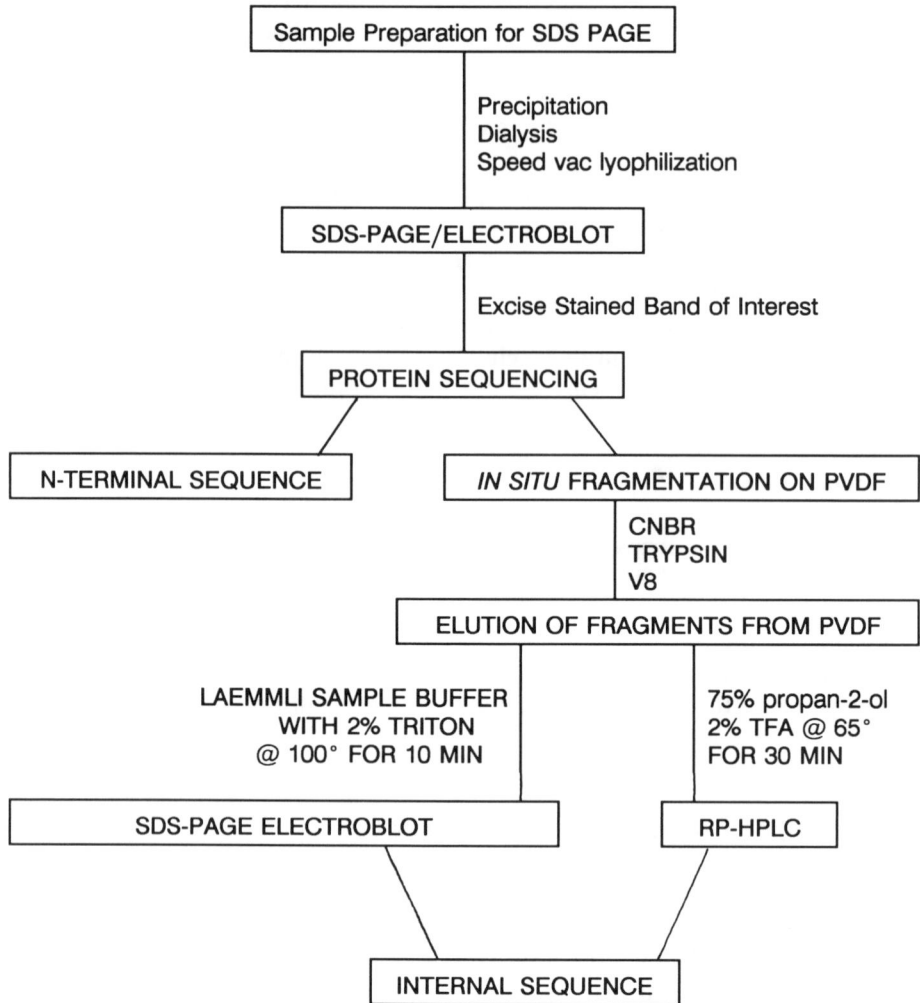

FIG 5.2. Summary of the procedure for using PVDF membrane for microsequencing. Reproduced with permission from Applied Biosystems Ltd.

bilon), including improved protein binding. We have also used both Immobilon and Pro-Blott in our laboratory with similar results. Yuen et al. (1988, 1990) have also reported the use of protein blotted on PVDF membranes for amino acid analysis and for chemical or enzymic cleavage prior to reseparation by HPLC or SDS-PAGE to determine internal sequences (Fig. 5.2).

A. Protocol

The following protocol is taken from Yuen et al. (1990), based on Matsudaira (1987):

- Prepare CAPS buffer.
  (a) *10× stock* (100 mM, pH 11). Dissolve 22.13 g of 3-[cyclohexylamino]-1-propanesulfonic acid in 900 ml of deionised water. Titrate with 2 M NaOH (approximately 15 ml) to pH 11, and add deionised water to a final volume of 1 litre. Store at room temperature.
  (b) *Electroblotting buffer*. Prepare 2 litres of buffer by mixing 200 ml of 10× stock buffer, 200 ml of methanol, and 1600 ml of deionised water.
- Wet the PVDF membranes (two sheets) with methanol for a few seconds, and place the membranes in a Petri dish containing blotting buffer.
- Remove the gel from the electrophoresis cell and soak in electroblotting buffer for 5 min.
- Assemble the transblotting sandwich, and electroblot at 90 V (300 mA), at room temperature for 10–30 min.
- Remove the membranes from transblotting sandwich and rinse with deionised water prior to staining.
- Visualise proteins by staining with Coomassie Blue R250.
  (a) Rinse membranes in distilled deionised water.
  (b) Saturate the membranes with 100% methanol for a few seconds.
  (c) Stain briefly with 0.1% (w/v) Coomassie Blue R250 in 1% acetic acid (v/v)/40% MeOH (v/v). Spots should appear within a minute.
  (d) Destain with 50% (v/v) aqueous MeOH.
- Rinse thoroughly with distilled deionised water and excise bands of interest.

VII. DETECTION OF DNA BINDING PROTEINS WITH 'SOUTHWESTERN BLOTTING'

Interactions of proteins with DNA are of considerable interest, particularly in relation to DNA replication and the control of transcription. 'southwestern blotting' was developed by Bowen *et al.* (1980) as a simple method of detecting DNA binding proteins in crude mixtures. The procedure is, as its name implies, a combination of protein and DNA blotting. Nuclear proteins are prepared and separated by SDS-PAGE. The proteins are then transferred to a nitrocellulose membrane and probed with DNA labelled either with ^{32}P or a non-radioactive label. Although simple in principle, the procedure can be difficult in practice, and requires considerable technical ability.

The most common application is to identify proteins that bind to a known DNA sequence, which can be an oligonucleotide or a gene fragment (Bowen *et al.*, 1980; Miskimins *et al.*, 1985). However, Keller and Maniatis (1991) have recently reported a novel preparative southwestern blotting procedure for selection of DNA sequences recognised by specific DNA binding proteins.

Relatively few studies of plant DNA binding proteins have been carried out (compared with studies of animal and microbial systems) and the use of southwestern blotting has consequently been restricted (but see for example the recent paper by So and Larkins, 1991). However, it will undoubtedly be used more widely in future studies.

A. Protocol

The following protocol is from J. M. Al-Rashdi and J. A. Bryant (personal communication), based on Miskimins *et al.* (1985).

- Separate nuclear proteins by SDS-PAGE using standard procedures.
- Transfer proteins to nitrocellulose membrane by electroblotting.
- Soak the nitrocellulose filter in prehybridisation solution at 22°C for 1 h. Prehybridisation solution is 10 mM HEPES buffer, pH 8.0 containing 5% (w/v) skimmed milk powder.
- Hybridise the DNA to the DNA-binding proteins by soaking the filter in hybridisation solution for 1 h at 22°C. Hybridisation solution is 10 mM HEPES, 50 mM NaCl, 10 mM $MgCl_2$, 0.1 mM Na_2EDTA, 1 mM dithiothreitol, 0.25% (w/v) skimmed milk powder and ^{32}P-end-labelled DNA (1 µg). The NaCl concentration and $MgCl_2$ concentration can be varied in order to determine optima.
- Wash filters for 2 × 1 h in hybridisation solution *minus labelled DNA* but plus 0.3 M NaCl.
- Autoradiograph the filter.

VIII. CONCLUSIONS

Protein blotting, in its various forms, has made major contributions to plant science, being used in studies of genetics and phylogeny, plant development, protein chemistry and gene structure and expression. The steady improvement of procedures and blotting membranes, coupled with developments in microsequencing and immunochemical labelling, has resulted in a degree of sensitivity which would have been regarded as impossible only a decade ago. There is no doubt that its use and influence will continue to expand.

REFERENCES

Aebersold, R. H., Teplow, D. B., Hood, L. E. and Kent, S. B. H. (1986). *J. Biol. Chem.* **261**, 4229–4238.
Alwine, J. C., Kemp, D. J. and Stark, G. R. (1977). *Proc. Natl. Acad. Sci. USA* **74**, 5350–5354.
Baker, C. S., Dunn, M. J. and Yacoub, M. H. (1991). *Electrophoresis* **12**, 342–348.
Beisiegel, U. (1986). *Electrophoresis* **7**, 1–18.
Bernstein, J. M., Stokes, C. E. and Fernie, B. (1987). *J. Clin. Microbiol.* **25**, 72–75.
Bird, C. R., Gearing, A. J. H. and Thorpe, R. (1988). *J. Immunol. Methods* **106**, 175–179.
Bjerrum, Ole J. and Hinnerfeldt, F. R. (1987). *Electrophoresis* **8**, 439–444.
Blake, M. S., Johnston, K. H., Russell-Jones, G. J. and Gotschlich, E. C. (1984). *Anal. Biochem.* **136**, 175–179.
Bowen, B., Steinberg, J., Laemmli, U. K. and Weintraub, H. (1980). *Nucl. Acids Res.* **8**, 1–20.
Brett, G. M., Mills, E. N. C., Tatham, A. S., Fido, R. J., Shewry, P. R. and Morgan, M. R. A. (1993) *Theor. Appl. Genet.* **86**, 442–448
Burnette, W. N. (1981). *Anal. Biochem.* **112**, 195–203.
Eckerskorn, C., Newes, W., Goretzki, H. and Lottspeich, F. (1988). *Eur. J. Biochem.* **176**, 509–519.

Gershoni, J. M. and Palade, G. E. (1982). *Anal. Biochem.* **124**, 396–405.
Gooderham, K. (1984). In "Methods in Molecular Biology", Vol. 1: Proteins (J. M. Walker, ed.), Ch. 20, pp. 165–178. Humana Press, Clifton, NJ.
Hames, B. D. (1990). In "Gel Electrophoresis of Proteins: A Practical Approach", 2nd edn (B. D. Hames and D. Rickwood, eds), pp. 1–147. IRL Press, Oxford.
Hancock, K. and Tsang, V. C. W. (1983). *Anal. Biochem.* **133**, 157–162.
Harper, D. R., Liu, K.-M. and Kangro, H. O. (1986). *Anal. Biochem.* **157**, 270–274.
Harper, D. R., Liu, K.-M. and Kangro, H. O. (1990). *J. Virol. Methods* **30**, 25–40.
Hauri, H.-P. and Bucher, K. (1986). *Anal. Biochem.* **159**, 386–389.
Jackson, P. and Thompson, R. J. (1984). *Electrophoresis* **5**, 35–42.
Johnson, D. A., Gautsch, J. W., Sportsman, J. R. and Elder, J. H. (1984). *Gene Anal. Tech.* **1**, 3–8.
Keller, A. D. and Maniatis, T. (1991). *Nucl. Acids Res.* **19**, 4675–4680.
Kyhse-Anderson, J. (1984). *J. Biochem. Biophys.* **10**, 203–209.
Li, K. W., Geraerts, W. P. M., Van Elk, R. and Joosse, J. (1989). *Anal. Biochem.* **182**, 44–47.
Lin, W. and Kasamatsu, H. (1983). *Anal. Biochem.* **128**, 302–311.
Matsudaira, P. (1987). *J. Biol. Chem.* **262**, 10035–10038.
McDonald M. B. Jr. (1991). *Seed Sci. & Technol.* **19**, 33–40.
Miskimins, W. K., Roberts, M. P., McClelland, A. and Ruddle, F. H. (1985). *Proc. Natl. Acad. Sci. USA* **82**, 6741–6744.
Moermans, M., Daneels, G. and De Mey, J. (1985). *Anal. Biochem.* **145**, 315–321.
Moermans, M., De Raeymaeker, M., Daneels, G. and De Mey, J. (1986). *Anal. Biochem.* **153**, 18–22.
Neilson, P. J., Manchester, K. L., Towbin, H., Gordon J. and Thomas, G. E. (1982). *J. Biol. Chem.* **257**, 12316–12321.
Olmsted, J. B. (1981). *J. Biol. Chem.* **256**, 11955–11957.
Ono, T. and Tuan, R. S. (1990). *Anal. Biochem.* **187**, 324–327.
Peferoen, M. (1988). In "Methods in Molecular Biology", Vol. 3: New Protein Techniques (J. M. Walker, ed.), Ch. 28, pp. 383–393. Humana Press, Clifton, NJ.
Peferoen, M., Huybrechts, R. and De Loof, A. (1982). *FEBS Lett.* **145**, 369–372.
Reinhart, M. P. and Malamud, D. (1982). *Anal. Biochem.* **123**, 229–235.
Renart, J., Reiser, J. and Stark, G. R. (1979). *Proc. Natl. Acad. Sci. USA* **76**, 3116–3120.
Salinovich, O. and Montelaro, R. C. (1986). *Anal. Biochem.* **156**, 341–347.
So, J.-S. and Larkins, B. A. (1991). *Plant Mol. Biol.* **17**, 309–319.
Southern, E. M. (1975). *J. Mol. Biol.* **98**, 503–517.
Talian, J. C., Olmsted, J. B. and Goldman, R. D. (1983). *J. Cell Biol.* **97**, 1277–1282.
Towbin, H. and Gordon, J. (1984). *J. Immunol. Methods* **72**, 313–340.
Towbin, H., Staehelin, T. and Gordon, J. (1979). *Proc. Natl. Acad. Sci. USA* **76**, 4350–4354.
Vandekerckhove, J., Bauw, G., Puybe, M., van Damme, J. and van Montagu, M. (1985). *Eur. J. Biochem.* **152**, 9–19.
Yuen, S. W., Chui, A. H., Wilson, K. J. and Yuan, P. M. (1988). *Applied Biosystems Protein Sequencer User's Bulletin*, No. 36.
Yuen, S., Sheer, D., Hsi, K.-L. and Mattaliano, R. (1990). *Research News*, Applied Biosystems Inc.

6 The Polymerase Chain Reaction

CHRIS THOMAS

Advanced Technologies (Cambridge) Ltd, Cambridge Science Park, Cambridge CB4 4WA, UK

I.	Introduction	118
II.	What is PCR?	118
III.	Setting up PCR	119
	A. A typical PCR protocol	119
	B. Template preparation	120
	C. Enzymes	121
	D. Reagents	121
	E. Equipment	122
	F. Cost	122
IV.	Applications of the PCR	123
	A. Detection of specific DNA sequences	123
	B. DNA fingerprinting with RAPDS	124
	C. Detecting mutations with PCR	125
	D. Cloning and PCR	126
	E. Sequencing and PCR	129
	F. Quantification and PCR	130
V.	Primer design	131
	A. General features of primers	131
	B. Calculating the annealing temperature	131
	C. Heterologous primers	132
VI.	Troubleshooting	133
	A. Use of controls	134
	B. Negative results	134
	C. Too many bands	134
	D. Contamination	135
	E. PCR machines	135
VII.	Conclusion	136
	Acknowledgements	136
	References	137

I. INTRODUCTION

The polymerase chain reaction (PCR; Mullis and Faloona, 1987) has rapidly become an indispensable tool for molecular biologists since its first use with a thermostable polymerase by Saiki *et al.* in 1988. The simple core process of amplifying DNA fragments *in vitro* has been adapted to a wide variety of uses ranging from diagnostic screening, fingerprinting and mutagenesis to cloning. The likelihood is, that wherever DNA is being used or required for a procedure, PCR can be applied in one form or another. This is reflected in the rapid rise of publications either on PCR or utilising PCR.

There are already a number of comprehensive publications on the principle of PCR and its application to a wide range of processes (Erlich, 1989; Innis *et al.*, 1990; McPherson *et al.*, 1991a; Sambrook *et al.*, 1989). These include detailed protocols for many of the procedures and are of great practical use.

The aim of this chapter is to give the plant molecular biologist an insight into the application of PCR to a variety of different work areas, based on our experience at Advanced Technologies (Cambridge) Ltd (ATC) with applying PCR in a research and development environment. The chapter also deals with some of the considerations on experimental design applicable to PCR in general and to specific areas. Particular emphasis is placed on primer design as experience has shown this to be a key factor in the success of any PCR experiment.

II. WHAT IS PCR?

PCR is a method for amplifying DNA exponentially *in vitro*. The process requires a suitable template DNA, a pair of oligonucleotide primers, a DNA polymerase with suitable reaction buffer and some method of incubating the PCR reaction at three different temperatures. The primers are specific to selected sequences on the template DNA. They are separated by a distance ranging from a few base pairs to over several thousand bases, depending on the requirements of the user. DNA synthesis primed by each primer will extend up to and beyond the region of the other primer; metaphorically, the primers face towards each other. Nowadays, the polymerases used are thermostable.

The reaction mixture containing all the components given above is heated to 94°C to denature the template DNA. The sample is then cooled until the primers anneal to the template. The temperature is raised to the optimum for the polymerase, e.g. 72°C for Taq polymerase and DNA synthesis proceeds from the primed regions along the DNA molecule. The original DNA and the newly synthesised strands are taken through further cycles of denaturation, annealing and synthesis. After the second cycle, products begin to emerge which only cover the region containing the two primers and intervening DNA. As the PCR progresses these double-stranded DNA molecules of a specific size become the predominant product. Amplification is exponential at first but after 20 or so cycles the number of product molecules exceeds the number of enzyme molecules and amplification slows down. By 25 to 30 cycles, there is enough product to be visualised by a variety of methods. These include electrophoresis and staining with ethidium bromide (Sambrook *et al.*, 1989), radiolabelling or the incorporation of fluorescent dyes (Edwards *et al.*, 1991; Mayrand *et al.*, 1992). When PCR was originally performed using Klenow enzyme (Saiki *et al.*, 1985; Scharf *et al.*, 1986), fresh enzyme had to be added after each PCR cycle. The introduction of thermostable polymerases has made this unnecessary and permits the automation of PCR (Saiki *et al.*, 1988).

III. SETTING UP PCR

This section gives a simple protocol for setting up a PCR with Taq polymerase as the thermostable enzyme. This is followed by some general considerations on the various components of the PCR amplification.

A. A Typical PCR Protocol

The following generalised protocol based on recommendations by Cetus should work in most cases with Taq polymerase.

(1) Prepare a master mix containing:
   1 × Taq buffer (10 mM Tris-Cl, pH 8.4 at room temperature, 50 mM KCl, 0.01% gelatin, 0.01% Nonidet P40 and 0.01% Tween 20)
   1.5 mM $MgCl_2$
   200 μM each dNTP
   0.25 μg of 20 mer primer 1 per 50 μl reaction
   0.25 μg of 20 mer primer 2 per 50 μl reaction
   1.25 units of Taq per 50 μl reaction.
(2) Mix gently and dispense the required number of 50 μl aliquots into 500 μl Eppendorfs that already contain 50 μl of light liquid paraffin. Some older machines may use 1.5 ml Eppendorf tubes whilst other models permit the use of microtitre plates.
(3) Add 1–2 μl of template (0.1 ng of plasmid or 20–500 ng of genomic DNA) to each reaction. The sample is always added last to prevent contamination of stock solutions or other samples.
(4) Always include a positive control if possible to confirm that the PCR amplification is working. Also include a negative control (no template). This checks whether any of the reagents are contaminated, giving false positives.
(5) Perform PCR with the following programme. All temperatures = sample temperature; all times = duration when sample achieves required temperature:
94°C for 90 s
T_a for 10 s
72°C for X min
3 cycles

(with some genomic DNA samples, the initial denaturation may not be adequate. In these cases denature the DNA for 5 min prior to the reaction.)

followed by:
94°C for 30 s
T_a for 10 s
72°C for X min, increasing to $2X$ min in increments during the next 27 cycles

finish with:
T_a for 10 s
72°C for $2X$ min
2 cycles

The last two cycles ensure that any residual single-stranded molecules are converted to double-stranded DNA.

T_a = the annealing temperature of the primers (see below for different methods of determining the T_a)

$X = 1$ for sequences up to 2 kb. For short PCR products of 200–500 bp use 30 s at 72°C throughout.

(6) Electrophorese 10 µl aliquots of the PCR on 1–2% agarose gels, depending on fragment size. Products should be clearly visible with ethidium bromide staining.

B. Template Preparation

DNA is the most common template used for PCR. RNA or mRNA can also be used with enzymes such as rTth polymerase that have reverse transcriptase activity. The purer the DNA, the better. However, PCR will function with crude nucleic acid preparations and even bacterial cells. The degree of purity required depends on whether the extract containing the template inhibits the polymerase.

When using genomic DNA, the best template is that purified on caesium chloride gradients or by other methods giving a high degree of purity. It is the method of choice when either using large amounts of template or repeated use of the same source material. Several rapid methods have been published which provide DNA suitable for PCR (Dellaporta et al., 1983; Lassner et al., 1989; McGarvey and Kaper, 1991; Hamill et al., 1990; Gawel and Jarret, 1991). These are useful where numerous different samples need to be assayed and only microgram quantities of DNA are needed. Many have been adapted to 1.5 ml microfuge tube scale extractions which also speeds up the extraction process. Examples of these methods range from the relatively pure CTAB-based methods of Dellaporta et al. (1983), to cruder ethanol precipitates (Lassner et al., 1989; McGarvey and Kaper, 1991), to squashes of plant tissue on nylon membranes (Langridge et al., 1991). Young or tissue-cultured leaf material from tobacco, tomato or potato provides ideal source material for DNA extractions. Material rich in polyphenols or oils such as eucalyptus can be less tractable, but the method of Gawel and Jarret (1991) has been used with success in our company (M. Maunders, personal communication).

Where bacterial colonies or phage plaques are to be screened, this can be done with a small stab of the culture direct from the agar plate or broth (Güssow and Clackson, 1989; Runnebaum et al., 1991).

With all these methods it is important to include controls to check that the substrate will amplify in a PCR. This can be done by spiking the reaction with a known functional positive sample, or by screening for a control sequence that is definitely present in the template. Often contaminants are carried over which inhibit the PCR; sometimes these can be diluted out. In our experience, if the DNA solution or pellet is clear or cream coloured, reactions tend to work. If the DNA is brown or black, you may experience problems.

C. Enzymes

The first thermostable polymerase to be used for PCR was Taq polymerase (Saiki *et al.*, 1988). The enzyme has been well characterised and described elsewhere (Gelfand, 1989). At present it is still the most economical enzyme. Cloned and modified versions of Taq polymerase are readily available, e.g. from Perkin Elmer, Promega and Pharmacia. Other enzymes that have come into use are Vent polymerase from New England Biolabs, Pfu DNA polymerase from Stratagene, and rTth polymerase from Perkin Elmer. More polymerases are bound to be discovered and exploited for PCR. Note that with many preparations, the enzyme contains small quantities of the host DNA from which it was isolated (Schmidt *et al.*, 1991). This may cause artefacts in some cases but can be counteracted if necessary (see Section VI.D below). Enzymes vary regarding their processivity, sensitivity to magnesium chloride concentration and stability at high temperatures.

The non-template specific addition of a deoxyribonucleotide at the 3'end of double-stranded molecules is a feature of many polymerases (Clark, 1988) which is countered by a 3'-5' exonuclease activity. The exonuclease also ensures that the DNA polymerases have a lower error rate for incorporation of mismatched bases. Taq polymerase does not have the exonuclease activity and is therefore more likely to introduce errors (Tindall and Kunkel, 1988; Krawczak *et al.*, 1989; Kwok *et al.*, 1990; Eckert and Kunkel, 1991). It is advisable to use an enzyme with a low error rate and with proofreading ability such as Vent DNA polymerase (Mattila *et al.*, 1991) or Pfu DNA polymerase (Bergseid *et al.*, 1991) in cases where this is important, such as the cloning of genes from one or only a very few template molecules. In most cases Taq polymerase is perfectly adequate.

The rTth polymerase has the added advantage of being able to function as a reverse transcriptase (Myers and Gelfand, 1991). This means that first strand cDNA synthesis and subsequent PCR reactions can be performed in the same reaction tube with minor buffer additions.

Thermostable ligase has also been used in chain reactions with primers. The process is called the ligase chain reaction (LCR) and is specifically designed to detect point mutations in sequences (Barany, 1991; Weiss, 1991).

D. Reagents

The standard buffer used for Taq polymerase contains 10 mM Tris-Cl, pH 8.4 at room temperature, 50 mM KCl, 1.5 mM $MgCl_2$, 200 mM each dNTP, 0.01% gelatin or bovine serum albumin (BSA), 0.01% NP40 and 0.01% Tween (Gelfand, 1989). As an alternative 0.1% Triton X-100 can be used instead of the NP40 and Tween. The potassium chloride concentration can be varied or reduced; this may be necessary for very long PCR fragments (Ponce and Micol, 1992). Because the magnesium chloride concentration is affected by the free deoxyribonucleotides, which mop up some of the ions, this must be compensated for at higher nucleotide concentrations (Gelfand, 1989). The magnesium chloride concentration can also be altered to optimise the individual PCR reaction but in most cases this is not necessary.

Although it is easy enough to prepare one's own buffer, many suppliers of PCR enzymes supply a basic buffer, with or without magnesium chloride and the non-ionic detergents. Stabilised solutions of nucleotides can also be purchased, e.g. from Pharmacia. The advantage of using their preparations is that they are quality controlled and are likely to be free of any contaminants commonly found in the laboratory in which they are to be used (see contamination below).

Primers and primer design will be dealt with separately below because of its importance to the successful outcome of a PCR.

A variety of methods and reagents are available for enhancing the specificity or amount of product produced in more recalcitrant PCRs. These include formamide (Sarkar et al., 1990), dimethylsulphoxide (DMSO) (Winship, 1989) and tetramethyl ammonium chloride (Hung et al., 1990). In our experience they have not proved essential, primer design often being far more critical.

E. Equipment

The equipment can fall into one of three categories. First, a student and three water baths. Second, home-made automated designs (e.g. Torgersen et al., 1989) and third, off the shelf PCR machines. There is now a wide range of commercially available PCR machines. Because Cetus has patented the PCR process, many machines are described under pseudonym such as Thermal cycler, etc.

Machines vary in their mechanism of heating and cooling the samples from Peltier devices to light bulbs and fans. The current minimum standard aspired to is a heating or cooling rate of $1°C\ s^{-1}$.

The following features may be relevant when choosing a machine. The ability to measure the sample temperature during a PCR run or even to control the PCR run by sample temperature can be reassuring and useful. This is primarily because the time taken for the sample to attain the required temperature is slightly longer than that taken by the heating block. How well and quickly the sample is heated is also dependent on the sample volume. The ability to record a PCR run by output to a chart recorder or computer terminal is of benefit. This allows a comparison between runs on the same machine under different conditions or between runs on different machines. For example, a PCR programme of 30 cycles at 94°C for 30 s, 50°C for 10 s and 72°C for 2 min as measured by sample temperature, could theoretically be complete in 1 h 20 min. In reality, older machines could take up to 4.5 h and newer machines take just under 3 h.

Another important feature is temperature variation across the heating block (Hoelzel, 1990; Van Leuven, 1991). It is worth doing a trial run with all wells filled with identical samples to check for uniformity. Other factors to take into account include serviceability, type of work to be done, flexibility and user-friendliness of the programming. A number of PCR machines now have heated lids or other mechanisms to reduce sample evaporation. This removes the need to overlayer the sample with liquid paraffin, though one should ensure that this cannot lead to cross-contamination between different samples.

F. Cost

The following considerations are based on prices at June 1991.

Programmable heating blocks are available at prices ranging from £3000 to £10 000.

It is worth bearing in mind that, once familiar with PCR, the machine is likely to be used to capacity and it is therefore worth anticipating the need for an additional machine within a short space of time.

The major cost with PCR is the consumption of the thermostable polymerase. Taq polymerase can be purchased at around £300 per 1000 units; this is equivalent to 800 × 50 µl reactions.

Primers are the other main cost. Depending on whether they are made in-house or purchased commercially, the cost can range from £2–6 per base, equivalent to £80–240 for 500 µg each of two 20 mer primers. This should last for 2000 × 50 µl reactions.

The other main consumable worth purchasing is a set of ultrapure dNTPs at about £80 for 4 × 100 µl of each dNTP at 100 mM concentration, enough for 1000 × 50 µl reactions.

The average cost for PCR can therefore average at around £0.50 per 50 µl PCR reaction. This assumes that the primers are completely used up. With primers only being used for say 10 PCRs, as in isolating a known gene, the price can reach £16 per reaction.

IV. APPLICATIONS OF THE PCR

PCR is such a versatile technique that it can be used in a wide variety of different ways. Below are some of the possible areas of application each with its own requirements. The purpose of the task influences the choice of parameters for PCR. Different objectives may also have their own specific problems.

A. Detection of Specific DNA Sequences

The rapid screening of large numbers of samples is perhaps the most obvious application of PCR. The simplest example is the detection of a specific sequence such as the confirmation that genes have been introduced into plants (Lassner *et al.*, 1989; Hamill *et al.*, 1990, 1991; McGarvey and Kaper, 1991). Progeny of transformants can also be screened for chosen PCR fragments in genetic analyses of crosses. The method can also be used to test for the presence of microbial or DNA contaminants (Nerenberg and Minor, 1991). Sequences of interest can be detected from RNA preparations by first synthesising cDNA and using this as a template for PCR (Goblet *et al.*, 1989; Sheardown, 1992; also see section below on cloning specific cDNAs). Several companies, e.g. Pharmacia, provide first-strand cDNA synthesis kits for the PCR user. PCR can also be used more specifically as a method of mapping transposon insertion sites (Barnes, 1990) or transposon tagging and genome characterisation (Earp *et al.*, 1990).

There is a rapidly growing list of publications on the screening of genomic or cDNA libraries for sequences of interest using PCR as an alternative or an aid to conventional methods. The number of clones that need to be screened to obtain a desired sequence can be dramatically reduced if there is sufficient sequence information to permit the detection of a diagnostic fragment by PCR. This is achieved by using the DNA from subsections of the library in PCR amplifications. Only those subsections that contain the diagnostic fragment need to be taken further for screening by hybridisation (Bloem and Yu, 1990; Kwiatkowski et al., 1990; Isola et al., 1991. The method can even be used to determine the clone frequency in a library (Alexander, 1990). Where there is not enough information to generate two primers for PCR of a specific fragment, one can either use an oligo(dT)-based oligonucleotide as one of the primers or an oligonucleotide to the vector sequence adjacent to the cloned sequence (Hovens and Wilks, 1989; Rasmussen et al., 1989; Rosenberg et al., 1991).

Alternatively, a more general screen may be performed. Testing transformed bacteria for the presence or absence of recombinant plasmids can be performed (Güssow and Clackson, 1989; Runnebaum et al., 1991). The primers are designed to complement the regions on either side of the cloning site in the vector. PCR products from individual clones are then characterised by size fractionation on an agarose gel. The method can be performed on crude bacteria and is a viable alternative to alkaline lysis. Several hundred colonies can be screened in one day.

Ideally, the size of the PCR product in screening experiments should be as small as possible, in the range of 100–500 bases. The exception to this is of course the general colony screen for recombinants. The advantages of small products are two-fold. First, small contaminants or spurious products of PCR will be amplified preferentially over a larger product, therefore, if the product is small it is less susceptible to this problem (Mullis, 1991). Second, the PCR reaction can be performed with a shorter extension time, reducing overall run time and allowing a faster throughput of samples.

Because PCR is particularly effective with short fragments, it is also possible to amplify several different fragments from the same template in the same reaction (multiplex PCR, Lassner et al., 1989; Chamberlain et al., 1990). This does not always work as some primer pairs interfere with one another, but prudent primer design (see below) and trial and error testing can result in substantial time savings by combining several screens. Fluorescent dye labelling of primers as an alternative to ethidium bromide is particularly attractive with this system (Edwards et al., 1991; Mayrand et al., 1992). PCR products of similar size or added internal controls can be monitored if the different primers are labelled with dyes fluorescing in different colours. The system is also more sensitive and reduces the number of cycles required for PCR.

The main disadvantage of routine screening is the risk of contamination. Repeated screening of a particular product results in the accumulation of enough DNA on work surfaces, pipetters, hands, etc. to begin giving false positives (Kwok and Higuchi, 1989; Kitchin et al., 1991). To avoid contamination, the working areas for PCR should be treated as sterile areas with dedicated pipetters, etc. The use of positive displacement pipetters may be an advantage. The preparation of 'sterile' batches of water, buffer stocks, reagents and plasticware or the purchase of preprepared material also minimises the risk of contamination. Running both positive and negative control reactions is particularly important to check for early signs of contamination.

B. DNA Fingerprinting with RAPDS

PCR is a method that relies on specific primers for the amplification of known sequences. It can however be applied to generate 'randomly' amplified polymorphic DNA sequences (RAPDS) which can be used as molecular markers for genome finger-

6. THE POLYMERASE CHAIN REACTION

printing (Welsh and McClelland, 1990, 1991; Williams *et al.*, 1990). Genomic DNA is used as a template for PCR with short primers under conditions where the specificity of annealing is reduced, such as low annealing temperature and high magnesium chloride concentration. The primers bind with a frequency that is sufficient to ensure that some are close enough to result in a PCR amplification. The patterns of bands can be polymorphic between different species and varieties and thus be used for fingerprinting.

At present, the choice and design of primers is arbitrary. Primers that result in polymorphisms are found by trial and error. Sets of RAPD primers that have been used successfully can now be obtained commercially from companies such as Genosys (Cambridge, UK) and Operon (California). Different groups working with RAPDS also exchange primers or primer sequences that are useful.

RAPDS have been used successfully in analyses of *Arabidopsis* (Reitner *et al.*, 1992), birch (Oliver, 1990), brassicas (Hu and Quiros, 1991), cocoa (Wilde *et al.*, 1992), maize (Welsh *et al.*, 1991), soybean (Williams *et al.*, 1990; Caetano-Anolles, 1991), tomato (Klein-Lankhorst *et al.*, 1991; Martin *et al.*, 1991) and white spruce (Carlson *et al.*, 1991; Tulsieram *et al.*, 1992).

C. Detecting Mutations with PCR

The specificity of a primer for PCR can be ensured by design and by tailoring the PCR run conditions. It is therefore possible to exploit the specificity in order to detect the presence or absence of mutations in a given sequence. Primers are most sensitive to mismatches with their template at their 3' end from which extension occurs. Generally, if there is a mismatch at the 3' end between primer and template, then extension from that primer does not occur and no PCR product is generated (Gibbs *et al.*, 1989; Newton *et al.*, 1989). This method has been used in studies on point mutations. The same principle has been used in a related amplification method, the ligase chain reaction (LCR) (Wu and Wallace, 1989; Weiss, 1991). Short primers of up to 18 bp in length are also very susceptible to single base differences internally and can also be used to detect point mutations under stringent PCR conditions (Gibbs *et al.*, 1989; Seyama *et al.*, 1992).

The only drawback with the above methods is that they rely on the absence of a response or PCR product to indicate a mutant sequence. One has to be sure that the negative result is real rather than a failure of the PCR to work for some other reason such as contamination with inhibitors.

Mutations can be detected by first amplifying two separate DNAs which are thought to differ by a point mutation and annealing one template to the other. Base changes can then be detected by difference in the melting points between homologous and heterologous combinations during denaturing gel electrophoresis (Farr *et al.*, 1988; Sheffield *et al.*, 1989; Roberts *et al.*, 1991; Seyama *et al.*, 1992). PCR can also be used to amplify DNA fragments that can then be tested for the presence or absence of a restriction site, avoiding the need for restriction and Southern blotting of genomic DNAs where the required locus has already been defined (Saiki *et al.*, 1985; Haliassos *et al.*, 1989).

D. Cloning and PCR

1. Amplification of the sequence

PCR is invaluable in cloning experiments where some sequence information is available. The sequence can be obtained either from prior publications (Saiki *et al.*, Scharf *et al.*, 1986; Scharf, 1990; Clackson *et al.*, 1991) or from backtranslated protein sequences (see Section V below).

Genes can be cloned from sources quite divergent from the material the original sequence was obtained from. In these cases, the primers are designed to take account of possible sequence variation. This is done on the basis of codon usage and codon preference tables for the species involved (see Section V). A similar process of primer design is undertaken when a known amino acid sequence is backtranslated into possible nucleic acid sequence (Lathe, 1985; Patil and Dekker, 1990). In both instances it is preferable to target either conserved regions within the gene, if known, or to choose areas containing codons with minimal redundancy.

The whole range of different nucleic acid templates can be utilised, according to availability or preference. Genomic DNA is perfectly adequate for small parts of genes, with the qualification that introns may give rise to bands of unexpected size. Obviously, it is the only template for attempts to obtain promoter sequences. cDNA is the preferred template for isolating genes. cDNA has the advantage that a much smaller number of different sequences are actually screened, reducing the risk of non-specific products arising. Only single-stranded cDNA is required and a number of companies provide first strand cDNA synthesis kits aimed at the PCR user.

cDNAs can be synthesised by conventional methods from purified poly(A)$^+$ RNA (Sambrook *et al.*, 1989) and then used for PCR (Frohman *et al.*, 1988; Sheardown, 1992). Total RNA can also be used as a source for oligo(dT) primed cDNA synthesis as the subsequent PCR amplifies the target molecule from the low levels of cDNA generated (Carothers *et al.*, 1989; Doherty *et al.*, 1989). If there is enough sequence information available, the target gene can be specifically synthesised from total RNA preparations with a specific primer rather than oligo(dT) and then amplified (Doherty *et al.*, 1989). Several methods have been published performing cDNA synthesis and PCR in the same reaction tube, simplifying the process even further (Goblet *et al.*, 1989; Myers and Gelfand, 1991). AMV reverse transcriptase has however been shown to inhibit Taq polymerase (Sellner *et al.*, 1992). The inhibition of the PCR amplification should not be encountered with rTth polymerase (Myers and Gelfand, 1991).

Often the amount of available sequence to design primers from is very short. A number of different methods have been devised to enable the cloning of sequence beyond the known region. Examples are the use of Inverse PCR (Ochman *et al.*, 1988, 1990; Huang *et al.*, 1990; Kandpal *et al.*, 1990), addition of linkers (Loh *et al.*, 1989) or the use of vector sequence (Hovens and Wilks, 1989; Rasmussen *et al.*, 1989; Rosenberg *et al.*, 1991). These methods also permit the cloning of the promoter regions upstream of the start codon for a coding sequence from genomic DNA.

PCR can be used in cases where it is difficult to obtain significant amounts of RNA for conventional cDNA synthesis and library generation. cDNA is first synthesised from the available RNA using oligo(dT). The first-strand cDNA can then either be dG tailed or linkers of approximately 20 bp or more are ligated onto the cDNAs after second strand synthesis. By using primers specific to either the poly(A) and poly(G) sequences of the cDNA (Belyavsky et al., 1989; Gurr and McPherson, 1991) or primers specific to the attached linkers (Edwards et al., 1991), it is then possible to amplify the cDNA population into clonable quantities. Size fractionation before and after PCR is advisable to reduce the quantities of small products that are preferentially amplified and cloned. Molecules recalcitrant to PCR will also be selected against. The ability to amplify cDNA generally is also of use for generating large amounts of material for subtractive hybridisation screening (Herfort and Garber, 1991).

The size of the gene to be cloned has to be taken into consideration. Whilst sequences up to 9 kb can be amplified (Schwartz et al., 1990; Maga and Richardson, 1991; Ponce and Micol, 1992), the greatest success in obtaining a product from a cDNA or genomic background is achieved with products no more than a few kb in length. This is in part due to reducing the chance of small PCR artefacts taking over a PCR reaction (Mullis, 1991). It is also a generally observed phenomenon that some DNA regions are recalcitrant or less than efficient templates in PCR reactions. Therefore, when attempting to clone a 3 kb gene, it may be advisable to try to amplify out overlapping subfragments of 1 to 1.6 kb as well as the full length product.

The two main problems with the use of PCR for isolating and cloning genes are obtaining a single product (or set of products for multigene families) and the reduction in yield due to spurious amplification of smaller PCR products (Mullis, 1991). Whilst a good primer design might minimise these difficulties, they should be anticipated. One solution is to design nested primers for each chosen sequence (Jackson et al., 1991). PCR is conducted with the first primer set for 25 cycles. An aliquot of one to two microlitres from the primary PCR is then used to seed another 15–25 cycles of PCR using the secondary primers. The only condition is that the secondary primers must initiate DNA synthesis from their 3' ends downstream of the 3' position of the primary primer end. This can be as little as 3 bases downstream. The secondary primers confer an additional specific selection on the templates that have been amplified during the first round of PCR. Single bands are thus readily obtained, often from primary PCR preparations that do not give any visible product in the expected size range.

The use of secondary primers is also of use when no PCR product is visible after 30 cycles of primary PCR and other sources of error like absence of template can be discounted. PCR with a second set of internal primers may reveal the presence of the required product whereas extended PCR with the primary primers is more likely to amplify small artefacts.

2. Cloning of PCR products

Before cloning PCR products, it is often advisable to remove the unused primers and any short 'junk' DNA that may also have been generated. Many PCR reactions give rise to very small products often referred to as 'primer-dimers'. The simplest way to remove these in our experience is to pass the PCR preparation through a Sephacryl 400 spun column which has been equilibrated with water or TE buffer. The theoretical exclusion limit for Sephacryl 400 is about 200 bp. There are numerous other proprietry methods which are equally as rapid which rely on selective adsorption or filtration.

Where the desired fragment is part of a mixture of larger DNA molecules, one has to use agarose gel electrophoresis (Sambrook *et al.*, 1989) or high performance liquid chromatography (HPLC) to separate out the different products (Kalnowski *et al.*, 1991). HPLC has the advantage of eluting the required DNA in a liquid media from which the DNA can be easily recovered. Bands can be isolated from agarose gels by any of the numerous methods ranging from electroelution into solution or onto supports (Girvitz *et al.*, 1980; Dretzen *et al.*, 1981), centrifugation (Ericson, 1990) filtration (Weichenhan, 1991) or liquefaction in low melting temperature agarose, to name but a few. Yields of 50% or more from the gel are considered to be good in our experience. Where yields are low, or where a large quantity of the desired product is required, secondary PCR may be the solution. A small agarose plug from the band can be used directly as a template in further PCRs (Zintz and Beebe, 1991). Alternatively, an aliquot of the purified but dilute product can be used. The secondary PCR products are checked on a gel before primer removal as above.

PCR fragments that have been synthesised with enzymes possessing proofreading ability can be cloned directly into blunt-ended vector by ligation (Lohff and Cease, 1992). Fragments synthesised with Taq polymerase need to be polished with Klenow or T4 polymerase. This step removes the additional dATP residue added at the 3′ ends of the fragments (Scharf, 1990). Most primers lack the 5′ phosphate and therefore have to be phosphorylated prior to blunt-end ligation. This is best done on the primers alone prior to PCR rather than on the product afterwards. The alternative to removing the dATP is to clone the fragment into vectors cut so that they have a single dT overhang at their 3′ ends (Kovalic *et al.*, 1991).

Primers can be designed to contain restriction sites at their 5′ ends (Scharf *et al.*, 1986; Jung *et al.*, 1990; Kaufman and Evans, 1990; Scharf, 1990). By ensuring that each primer contains a different restriction site, fragments can be cloned directionally into a suitable double-digested vector. This method is particularly attractive as the digested vector does not have to be dephosphorylated prior to ligation and yet a large proportion of the obtained clones will contain the insert. With increasing fragment size, the choice of restriction sites that will cut the primer but not the fragment decreases. In such cases it is better to use some of the sites for enzymes such as Not I which recognise eight or more base pair palindromes or sequences. The only stricture with making primers with restriction sites, is that a minimum of four additional bases should be tagged onto the 5′ end beyond the restriction site. Some enzymes do not digest sites that are at the 5′ termini and base addition alleviates this problem (Kaufman and Evans, 1990). Alternatively, the PCR products can be ligated into concatamers and then digested, although this may increase the risk of obtaining ligation artefacts.

The quality of the primers used can be quite critical in cloning experiments, especially if the sequence at the termini of the fragments is essential. Oligonucleotides are synthesised by the sequential addition of single bases from the 3′ to the 5′ end. During each cycle of addition, some molecules are not extended. Nowadays, base additions can be made with 99% or more efficiency at each step. At a 99% efficiency, approximately 80% of the final product will be full length in a 20 mer synthesis, 74% will be full length for a 30 mer. Where PCR is conducted with two 30 mer primers made at 99% efficiency per base addition, only 55% of the PCR product will contain both full length primers. Where primer length is critical, it is therefore best to obtain purified full length primer.

3. Alteration and manipulation of sequences by PCR

The PCR can also be used add sequences, join different fragments of DNA together and to introduce specific alterations (Clackson *et al.*, 1991).

The simplest way of adding short sequences up to 200 bp such as restriction sites or transit peptide signals to fragments of DNA is by primer design. An oligonucleotide is made that contains the required sequence plus an additional 18–20 bp which overlap the 5′ end of the product to which they are to be attached. PCR is then performed with the 5′ primer and a primer at the 3′ end of the desired molecule to generate the fusion product (Scharf *et al.*, 1986; Higuchi, 1989; Mensa-Wilmot and Englund, 1992).

Several fragments can be fused together by generating PCR products that overlap (Jayaraman *et al.*, 1989; Yon and Fried, 1989). The products are mixed together, annealed and extended for a few cycles. Primer specific to the 5′ and 3′ ends of the final full length molecule are then added and the full length molecule is recovered by PCR. Large insertions can be made into sequences by a method of 'sticky feet PCR' (Clackson and Winter, 1989). The product to be inserted is amplified using primers that contain homologies to the area of insertion at their termini. The product is annealed to the sequence to be modified, which ideally is in a plasmid vector capable of going through a single-stranded phase of replication. The vector has also been grown in a host which introduces uracil instead of dTTP into the plasmid. The annealed molecules are extended to make a complete circuit of the vector and the resultant junction between the 5′ end of the primer and the 3′ end of the extended molecule is closed by ligation. The chimaeric plasmids thus created are transferred into a suitable host which digests the original uracil containing template molecules but not the newly synthesised strands containing the required insertion. Double-stranded modified plasmid is then synthesised by the host DNA polymerases and can be isolated.

Point mutations can be introduced into sequences by PCR (Higuchi, 1989; Kuipers *et al.*, 1991; Sharrocks and Shaw, 1992). The mutation is incorporated into a primer sequence covering the region to be altered and PCR amplification is performed with the addition of the primer for the 5′ or 3′ end of the gene to be altered. The resulting fragment is then used in turn as a primer with a suitable complementary primer to achieve the amplification of the required full length mutated product. The amplified mutated product can then be cloned conventionally or be incorporated back into the target sequence by any of the above fusion methods.

E. Sequencing and PCR

The thermostable polymerases can be used in sequencing reactions (Innis *et al.*, 1988; Peterson, 1988; Brow, 1990). The advantages are two-fold: first, problems of secondary structure in a sequence, which may block sequencing with Klenow, reverse transcriptase or Sequenase, may be overcome because the sequencing reaction is conducted at higher temperature; second, the sequencing reactions can be repeated several times on the same template (Voss *et al.*, 1989). This generates a detectable amount of sequence product from small quantities of template that could otherwise not give results. The technique is well established and again Taq sequencing kits are available commercially for sequencing with or without amplification.

PCR products can be sequenced by any of the usual methods once they have been cloned in a vector. Sequencing the PCR product directly is also possible (Wrischnik et al., 1987; Innis et al., 1988) but first the primers used for the PCR have to be removed from the DNA preparation as for cloning described above. Straight double-stranded PCR products do not sequence very well as the two linear strands can reanneal fairly rapidly, reducing the efficiency of the sequencing reaction. The sample can be treated so that the strand to be sequenced is present in a greater proportion than the complementary strand using assymetric PCR (Gyllensten and Erlich, 1988; McCabe, 1990). In asymmetric PCR, one of the primers used in the PCR reaction is present at limiting amounts. Initially the PCR will generate double-stranded product exponentially. During later cycles, when one primer has been depleted, one strand is amplified linearly, resulting in a single predominantly single-stranded preparation.

If one of the primers in a normal PCR amplification is biotinylated, then the two DNA strands can be separated after denaturation by binding one strand to Streptavidin coated support such as magnetic beads. The method is particularly attractive as both strands of a PCR product can be separated from each other, one on the beads, the other in solution. These strands can act as highly efficient templates in any single-stranded sequencing protocol (Green et al., 1990).

Where PCR is going to be part of a large scale sequencing programme, consideration ought to be given to using fluorescent dye-labelled sequencing (Smith et al., 1986) in conjunction with a thermocycling-based sequencing system (Connell et al., 1987; Gocayne et al., 1987). This confers the advantages of single-stranded sequencing from small amounts of template with direct processing of results on an automatic gel reader.

F. Quantification and PCR

PCR products can be measured as any normal DNA by either ultraviolet absorption at 260 nm or a variety of staining methods, though the unused primers may have to be removed, e.g. by gel fitration.

Measuring the amount of original template from the amount of PCR product obtained requires a more rigorous approach. This involves performing a series of amplifications in the presence of a serial dilution of an internal standard which co-amplifies with the target DNA in the reaction (Gilliland et al., 1990). Internal standards can take the form of naturally present members of gene families (Buck et al., 1991) or a mutated and purified form of the target sequence as in DIANA (the *det*ection of *i*mmobilised *a*mplified *n*ucleic *a*cids: Lundeberg et al., 1990; Wahlberg et al., 1990).

Another method is the use of fluorescent dye-labelled dNTPs (Edwards et al., 1991). With fluorescent dyes it is now possible to quantify the amount of PCR product much more accurately and at lower DNA concentrations than with ethidium bromide. PCR can be conducted for a smaller number of cycles and DNA concentrations determined during the linear amplification range. The added advantage of performing fewer amplification cycles is also that the product is often much cleaner and less contaminated with low molecular weight artefacts.

6. THE POLYMERASE CHAIN REACTION

V. PRIMER DESIGN

A. General Features of Primers

The three main factors in primer design are specificity, annealing temperature and cost. Cost is wholly dependent on the length of the oligonucleotide, whilst the other two factors are also affected by the sequence chosen. Primers as short as nine bases can be used as in RAPDS. They are used at low annealing temperatures such as 30°C and in most cases they prime at a number of sites in a genome and therefore have a relatively low specificity.

Specificity increases with length. With a 16 mer primer one would expect to find only one target copy in a genomic template of 10^9 base pairs (Sambrook et al., 1989). Up to about 20 bases the primer specificity is sensitive to single base changes under stringent annealing conditions. Above 20 bases, single base changes will be tolerated (Lathe, 1985), especially if they are at the 5' end of the primer, however specificity does increase as the annealing temperature rises. Primer length can in most cases be kept within a maximum length of 30 bases both in terms of specificity (Lathe, 1985) and cost. This maximum includes the addition of non-template sequence such as restriction sites.

The actual sequence chosen for the primer is also very important. Most workers recommend a 50–60% G+C ratio as well as the avoidance of GC-rich sequences and repetitive sequences (Innis and Gelfand, 1990; Saiki, 1990). The main aim is to find a unique primer. The primer sequence should also be checked for the ability of the primer to fold on itself or to hybridise with the other primer in a PCR reaction (Saiki, 1990) as this can lead to reduced efficiency. Primers are short and present in great abundance relative to the target template and any artefacts created by primer interactions can rapidly take over a PCR and reduce the yield of the required product (Mullis, 1991). Often referred to as primer-dimers, these will be present to a greater or lesser degree in many PCRs, depending on primer design.

The specificity of annealing by an oligonucleotide to a template has been shown by Lathe (1985) to be influenced by binding via a short run of bases within the oligo. Greater importance is therefore attached to the sequence at the 3' end of the printer from which extension occurs to ensure that most of the specificity of binding resides in this part of the molecule. Special attention is also given to the 3' terminal base. Studies on Taq and other polymerases indicate that extension can occur from primers having a single base mismatch at the 3' primer end to varying degrees. This is particularly the case with primers ending in a terminal dTTP, which extend as if there were no mismatch at all (Mendelman et al., 1989; Kwok et al., 1990). Primers ending in CC, AG and GA are 100 times less likely to extend at mismatched termini (Mendelman et al., 1989).

B. Calculating the Annealing Temperature

The annealing temperature (T_a) of an oligonucleotide is generally chosen to be 15°C below the melting temperature (T_m) of the oligonucleotide when bound to the template (Sambrook et al., 1989). The T_m is not only dependent on the length of the molecule but also on the sequence. Several equations are available for determining the

T_m of an oligonucleotide. The simplest ascribes a value of 2°C for every T or A and 4°C for every G or C in a primer (Sambrook *et al.*, 1989). A 20 mer with a 50% G+C content would then have a T_m of 60°C. A more accurate method for determining T_m can also be found in Sambrook *et al.* (1989) for oligos up to 60 nucleotides in length:

$$T_m = 81.5 + 16.6 \log(M) + 0.41(P_{gc}) - (B/L) - (P_m) - 0.65(P_f) \qquad (1)$$

where M is the molar concentration of sodium ions to a maximum of 0.5 M; P_{gc} is the percentage of G+C bases in the oligo; P_m is the percentage of mismatched bases if known; P_f is the percentage of formamide in the buffer; B is 675; L is the probe length in bases.

Perhaps the most precise method is that of Rychlik *et al.* (1990), which calculates T_m on the basis of duplex formation in the equation

$$T_m = \frac{\Delta H}{\Delta S + R\ln(c/4)} - 273.15 + 16.6 \log[K^+] \qquad (2)$$

where ΔH and ΔS are the enthalpy and entropy for duplex formation, respectively; R is the molar gas constant at 8.314 J K^{-1} mol^{-1} and c is the concentration of template, set at 250 pM.

Rychlik then calculates the optimum T_a from T_m values calculated for the primer and template from the above equations.

$$T_a^{OPT} = 0.3 T_m^{primer} + 0.7 T_m^{product} - 14.9 \qquad (3)$$

Fortunately, Rychlik's method has been developed as a computer program and is available from National Biosciences (Plymouth, MN, USA). The advantage of the program is that it gives a graphical illustration of the stability of any part of the sequence (if known), allowing primers to be designed to more stable regions of the template. We have found that primers designed and used according to Rychlik on the basis of T_m^{primer} alone or T_a^{OPT} work well. Other software aiding primer design is also available (e.g. Lowe *et al.*, 1990).

Calculations of annealing temperature are at best a guide to the best starting point for setting up PCRs. The most suitable annealing temperature should be determined empirically where conditions are critical or problems in PCR have been encountered. A comparison of calculated and actual annealing temperatures was conducted by Hamill *et al.* (1991) using three methods for determining the optimum annealing temperature for a variety of primers for use on transformed plants. Their study revealed discrepancies of varying degrees between theoretical and empirically determined values. The general aim should be to find an annealing temperature that is as high as practicably possible. Ideally, the T_a should be the same as the extension temperature, 72°C. This removes one temperature incubation step in the PCR cycle and reduces the total run time for PCRs.

C. Heterologous Primers

Primers can be designed to function specifically in situations where there is a known or anticipated sequence diversity, e.g. when cloning genes from different sources or working from a nucleic acid sequence derived from an amino acid sequence. The following approach to primer design is essentially that of Sambrook *et al.* (1989) for designing degenerate oligonucleotides for hybridisation, adapted to PCR (see also McPherson et al., 1991b).

The first aim is to limit the number of possible base changes that could affect the primer. Conserved regions with little or no variation can be found by comparison where sequence information exists from a number of different organisms for the gene of interest.

Backtranslation of amino acid sequences can result in a high degree of redundancy. This can be reduced by selecting areas of sequence which do not contain highly redundant amino acids such as serine, arginine or leucine at high abundance. The next step is to write down the selected backtranslated amino acid sequence, highlighting the variable bases. Codon usage varies between different organisms and some codons are used more frequently than others. By consulting a codon usage table such as that published by Murray et al. (1989) or one prepared from a selection of sequences already determined from the organism of interest, it is possible to omit codons such as AAT for asparagine in monocots or ATA for isoleucine in plants in general. Murray et al. (1989) also give other base preferences in codon usage between dicotyledons and monocotyledons.

Having reduced redundancy as far as possible, there are several approaches to primer design. The simplest is to synthesise a mixture of primers encompassing all potential variants of the sequence. Theoretically, only the correct primers amplify a product that is still visible on an ethidium bromide-stained gel, as the primers are present at high concentration in PCR reactions. The second approach is to minimise the sequence differences by using deoxyinosine where three or four bases can encode for a particular position. Deoxyinosine will base pair with any base. An additional more controversial method is to code any wobble comprised of an A or G as G and any C or T wobble as T. This is a transfer to DNA of the observation by Uhlenbeck et al. (1971) that uracil can base pair with both adenosine and guanosine in RNA. These measures are all aimed at providing a continuous sequence which will enhance the binding of the primer to the template (Lathe, 1985). Any sequence divergence at the last four or five 3' bases is retained to ensure maximum specificity of the primers (Sommer and Tautz, 1989). Where, possible, the 3' terminus of the primer should be at a non-redundant base to ensure extension.

With heterologous primers, the annealing temperature should be determined empirically. Two sets of nested primers can also be used (or at least designed) to maximise both specificity and yield of PCR product.

The design of primers for RAPDS is much more empirical. The concensus appears to be that primer length can vary from 8 to 12 bases, with 10 bases being used most commonly, and that the G+C content should be 50% or greater.

VI. TROUBLESHOOTING

Whilst PCR has been hailed as the answer to all molecular biologists problems, the novice soon finds that the method is not without its own pitfalls. The most obvious and damaging phenomenon is the apparent variability in obtaining results which can rapidly lead to a discrediting of the technique (or the person using it!). The following points may be of use when the inevitable snag is encountered in a PCR.

A. Use of Controls

ALWAYS include a positive and negative control in the PCRs. The former will indicate whether the system is working, and the latter whether contamination has crept in. It is also advisable to perform control reactions in the presence or absence of template with each primer used on its own as single primers can give rise to bands that could be misinterpreted in a PCR with two primers. These steps alone should guard against the major pitfalls.

B. Negative Results

Negative results can take two forms. In the first case, no PCR product whatsoever is seen when run on a gel. The most likely cause is a poisoning of the PCR by the template, especially if some of the cruder DNA purification methods are used. This can be checked for by spiking a series of PCRs that are known to function reliably with a serial dilution of the suspect template.

The second case is where the PCR is apparently functional but the expected band does not appear. A wide variety of reasons may account for this. Try out a range of different annealing temperatures from 35°C upwards in 5°C increments. Check that the extension time is adequate. If using genomic DNA, then ensure that the denaturation time during the first few cycles is sufficient, e.g. 5 min before the first cycle, 90 s for the next two cycles.

The appearance of the PCR on a gel may also give an indication of possible causes for failure. A smear of DNA rather than discrete bands suggests either template overload at the start of amplification or excessive amplification (Bell and DeMarini 1991). Reducing the number of cycles can alleviate this. Large amounts of products smaller than that expected will have a significant effect on the yield of the desired product as they will be amplified preferentially. Secondary PCR with nested primers should be tried as this increases specificity and often reveals products amplified suboptimally during the first round.

If the above do not work, then one should try a new set of primers — sometimes a shift of even a few bases can make the difference.

Finally, it may just be that the template is recalcitrant to PCR. In our experience there are certain genes or regions of genes that do not amplify, irrespective of all alterations tried (A.F.Weir, personal communication).

C. Too Many Bands

The first step is to try higher annealing temperatures and secondary PCR. In some cases the primers may be annealing to the template and are being extended in the period after setting up the PCR but before the very first denaturation. This can be avoided with a 'hot start' (D'Aquila et al., 1991). The PCR with the template but without enzyme is allowed to go through the first denaturation before the enzyme is added at or above the annealing temperature.

Individual primers should be checked to see if one is giving rise to spurious products. The reactions may have become contaminated. The magnesium chloride in the reaction

6. THE POLYMERASE CHAIN REACTION

buffer can also lead to reduced specificity if too high. If all else fails, try different primers and primer combinations.

D. Contamination

PCR also stands for *p*reviously *c*ontaminated *r*eagents, especially where the same product is regularly screened for. Contamination can arise either from the work area, the solutions and reagents or the equipment such as pipetters and tips used (Kwok and Higuchi, 1989; Kitchin *et al.*, 1991). In all cases it is advisable to buy in the basic reagents for PCR such as the enzyme, the enzyme buffer (generally supplied as 10-fold concentrate as a matter of course), and the deoxynucleotides. Primers should also be purchased or at least made at a separate location to the main DNA or PCR work. Where possible dispense the reagents into smaller aliquots before use and minimise the number of times a solution may be reopened for use.

Store PCR solutions and DNAs at separate locations. Always add DNA last to reaction mix. The only exception to this rule is when a 'hot start' PCR is being performed. In this case prepare a fresh aliquot of the enzyme to be added to the reaction in a separate container and pipette from this material rather than directly from the stock.

Where the PCR work is varied enough, the above should be sufficient for work at the laboratory bench. For repetitive screening, PCR should be treated as a sterile technique with its own demarcated work area, special designated pipetters and other materials. Kitchin *et al.* (1991) give a cautionary tale on how even a protective hat may have to be worn! Primer sequences which can detect most common plasmid contaminations have also been designed and published (Nerenberg and Minor, 1991). The enzymes used in PCR are obtained from bacteria and therefore contain trace quantities of bacterial RNA and DNA. This can be checked by using another set of primers designed by Schmidt *et al.* (1991). In cases where this material turns out to be a major problem, the DNA can be removed, or at least made unsuitable for PCR, by a variety of methods such as UV irradiation (Sarkar and Sommer, 1990), enzymic digestion (Furrer *et al.*, 1990; Li *et al.*, 1991) or chemical modification (Jinno *et al.*, 1990).

Contamination of the template before PCR can be reduced by using disposable plasticware where possible and treating reusable materials with 0.2 M sodium hydroxide for 1 h followed by 0.2 M hydrochloric acid for a further hour and rinsing with sterile water.

Where contamination is detected, it is better to start afresh with fresh reagents, a clean work area and new template. Nine times out of ten this will be sufficient to remove the problem and is also less time consuming than hunting for a source of contamination.

E. PCR Machines

There are many different PCR machines available. Problems may therefore sometimes be encountered when trying to repeat a PCR on two different machines. In our experience, this can even be observed between machines from the same manufacturer.

Check whether the original PCR was controlled by the block temperature or the original sample temperature. Alteration of the duration of denaturation, annealing and extension may be needed. Alterations in sample volume can also affect cycling time, as can the positioning of sample temperature probes controlling an amplification in the heating block.

It is always worth checking for PCR reproducibility across a heating block with a standard PCR to see if certain positions give reproducibly poor results. A number of such studies have been published (Hoelzel, 1990; Van Leuven, 1990) but fortunately the manufacturers are addressing the problem with improved block designs. A temperature probe linked to a chart recorder or other recording device can be useful. Using the probe, one can either monitor an entire PCR run or check the temperature profiles of individual sample wells. A good PCR machine will give square temperature profiles (see Hoelzel, 1990), indicating rapid heating to set temperature without overheating or undercooling.

VII. CONCLUSION

The polymerase chain reaction is a deceptively simple concept, the amplification of a piece of nucleic acid *in vitro* using two primers and a polymerase. The virtue of the simplicity is that the principle is so versatile. PCR has been applied in many ingenious ways and this chapter can only provide an outline of some of the applications of PCR and considerations on its use. For collections of such methods, the reader is again referred to the references given in the introduction. More specific methods and PCR tips are frequently published in the methods and new techniques sections of *Nucleic Acids Research* and in *Biotechniques*. There is also a journal devoted entirely to PCR, *PCR Methods and Applications*, published by the Cold Spring Harbor Laboratory Press.

The disadvantage of the simplicity of the concept of PCR is that the importance of experimental design for achieving reproducible and valid results is often underestimated. Primer design and the inclusion of suitable controls are the key elements to ensure success.

The future will no doubt bring many further novel applications of PCR. The likely trends apart from experimental methodology are towards faster cycling times (reducing the time of transition from one temperature incubation to another), greater automation of sample handling and analysis, and higher sample throughput.

Finally, a note of caution — PCR is compulsive. It takes just one successful application and you are hooked!

ACKNOWLEDGEMENTS

I would like to thank Dr M. Maunders for his help with the section on RAPDS and Dr K. S. Blundy for his comments during the writing of this chapter. I am indebted to all my colleagues at ATC for adopting and adapting PCR to great effect, thus making the writing of this chapter possible.

REFERENCES

Alexander, D. C. (1990). *Nucl. Acids Res.* **18**, 7453-7454.
Barany, F. (1991). *Proc. Natl. Acad. Sci. USA* **88**, 189-193.
Barnes, T. M. (1990). *Nucl. Acids Res.* **18**, 6741-6742.
Bell, D. A. and DeMarini, D. M. (1991). *Nucl. Acids Res.* **19**, 5079.
Belyavsky, A., Vinogradov, T. and Rajewsky, K. (1989). *Nucl. Acids Res.* **17**, 2919-2932.
Bergseid, M., Scott, B., Mathur, S., Nielson, K., Shoemaker, D. and Mathur, E. (1991). *Strategies* **4**, 34-35.
Bloem, L. J. and Yu, L. (1990). *Nucl. Acids Res.* **18**, 2830.
Brow, M. A. D. (1990). *In* "PCR Protocols: A Guide to Methods and Applications" (M. A. Innis, D. H. Gelfand, J. J. Sninsky and T. J. White, eds), pp. 189-196. Academic Press, San Diego.
Buck, K. J., Adron, H. and Sikela, J. M. (1991). *Biotechniques* **11**, 636-641.
Caetano-Anolles, G., Bassam, B. J. and Gresshoff, P. M. (1991). *BioTechnology* **9**, 553-557.
Carlson J. E., Tulsieram, L. K., Glaubitz, J. C., Luk, V. W. K., Kauffeldt, C. and Rutledge R. (1991). *Theor. Appl. Genet.* **83**, 194-200.
Carothers A. M., Urlaub, G., Mucha, J., Grunberger, D. and Chasin, L. A. (1989). *Biotechniques* **7**, 494-499.
Chamberlain, J. S., Gibbs, R. A., Ranier, J. E. and Laskey, C. T. (1990). *In* "PCR Protocols: A Guide to Methods and Applications" (M. A. Innis, D. H. Gelfand, J. J. Sninsky and T. J. White, eds), pp. 272-281. Academic Press, San Diego.
Clackson, T. and Winter, G. (1989). *Nucl. Acids Res.* **17**, 10163-10170.
Clackson, T., Güssow, D. and Jones, P. T. (1991). *In* "PCR: A Practical Approach" (M. J. McPherson, P. Quirke and G. R. Taylor, eds), pp. 187-214. Oxford University Press, Oxford.
Clark, J. M. (1988). *Nucl. Acids Res.* **16**, 9677-9686.
Connell, C., Fung, S., Heiner, C., Bridgham, J., Chakerain, V., Heron, E., Jones, B., Menchen, S., Mordan, W., Raff, M., Recknor, M., Smith, L., Springer, J., Woo, S. and Hunkapiller, M. (1987). *Biotechniques* **5**, 348-362.
D'Aquila, R. T. D., Bechtel, L. J., Videler, J. A., Eron, J. J., Gorczyca, P. and Kaplan, J. C. (1991). *Nucl. Acids Res.* **19**, 3749.
Dellaporta, S. L., Wood, J. and Hicks, J. B. (1983). *Plant Mol. Biol. Rep.* **1**, 19-21
Doherty, P. J., Huesla-Contreras, M., Dosch, H. M. and Pan, S. (1989). *Anal. Biochem.* **177**, 7-10.
Dretzen, G., Bellard, M., Sassone-Corsi, P. and Chambon, P. (1981). *Anal. Biochem.* **112**, 295-298.
Earp, D. J., Lowe, B. and Baker, B. (1990). *Nucl. Acids Res.* **18**, 3271-3279.
Eckert, K. A. and Kunkel, T. A. (1991). *In* "PCR: A Practical Approach" (M. J. McPherson, P. Quirke, and G. R. Taylor, eds), pp. 225-244. Oxford University Press, Oxford.
Edwards, J. B. D. M., Delort, J. and Mallet, J. (1991). *Nucl. Acids Res.* **19**, 5227-5232.
Ericson, M. L. (1990). *Trends Genet.* **6**, 278.
Erlich, H. A. (ed.) (1989). "PCR Technology: Principles and Applications for DNA Amplification", 246 pp. Stockton Press, New York.
Farr, C. J., Saiki, R. K., Erlich, H. A., McCormick, F. and Marshall, C. J. (1988). *Proc. Natl. Acad. Sci. USA* **85**, 1629-1633.
Frohman, M. A., Dush, M. K. and Martin, G. R. (1988). *Proc. Natl. Acad. Sci. USA* **85**, 8998-9002.
Furrer, B., Candrian, U., Wieland, P. and Luthy, J. (1990). *Nature* **346**, 324.
Gawel, N. J. and Jarret, R. L. (1991). *Plant. Mol. Biol. Rep.* **9**, 262-266.
Gelfand, D. (1989). *In* "PCR Technology: Principles and Applications for DNA Amplification" (H. A. Erlich, ed.), pp. 17-22. Stockton Press, New York.
Gibbs, R. A., Nguyen, P.-N. and Caskey, C. T. (1989). *Nucl. Acids Res.* **17**, 2437-2448.
Gilliland, G., Perrin, S. and Bunn, H. F. (1990). *In* "PCR Protocols: A Guide to Methods and Applications" (M. A. Innis, D. H. Gelfand, J. J. Sninsky and T. J. White, eds), pp. 60-75. Academic Press, San Diego.

Girvitz, S. C., Bacchetti, S., Rainbow, A. J. and Graham, F. L. (1980). *Anal. Biochem.* **106**, 492-496.
Goblet, C., Prost, E. and Whalen, R. G. (1989). *Nucl. Acids Res.* **17**, 2144.
Gocayne, J., Robinson, D. A., FitzGerald, M. G., Chung, F.-Z., Kerlavage, A. R., Lentes, K.-U., Lai, J., Wang, C.-D., Fraser, C. M. and Venter, J. C. (1987). *Proc. Natl. Acad. Sci. USA* **84**, 8296-8300.
Green, A., Roopra, A. and Vaudin, M. (1990). *Nucl. Acids Res.* **18**, 6163-6164.
Gurr, S. J. and McPherson, M. J. (1991). In "PCR: A Practical Approach" (M. J. McPherson, P. Quirke, and G. R. Taylor, eds), pp. 147-170. Oxford University Press, Oxford.
Güssow, D. and Clackson, T. (1989). *Nucl. Acids Res.* **17**, 4000.
Gyllensten, U. B. and Erlich, H. A. (1988). *Proc. Natl. Acad. Sci. USA* **85**, 7652-7656.
Haliassos, A., Chomel, J. C., Tesson, L., Baudis, M., Kruh, J., Kaplan, J. C. and Kitzis, A. (1989). *Nucl. Acids Res.* **17**, 3606.
Hamill, J., Rounsley, S., Spencer, A., Todd, G. and Rhodes, M.J.C. (1990). In "Progress in Plant Cellular and Molecular Biology" (H. J. J. Nijkamp, L. H. W. Van der Plas and J. Van Aartrijk, eds), pp. 183-188. Kluwer Academic Publishers, Dordrecht.
Hamill, J. D., Rounsley, S., Spencer, A., Todd, G. and Rhodes, M. J. C. (1991). *Plant Cell Rep.* **10**, 221-224.
Herfort, M. R. and Garber, A. T. (1991). *Biotechniques* **11**, 598-603.
Higuchi, R. (1989). In "PCR Technology: Principles and Applications for DNA Amplification" (H. A. Erlich, ed.), pp. 61-70. Stockton Press, New York.
Hoelzel, R. (1990). *Trend. Genet.* **6**, 237.
Hovens, C. M. and Wilks, A. F. (1989). *Nucl. Acids Res.* **17**, 4415-4416.
Hu, J. and Quiros, C. F. (1991). *Plant Cell Rep.* **10**, 505-511.
Huang, S., Yiyuan, H., Wu, C. and Holcenberg, J. (1990). *Nucl. Acids Res.* **18**, 1922.
Hung, T., Mak, K. and Fong, K. (1990). *Nucl. Acids Res.* **18**, 4953.
Innis, M. A. and Gelfand, D. H. (1990). In "PCR Protocols: A Guide to Methods and Applications" (M. A. Innis, D. H. Gelfand, J. J. Sninsky and T. J. White, eds), pp. 3-12. Academic Press, San Diego.
Innis, M. A., Myambo, K. B., Gelfand, D. H. and Brow, M. A. D. (1988). *Proc. Natl. Acad. Sci. USA* **85**, 9436-9440.
Innis, M. A., Gelfand, D. H., Sninsky, J. J. and White, T. J. (eds) (1990). "PCR Protocols: A Guide to Methods and Applications", 482 pp. Academic Press, San Diego.
Isola, N. R., Harn, H. J. and Cooper, D. L. (1991). *Biotechniques* **11**, 580-582.
Ivinson, A. J. and Taylor, G. R. (1991). In "PCR: A Practical Approach" (M. J. McPherson, P. Quirke, and G. R. Taylor, eds), pp. 15-27. Oxford University Press, Oxford.
Jackson, D. P., Hayden, J. D. and Quirke, P. (1991). In "PCR: A Practical Approach" (M. J. McPherson, P. Quirke and G. R. Taylor, eds), pp. 29-50. Oxford University Press, Oxford.
Jayaraman, K., Shah, J. and Fyles, J. (1989). *Nucl. Acids Res.* **17**, 4403.
Jinno, Y., Yoshiura, K. and Niikawa, N. (1990). *Nucl. Acids Res.* **18**, 6739.
Jung, V., Pestka, S. B. and Pestka, S. (1990). *Nucl. Acids Res.* **18**, 6156.
Kalnowski, M. H., McCoy-Haman, M. F. and Hollis, G. F. (1991). *Biotechniques* **11**, 246-249.
Kandpal, R. P., Shukla, H., Ward, D. C. and Weissman, S. M. (1990). *Nucl. Acids Res.* **18**, 3081.
Kaufman, D. L. and Evans, G. A. (1990). *Biotechniques* **9**, 304-306.
Kitchin, P. A., Szotyori, Z., Fromholc, C. and Almond, N. (1991). *Nature* **344**, 201.
Klein-Lankhorst, R. M., Vermunt, A., Weide, R., Liharska, T. and Zabel, P. (1991). *Theor. Appl. Genet.* **83**, 108-114.
Kovalic, D., Kwak, J. H. and Weisblum, B. (1991). *Nucl. Acids Res.* **19**, 4560.
Krawczak, M., Reiss, J., Schmidke, J. Rosler, U. (1989). *Nucl. Acids Res.* **19**, 2197-2201.
Kuipers, O. P., Boot, H. J. and deVos, W. M. (1991). *Nucl. Acids Res.* **19**, 4558.
Kwiatkowski, T. J., Zoghbi, H. Y., Ledbetter, S. A., Ellison, K. A. and Chinault, A. C. (1990). *Nucl. Acids Res.* **18**, 7191.
Kwok, S. and Higuchi, R. (1989). *Nature* **339**, 237-238.
Kwok, S., Kellog, D. E., Spasic, D., Goda, L., Levenson, C. and Sninsky, J. J. (1990). *Nucl. Acids Res.* **18**, 999-1005.

Langridge, U., Schwall, M. and Langridge, P. (1991). *Nucl. Acids Res.* **19**, 4954.
Lassner, M. W., Peterson, P. and Yoder, J. I. (1989). *Plant Mol. Biol. Rep.* **7**, 116-128.
Lathe, R. (1985). *J. Mol. Biol.* **183**, 1-12.
Li, H., Cui, X. and Arnheim, N. (1991). *Nucl. Acids Res.* 3139-3141.
Loh, E., Elliot, J. F., Cwiria, S., Lanier, L. L. and Davis, M. M. (1989). *Science* **243**, 217-220.
Lohff, C. J. and Cease, K. B. (1992). *Nucl. Acids Res.* **20**, 144.
Lowe, T., Sharefkin, J., Yang, S. Q. and Dieffenbach, C. W. (1990). *Nucl. Acids Res.* **18**, 1757-1761.
Lundeberg, J., Wahlberg, J. and Uhlen, M. (1990). *Biotechniques* **10**, 68-75.
Maga, E. A. and Richardson, T. (1991). *Biotechniques* **11**, 186-187.
Martin, G. B., Williams, J. G. K. and Tanksley, S. D. (1991). *Proc. Natl. Acad. Sci. USA* **88**, 2336-2340.
Mattila, P., Korpel, J., Tenkanen, T. and Pitkanen, K. (1991). *Nucl. Acids Res.* **19**, 4967-4973.
Mayrand, P. E., Corcoran, K. P., Ziegle, J. S., Robertson, J. M., Hoff, L. B. and Kronick, M. N. (1992). *Appl. Theor. Electrophoresis* **3**, 1-11.
McCabe, P. C. (1990). In "PCR Protocols: A Guide to Methods and Applications" (M. A. Innis, D. H. Gelfand, J. J. Sninsky and T. J. White, eds), pp. 76-83. Academic Press, San Diego.
McGarvey, P. and Kaper, J. M. (1991). *Biotechniques* **11**, 428-432.
McPherson, M. J., Quirke, P. and Taylor, G. R. (eds) (1991a). "PCR: A Practical Approach", 253 pp. Oxford University Press, Oxford.
McPherson, M. J., Jones, K. M. and Gurr, S. J. (1991b). *In* "PCR: A Practical Approach" (M. J. McPherson, P. Quirke, and G. R. Taylor, eds), pp. 171-186. Oxford University Press, Oxford.
Mendelman, L. V., Petruska, J. and Goodman, M. F. (1989). *J. Biol Chem.* **265**, 2338-2346.
Mensa-Wilmot, K. and Englund, P. T. (1992). *Nucl. Acids Res.* **20**, 143.
Mullis, K. B. (1991). *PCR Methods and Applications* **1**, 1-4.
Mullis, K. B. and Faloona, F. A. (1987). *In* "Methods in Enzymology", Vol. 155 (R. Wu, ed.), pp. 335-350. Academic Press, San Diego.
Murray, E. E., Lotzer, J. and Eberle, M. (1989). *Nucl. Acids Res.* **17**, 477-498.
Myers, T. W. and Gelfand, D. H. (1991). *Biochemistry* **30**, 7661-7666.
Nerenberg, M. I. and Minor, T. (1991). *Biotechniques* **11**, 332-333.
Newton, C. R., Graham, A., Heptinstall, L. E., Powell, S. J., Summers, C., Kalsheker, N., Smith, J. C. and Markham, A. F. (1989). *Nucl. Acids Res.* **17**, 2503-2516.
Ochman, H., Gerber, A. S. and Hartl, D. L. (1988). *Genetics* **120**, 621-623.
Ochman, H., Medhora, M. M., Garza, D. Hartl, D. L. (1990). *In* "PCR Protocols: A Guide to Methods and Applications" (M. A. Innis, D. H. Gelfand, J. J. Sninsky and T. J. White, eds), pp. 219-227. Academic Press, San Diego.
Oliver, R. P. (1990). *Biol. Res. Norwich* **5**, 4.
Patil, R. V. and Dekker, E. E. (1990). *Nucl. Acids Res.* **18**, 3080.
Peterson, M. G. (1988). *Nucl. Acids Res.* **16**, 10915.
Ponce, M. R. and Micol, J. L. (1992). *Nucl. Acids Res.* **20**, 623.
Rasmussen, U. B., Basset, P. and Daniel, J.-Y. (1989). *Nucl. Acids Res.* **17**, 3308.
Reitner, R. S., Williams, J. G. K., Feldman, K. A., Rafalski, J. A., Tingey, S. V. and Scolnik, P. A. (1992). *Proc. Natl. Acad. Sci. USA* **89**, 1477-1481.
Roberts, R. G., Montandon, A. J., Green, P. M. and Bentley, D. R. (1991). *In* "PCR: A Practical Approach" (M. J. McPherson, P. Quirke and G. R. Taylor, eds), pp. 51-66. Oxford University Press, Oxford.
Rosenberg, H. F., Corrette, S. E., Tenen, D. G. and Ackerman, S. J. (1991). *Biotechniques* **10**, 53-54.
Runnebaum, I. R., Syka, P. and Sukumar, S. (1991). *Biotechniques* **11**, 446-447.
Rychlik, W., Spencer, W. J. and Rhoads, R. E. (1990). *Nucl. Acids Res.* **18**, 6409-6412.
Saiki, R. K. (1990). *In* "PCR Protocols: A Guide to Methods and Applications" (M. A. Innis, D. H. Gelfand, J. J. Sninsky and T. J. White, eds), pp. 13-20. Academic Press, San Diego.
Saiki, R. K., Scharf, S., Faloona, F., Mullis, K. B., Horn, G. T., Erlich, H. A. and Arnheim, N. (1985). *Science* **230**, 1350-1354.

Saiki, R. K., Gelfand, D. H., Stoffel, S., Scharf, S. J., Higuchi, R., Horn, G. T., Mullis, K. B. and Erlich, H. A. (1988). *Science* **239**, 487-491.
Sambrook, J., Fritsch, E. F. and Maniatis, T. (eds) (1989). "Molecular Cloning: A Laboratory Manual", 2nd edn. Cold Spring Harbor Laboratory Press, NY.
Sarkar. G. and Sommer, S. S. (1990). *Nature* **343**, 27.
Sarkar, G., Kapelner, S. and Sommer, S. S. (1990). *Nucl. Acids Res.* **18**, 7465.
Scharf, S. J. (1990). *In* "PCR Protocols: A Guide to Methods and Applications" (M. A. Innis, D. H. Gelfand, J. J. Sninsky and T. J. White, eds), pp. 84-91, Academic Press, San Diego.
Scharf, S. J., Horn, G. T. and Erlich, H. A. (1986). *Science* **233**, 1076-1078.
Schmidt, T. M., Pace, B. and Pace, N. R. (1991). *Biotechniques* **11**, 176-177.
Schwartz, K., Hansen-Hagge, T. and Bartram, C. (1990). *Nucl. Acids Res.* **18**, 1079.
Sellner, L. N., Coelen, R. J. and Mackenzie, J. S. (1992). *Nucl. Acids Res.* **20**, 1487-1490.
Seyama, T., Ito, T., Hayashi, T., Mizuno, T., Nakamura, N. and Akiyama, M. (1992). *Nucl. Acids Res.*, **20**, 2493-2496.
Sharrocks, A. D. and Shaw, P. E. (1992). *Nucl. Acids Res.* **20**, 1147.
Sheardown, S. A. (1992). *Trends Genet.* **8**, 121.
Sheffield, V. C., Cox, D. R., Lerman, L. S. and Myers, R. M. (1989). *Proc. Natl. Acad. Sci. USA* **86**, 232-236.
Smith, L. M., Sanders, J. Z., Kaiser, R. J., Hughes, P., Dodd, C., Connell, C.R., Heiner, C., Kent, S. B. H. and Hood, L. E. (1986). *Nature* **321**, 674-679.
Sommer, R. and Tautz, D. (1989). *Nucl. Acids Res.* **17**, 6749.
Tindall, K. R. and Kunkel, T. A. (1988). *Biochemistry* **27**, 6008-6013.
Tulsieram, L. K., Glaubitz, J. C., Kiss, G. and Carlson, J. E. (1992). *BioTechnology* **10**, 686-690.
Torgersen, H., Blaas, D. and Skern, T. (1989). *Anal. Biochem.* **176**, 33-35.
Uhlenbeck, O. C., Martin, F. H. and Doty, P. (1971). *J. Mol. Biol.* **57**, 217.
Van Leuven (1991). *Trends Genet.* **7**, 142.
Voss, H., Schwager, C., Wirkner, U., Sproat, B., Zimmermann, J., Rosenthal, A., Erfle, H., Stegemann, J. and Ansorge, W. (1989). *Nucl. Acids Res.* **17**, 2517-2527.
Wahlberg, J., Lundeberg, J., Hultman, T. and Uhlen, M. (1990). *Proc. Natl. Acad. Sci. USA* **87**, 6569-6573.
Weichenhan, D. (1991). *Trends. Genet.* **7**, 109.
Weiss, R. (1991). *Science* **254**, 1292-1293.
Welsh, J. and McClelland, M. (1990). *Nucl. Acids Res.* **18**, 7213-7218.
Welsh, J. and McClelland, M. (1991). *Nucl. Acids Res.* **19**, 5275-5279.
Welsh, J., Honeycutt, R. J., McClelland, M. and Sobral, B. W. S. (1991). *Theor. Appl. Genet.* **82**, 473-476.
Wilde, J., Waugh, R. and Powell, W. (1992). *Theor. Appl. Genet.* **83**, 871-877.
Williams, J. G. K., Kubelik, A. R., Livak, K. J., Rafalski, J. A. and Tingey, S. V. (1990). *Nucl. Acids. Res.* **18**, 6532-6535.
Winship, P. R. (1989). *Nucl. Acids Res.* **17**, 1266.
Wrischnik, L. A., Higuchi, R. G., Stonekink, M., Erlich, H. A., Arnheim, N. and Wilson, A. C. (1987). *Nucl. Acids Res.* **15**, 529-542.
Wu, D. Y. and Wallace, R. B. (1989). *Genomics* **4**, 560-569.
Yon, J. and Fried, M. (1989). *Nucl. Acids Res.* **17**, 4895.
Zintz, C. B. and Beebe, D. C. (1991). *Biotechniques* **11**, 158-162.

7 Non-radioactive *In Situ* RNA Hybridisation Using Digoxigenin and an Application for Co-localisation Studies with Radioisotopes

EUGENE Y. TANIMOTO and THOMAS L. ROST

Section of Botany, University of California, Davis, CA 95616–8537, USA

I.	Introduction...	142
II.	Background discussion and protocols	143
	A. General precautions..	143
	B. Tissue selection...	143
	C. Microtechnique..	143
	D. Embedding	145
	E. Sectioning and mounting...	145
	F. Microtechnique protocol..	145
	G. Probe preparation	146
	H. Protocol for linearisation of the vector..............................	147
	I. Protocol for preparation of DIG-labelled transcripts	147
	J. Testing probe	148
	K. Protocol for preparing dot blots to test probe	148
	L. Prehybridisation..	149
	M. Prehybridisation protocol..	149
	N. Hybridisation	150
	O. Hybridisation protocol..	151
	P. Posthybridisation treatment	151
	Q. Posthybridisation protocol..	152
	R. Immunodetection	152

	S. Immunodetection and staining protocol	153
III.	Co-localisation protocol	154
	A. Fixation, embedding and sectioning	154
	B. Autoradiography	154
	C. Prehybridisation	154
	D. Hybridisation	155
	E. Posthybridisation	155
	F. Immunological detection	155
IV.	Results and discussion	156
	References	157

I. INTRODUCTION

In situ hybridisation technology is a tool to localise nucleic acid molecules in plant (Meyerowitz, 1987; Smith *et al.*, 1987; Cox and Goldberg, 1988; Raikhel *et al.*, 1989; Bochenek and Hirsch, 1990) and animal cells. Several reviews and comparison studies have been written which outline the general methods (Singer *et al.*, 1986; Raikhel *et al.*, 1989; Larsson, 1989; Kiyama *et al.*, 1990; Clavel *et al.*, 1991). The two main approaches involve either radioactive or non-radioactive labelled probes. The choice of which procedure to use is usually a compromise decision weighing the good and bad characteristics of each. The use of radioactive probes is increasingly more bothersome because of the monitoring equipment, special procedures and the expense of handling, storage and waste disposal. Radioactive procedures will probably never be abandoned completely, but when starting an *in situ* study for the first time, non-radioactive methods should probably be considered carefully as a first approach.

The most widely used non-radioactive methods use probes labelled with biotin (BIO) or digoxigenin (DIG). Both methods are safe and stable, but neither is considered to have the high sensitivity of radioactive probes (Negro *et al.*, 1985; Kiyama *et al.*, 1990). Results with BIO and DIG are faster to obtain than radioactive methods, since there are short exposure and stain reaction times. The reaction products produced in the non-radioactive method show greater resolution because there is no silver grain scatter to deal with (Larsson, 1989).

There are a few reports of the use of DIG on animal (Kiyama *et al.*, 1990; Tsukamoto *et al.*, 1991) and plant tissues (Bochenek and Hirsch, 1990; Tanimoto *et al.*, 1992). DIG is a plant aglycone which is conjugated to nucleotides incorporated into antisense RNA. The DIG method is about as sensitive as the BIO method (Kiyama *et al.*, 1990; Clavel *et al.*, 1991) and since the overall procedures used for both are similar, there is no obvious advantage in the use of one over the other.

In this chapter we will describe the use of DIG-labelled probes on paraffin-embedded plant tissues. We will also discuss the use of the DIG method along with radioactive probes as a means to double label for two components simultaneously. The example we will describe involves localising an mRNA for histone H2A and the co-localisation of cells which are progressing through DNA synthesis as measured by ^{3}H-thymidine incorporation and autoradiography. This general method should be equally useful for other double label combinations following the same procedural logic.

The following steps are generally followed regardless of the method used for probe labelling (Larsson, 1989, and others):

(1) Tissue preparation and microtechnique: fixation (selection of fixation method); embedding (selection of medium — cryosectioning, paraffin or plastic resin).
(2) Prehybridisation of tissues to enhance probe penetration (proteinase K or other enzymes).
(3) Hybridisation of labelled probe.
(4) Posthybridisation washes to eliminate background.
(5) Detection of labelled probe.

II. BACKGROUND DISCUSSION AND PROTOCOLS

A. General Precautions

The handling of RNA, whether in tissues or in Eppendorf tubes, requires special attention. Degradation can be caused by endogenous or contaminating RNases from a variety of origins. Collected tissues should be quickly placed in fixative and stably stored. Whenever possible, all aqueous solutions should be made with distilled H_2O and treated with diethyl pyrocarbonate (DEPC) to a concentration of 0.05% (v/v) prior to autoclaving. DEPC should be handled with care because it can explode at temperatures above 4°C (Scofield et al., 1992). Except when noted, all reagents should be of the highest grade. The use of sterile disposable laboratory ware is recommended, and all glassware should be washed with a 0.1% (v/v) DEPC solution in H_2O, or a 0.5% (v/v) sodium dodecyl sulphate solution in H_2O, or baked in an oven at 150°C for 30 min. Gloves should be worn when handling any objects where RNases could become a problem.

B. Tissue Selection

The practical limits of *in situ* hybridisation should be taken into consideration when choosing tissues to study. The abundance of target mRNAs not only affects the signal-to-noise ratio, but also the amount of time required for detection. Detection time varies with the mRNA abundance, where low abundance mRNAs can require lengthy detection periods. Other considerations for tissue selection include its physical dimension and how this affects handling during sectioning. Very small material, although microscopically viewable, presents problems in orientation and acquisition of the proper planes of section required to view expression patterns.

C. Microtechnique

Cryosectioning of unfixed material is the best procedure for difficult to preserve tissues or for samples which must be examined rapidly. The disadvantage of frozen sections is the difficulty of handling the sections, adhering sections to slides, and problems of poor preservation due to ice damage. Still, cryosectioning is a method of choice in some instances (Bochenek and Hirsch, 1990).
 Plant cells present a unique challenge to the microscopist because of their structure — a rigid cell wall surrounding variably dense cytoplasm and often a large vacuole.

Sometimes a compromise solution is best where good fixation of a cytoplasmic component is exchanged for slight shrinkage or some other less than perfect result.

Fixatives are either coagulants or non-coagulants. Coagulant substances, such as hydrochloric acid, picric acid, mercuric chloride and chromium trioxide, will precipitate proteins and have been traditional components for paraffin-embedding techniques. Non-coagulants, such as formaldehyde, glutaraldehyde, acrolein, osmium tetroxide, potassium dichromate and acetic acid, will cross-link proteins without causing extensive denaturation and are preferred for mRNA detection. The aldehydes, especially formaldehyde alone or in combination with glutaraldehyde, are especially useful for immunolocalisation and *in situ* hybridisation studies.

The procedure for preparing formaldehyde is as follows (R. Rasmussen and C. Houck, Calgene, Davis, CA, personal communication):

Preparation of a 4% stock solution

(1) Add 4 g of paraformaldehyde to 100 ml of phosphate buffered saline (PBS). Care must be taken in the preparation of formaldehyde — always use a fume hood, never breathe the vapours and wear gloves. Restrictions of the use of this excellent fixative will no doubt be installed in the near future for health and safety reasons. [PBS: 10 mM phosphate buffer + 150 mM NaCl. Make up stocks of 20 mM sodium phosphate mono- and dibasic, start with the dibasic and titrate to pH 7.4 with the monobasic solution (the final solution is approximately 80% dibasic and 20% monobasic). Add NaCl to make 300 mM. This makes 2 × PBS. Filter sterilise and store in the refrigerator. To use bring to 1 × with sterile water.]

(2) Stir on low heat for 2–4 h in the fume hood. After the paraformaldehyde has dissolved add 0.10165 g $MgCl_2$ (5 mM $MgCl_2$). Filter-sterilise and use. May be stored in the refrigerator for a few days.

Note that commercial formalin is about 37% formaldehyde. It is permissible to use formalin for paraffin level work, but generally paraformaldehyde is preferred.

Other problems associated with fixation include access (size of tissue piece), mobility of the fixative (e.g. aldeyhydes penetrate tissues quite slowly, acetic acid penetrates very fast), density of tissue, amount of lipid, presence of the cuticle, thick cell walls, and so on — unfortunately most of these problems have to be overcome by trial and error. You cannot assume that one fixation schedule will be the best for everything. Many other physical variables — concentration, pH, temperature, osmolarity and time of exposure — must also be considered if optimum preservation is expected. Several books discuss these problems fully (e.g. O'Brien and McCully, 1981).

The primary functions of chemical fixation for *in situ* hybridisation are to retain hybridisable mRNAs and preserve recognisable tissue morphology. It has been found that of all fixatives tested, freshly prepared paraformaldehyde solutions worked best for retention of mRNAs, efficiency of hybridisation, and maintenance of tissue morphology (Hafen and Levine, 1986; Singer *et al.*, 1986; Raikhel *et al.*, 1989). Paraformaldehyde did not excessively cross-link proteins to RNA and allowed good cellular penetration of long probes (Singer *et al.*, 1986). Singer *et al.* also reported that fixative made from bottled formalin was unpredictable in the retention of mRNAs and that precipitating fixatives such as alcohol or alcohol–acetic acid provided insufficient retention. Contrary to that report, others have successfully used alcohol fixatives on plant material (Meyerowitz, 1987; Bochenek and Hirsch, 1990; E. Y. Tanimoto and Lassner, unpublished). Cox and Goldberg (1988) recommended glutaraldehyde for general use in plant *in situ* hybridisation and formalin–acetic acid–ethanol (FAA) when other fixatives could not penetrate the tissue. All these formulas have unique advantages regarding the 'fixation image' they provide in the different tissues that have been tested. We do not recommend lengthy testing of fixatives unless the sensitivity of detection falls below expectations predicted by Northern analysis.

D. Embedding

We prefer to embed plant tissue in paraffin for ease of tissue preparation (Raikhel *et al.*, 1989) and storage. In the studies that are described in this chapter, all fixed tissue was embedded in melted Paraplast® (Monoject Scientific; St Louis, MO, USA), a blend of paraffin and synthetic polymers, in a constant temperature oven at 60°C. Care must be taken not to exceed 62°C because this will damage the synthetic polymers in Paraplast® and cause crumbling.

E. Sectioning and Mounting

Glass slides should be washed, coated (subbed) with an adhesive and dried before tissues are mounted. We used a 5% (v/v) glacial acetic acid solution in 95% ethanol to clean the slides. After 10 min in acidified ethanol, the slides were rinsed thoroughly in distilled water and oven dried. The dried slides were coated by soaking in 50 μg ml^{-1} poly L or D lysine (Sigma) in 10 mM Tris-HCl, pH 8.0, for 30 min and air-dried. We found that subbed slides were good for several months if stored in a dry place. In our hands, gelatin–chrome alum adhesives did not provide secure attachment of sections.

Embedded material can be trimmed and mounted on stubs for sectioning. Ribbons of serial sections can be cut at any desired thickness above 5 μm on a rotary microtome. To facilitate sectioning and lessen distortion of cellular structure in fragile tissue, the mounted stubs can be chilled in ice water immediately before use (Johansen, 1940).

Serial sections of interest were located within the ribbons under a dissecting microscope. Subbed slides were placed on a slide-warming tray set to 45°C and drops of clean distilled water were applied on top of the slide. A ribbon of sections was floated on top of the warm water and allowed to expand for a few minutes. The number of sections that could be mounted on the slide was dependent on the expansion of ribbons, and was adjusted to the coverslip size. The ribbon was centred with forceps, and the water drained off with a paper towel. All traces of water were removed with the edge of a paper towel, to minimise the formation of disruptive bubbles under the ribbon. Complete drying was essential for good adhesion of sections to slides; this was done overnight at 45°C on the warming tray. Mounted slides can be kept indefinitely for later use.

F. Microtechnique Protocol

(1) Pea roots (usually about 10 mm long segments) were aspirated for 15 min in 4% paraformaldehyde in PBS then fixed for an additional 12 h. A 1:10 ratio

of tissue to fixative volume was always used. Larger tissues would require longer periods for fixation periods, dehydration time, and infiltration by embedding medium.

(2) Roots that were to be embedded in paraffin were dehydrated in a seven-step tertiary butyl alcohol (TBA) series up to 100% TBA (Johansen, 1940). Solutions were changed twice each day.

- (a) 50% step (500 ml dH_2O, 400 ml 95% ethanol, 100 ml TBA). Absolute TBA must be stored in a warm place; freezing point is 26°C.
- (b) 70% step (300 ml dH_2O, 500 ml 95% ethanol, 200 ml TBA).
- (c) 85% step (150 ml dH_2O, 500 ml 95% ethanol, 350 ml TBA).
- (d) 95% step (450 ml 95% ethanol, 550 ml TBA).
- (e) 100% step (250 ml 100% ethanol, 750 ml TBA).
- (f) Absolute TBA.
- (g) Absolute TBA.

(3) Pea roots in absolute TBA were placed in a container half filled with melted and re-solidified Paraplast® (20 ml container) and covered by 100% TBA. The container was sealed and placed in the oven (2-3 h) until the Paraplast® melted. The seal was removed and the container placed in the oven overnight to allow the TBA to evaporate. The liquified Paraplast® was then changed at least three times (12 h between changes). The tissue in melted Paraplast® was oriented in a prewarmed embedding mould using prewarmed forceps. The mould was cooled rapidly on ice once the tissue was immobilised in the solidifying Paraplast®.

(4) Paraffin blocks were trimmed and cut at 10 μm and serial sections were mounted on subbed slides as described above.

G. Probe Preparation

A variety of *in vitro* transcription/cloning vectors are commercially available that contain different RNA polymerase promoters (SP6, T3 or T7) at opposite ends of the multiple cloning site. Any one of these can be used to subclone cDNA in any manner where its orientation is known. Transcription from the promoter at the 3' end of the cDNA will result in the 'antisense' RNA probe that is complementary to the mRNA of interest. Transcription from the 5' end of the cDNA will result in 'sense' strand RNA that should not be complementary to any endogenous RNA and can therefore be used as a control. Extensive purification of a cDNA-containing vector during isolation is not necessary — crude minipreps will work as long as restriction digestions completely cut the vector.

One μg of linear DNA template is used for *in vitro* transcription and normally yields enough RNA probe (10 μg) for over 50 slides. For synthesis of the antisense RNA from the 3' end of the cDNA, the vector is linearised by restriction digestion at the unique site on the 5' end of the cDNA where transcripts will run off the template and terminate. Sense RNA is synthesised from template that is cut at the 3' end of the cDNA. Vectors with unique restriction sites for enzymes that cut and leave either a 5' overhang or blunt end at the cut end of the cDNA are preferred because *in vitro* run-off transcription is more efficient with these DNA templates.

H. Protocol for Linearisation of the Vector

(1) *Linearization*. The following were added to a microcentrifuge tube for a 100 μl final reaction volume:

  (a) Plasmid DNA 10 μg.
  (b) 10× digestion buffer 10 μl.
  (c) Autoclaved distilled H_2O fill to final volume of 100 μl.
  (d) Restriction enzyme, 20 units. The vectors were digested for 2 h. Complete linearisation was checked by electrophoresis of a sample of the digest on an agarose gel.

(2) *Postlinearisation*. The following steps were done in the microcentrifuge tube containing the restriction digest:

  (a) Twenty μg of proteinase K was added (Sigma Chemical Co., St Louis, MO, USA), and incubated at 37°C for 30 min.
  (b) The proteinase K digest was extracted once with an equal volume of phenol by vortexing the mixture and then centrifugation, 11 000 × g for 5 min.
  (c) The aqueous phase was transferred to a new microcentrifuge tube and extracted with an equal volume of phenol:chloroform (1:1). The aqueous phase was transferred to a new tube and extracted with an equal volume of chloroform as above.
  (d) The aqueous phase was collected and the DNA was precipitated by addition of a 0.1 vol of 3 M sodium acetate, pH 5.0, and 2 vols of 100% ethanol. The DNA was pelleted by centrifugation, 11 000 × g for 20 min. The pellet was washed with 70% ethanol and dried. The linearised DNA was resuspended to a concentration of 1 μg μl^{-1} in DEPC-treated H_2O.

I. Protocol for Preparation of DIG-Labelled Transcripts

The protocol from the SP6/T7 RNA labelling kit from Boehringer Mannheim Biochemical was used for this procedure.

(1) The following were added to a microcentrifuge tube:

  (a) Linearised DNA template, 1 μl.
  (b) DEPC H_2O, 12 μl.
  (c) NTP labelling mix (w/DIG-UTP), 2 μl.
  (d) 10× transcription buffer, 2 μl.
  (e) RNA polymerase (20 U μl^{-1}), 2 μl.
  (f) RNase inhibitor (20 U μl^{-1}), 1 μl.

The tube was mixed, then centrifuged briefly and incubated for 2 h at 37°C.

  (g) RNase-free DNase, 2 μl.

(2) The tube was incubated for 15 min at 37°C. The reaction was stopped by addition of 2 μl of 0.2 M EDTA. The RNA was precipitated by addition of 2.5 μl of 4 M LiCl and 75 μl 100% ethanol (overnight at −20°C). The RNA was collected by centrifugation at 11 000 × g for 20 min. The RNA pellet

was washed with clean 70% ethanol and dried. The pellet, 10 μg of RNA (expected yield), was resuspended in 50 μl DEPC H$_2$O containing 20 units of RNase inhibitor. The probe was stable at −80°C for a least 6 months.

Contrary to published reports, no alkaline hydrolysis of transcripts, with lengths up to 1200 bases, was necessary prior to hybridisation. Normally, transcripts are hydrolysed to approximately 200 bases to facilitate probe penetration. One can utilise the published hydrolysis formula to fragment the transcript (Cox et al., 1984).

J. Testing Probe

There are two factors which will affect the performance of the DIG-labelled transcripts. The first is the specific detectable activity of the probe, and the second is whether the probe will hybridise to non-targeted RNA. Boehringer Mannheim Biochemical (BMB) claimed that 'the non-radioactive DIG RNA labelling and detection system allowed the detection of 0.3 pg of homologous DNA or RNA on membranes. From a calculation of molar ratios obtained from BMB, 1/12 of the total sequence in the transcripts synthesised using the BMB RNA labelling kit could be labelled with digoxigenin (UTP:DIG-UTP ratio was 2:1) We assumed that transcripts have normal base composition and that the RNA polymerase did not discriminate against the use of the labelled nucleotide. In theory, however, the highest specific activity of usable probe would occur if all the nucleotides of the probe were labelled. The exact specific detectable activity of the DIG-labelled probes is not as easily calculated as that of isotopically labelled probes, which can be directly measured as dpm per μg of probe. DIG-labelled probes are detected indirectly by the signal-amplifying reactions of antibody–enzyme conjugates; this causes uncertainty in calculations of specific activity. Comparisons with detectable activity of known transcripts can give rough estimates of the quality of the transcripts that are synthesised.

Determination of the specific detectable activity of a transcript can be done by comparison with a dilution series of labelled factory-prepared RNA on dot blots. We utilised a more comprehensive technique that determines the specific activity of hybridisable probe.

The second factor, unintended hybridisation, affects the probe performance and shows up by the creation of background. Hybridisation to non-targeted RNA can be discovered in preliminary Northern analysis of total RNA isolated from the tissue(s). Strand-specific RNA probes make it easier to resolve problems caused by the appearance of multiple bands on Northerns blots. The unwanted hybridisation of sense or antisense strand may be eliminated by deletions of homologous sequences through a process of trial and error.

K. Protocol for Preparing Dot Blots to Test Probe

(1) A dilution series of template cDNA was prepared on two dot blots (1 ng to 10 pg).
(2) Dots blots were prepared for hybridisation with the sense and antisense RNA probes in two separate bags according to BMB's recommendations. The blots

were prehybridised in 5 ml of fluid, and 10% of the synthesised probes (1 μg) was used in the hybridisation.

(3) After the blots were washed at high stringency, the probe was immunodetected with the BMB nucleic acid detection kit. We used only probes that could detect a minimum of 10 pg of cDNA.

L. Prehybridisation

The purpose of prehybridisation is to reduce background noise and facilitate hybridisation. Current protocols have been able to reduce background levels so that signal-to-noise ratios during the detection of moderate to high abundance mRNAs remains favourable, but the signal-to-noise ratio for low abundance mRNAs is often very poor.

Probe penetration and hybridisation can be impeded by cross-linked proteins and cellular debris attached to mRNA that were formed by aldehyde fixatives (Singer et al., 1986; Raikhel et al., 1989; Tsukamoto et al., 1991). In general, paraformaldehyde fixatives worked best at optimising hybridisation (Singer et al., 1986; Raikhel et al., 1989). Paraformaldehyde did not overly cross-link protein and RNA, while retention remained high. Singer et al. (1986) found that any type of cell pretreatment, including protein hydrolysis, did not improve hybridisation in tissue fixed with paraformaldehyde. They advised against the use of protein hydrolysis because of excessive loss of mRNA in muscle and fibroblast cells. In contrast, Tsukamoto et al. (1991) found that proteinase K treatment of paraformaldehyde-fixed embryonic mouse cells was essential for optimum hybridisation and that the level of necessary proteinase K treatments varied with different tissues. Favourable results were obtained on paraformaldehyde-fixed pea root sections whether or not protein hydrolysis was done. We presume that the specific benefit from protein hydrolysis will need to be determined for each individual case.

M. Prehybridisation Protocol

The following prehybridisation protocol was modified from Raikhel et al. (1989).

(1) Prehybridisation of paraffin-embedded material began with removal of the paraffin with xylene or a non-toxic xylene substitute made for histological use from organic citrus derivatives (e.g. Hemo-di, Fisher Biochemical, Pittsburgh, PA 15219, USA). Slides were hydrated through a graded alcohol series of 1:1 (xylene: 100% ethanol), 100%, 95%, 70%, 30% ethanol and then into DEPC H_2O, 2 min each.

(2) Proteins that impede hybridisation were hydrolysed in a two-step process.

(a) *Hydrochloric acid hydrolysis*. Slides were immersed in 0.2 M HCl for 20 min, then immersed twice in 2× standard saline phosphate EDTA (SSPE), 5 min each. The slides were rinsed in H_2O twice for 5 min each. Heat treatment between these steps was not done because sections would become detached from the slides at 70°C.

(b) *Proteinase K digestion*. Slides were placed in a solution containing 0.5 mg ml^{-1} proteinase K in 10 mM Tris-HCl (pH 8.0) and 5 mM EDTA at

37°C for 30 min. Slides were rinsed with H_2O twice, 5 min each.

(3) Background noise was reduced by neutralisation of positively charged cellular material that electrostatically attract the negatively charged probe. Acetylation has been the preferred method to reduce electrostatic interference (Hayashi *et al.*, 1978). Slides were dipped in 100 mM triethanolamine and removed. A 0.5% acetic anhydride solution (v/v) was made with the triethanolamine solution and the slides were returned for 10 min. The slides were rinsed twice in 2 × SSC for 5 min each and twice in H_2O for 5 min each. The slides were dehydrated in an ethanol series of 30%, 70%, 95% and 100% ethanol for 2 min each and then air-dried.

N. Hybridisation

The hybridisation conditions were chosen to maximise the extent of hybridisation (modified from Raikhel *et al.*, 1989). The mixture of formamide, buffers and salts included dextran sulphate, which reduced the 'apparent hybridisation volume' by exclusion of water and probe and thereby reduced the time necessary for maximum hybridisation. The addition of dithiothreitol (DTT) is optional when using DIG probes, but it was always included when ^{35}S-labelled probes were used to prevent random disulphide cross-bridges with the tissue. We elected to include it in our hybridisation as a precautionary measure. Excess hybridisation time or probe concentration can increase background noise. Probe concentrations in 100-fold excess over target mRNAs required only a few hours for proper levels of hybridisation (Singer *et al.*, 1986). Background noise was not a problem and so we did no further calculation for optimal probe concentration or length of hybridisation. Cox and Goldberg (1988), Cox *et al.* (1984) and Cox *et al.* (1986) briefly discussed the rationale for proper probe concentration.

Calculations of the volume of probe and buffers can be done in the following manner:

(1) Volume of DIG-labelled transcript (TV):

$$\text{Volume } (\mu l) = 0.5 \, \mu l \text{ per slide} \times (\text{no. of slides} + 1)$$

(2) Final volume (FV) of hybridisation solution (Hyb-sol):

$$\text{FV } (\mu l) = (\text{no. of slides} + 1) \times 50 \, \mu l \text{ per slide}$$

(3) Volume of transcript dilution buffer (DB):

$$\text{Volume } (\mu l) = (0.2 \times \text{FV}) - \text{volume of transcript}$$

(4) Volume of hybridisation buffer (HB):

$$\text{Volume } (\mu l) = 0.8 \times \text{FV}$$

7. NON-RADIOACTIVE *IN SITU* RNA HYBRIDISATION

O. Hybridisation Protocol

(1) Transcript volume (TV) was diluted with the calculated DB volume. The mixture was heated at 70°C for 5 min to remove potential secondary structure in the RNA.

(2) The diluted transcript was added to the calculated volume of HB to make the final hyb-sol volume. Fifty μl of the mixed hyb-sol was applied to the prehybridised sections. Bubbles were avoided by gently lowering the untreated 22 × 40 mm coverslip (Corning) from an inclined angle over the slide. Other manufactured coverslips can be used, but they sometimes leach and make the hyb-sol alkaline. These can be checked with indicator solutions prior to use, or coated by dipping into Sigmacoat and air-dried (Raikhel *et al.*, 1989).

(3) The edges of the coverslip were sealed with rubber cement (Carters). A 10 ml disposable syringe and 18 gauge needle made a good dispenser. Seeping hyb-sol from under the coverslips did not prevent the rubber cement from sealing.

(4) The sections were incubated in a constant temperature oven at 50–60°C overnight in a humid chamber (humidity provided by a bowl of H_2O). Optimum temperature for hybridisation, assuming 50% GC content of the probe, is approximately 50°C (Cox *et al.*, 1984; Cox and Goldberg, 1988). For some DIG-RNA probes, substantially higher background noise was generated at 50°C than 60°C.

Solutions

Transcript dilution buffer: 50% formamide, 10 mM DTT.

Hybridisation buffer. To make 1 ml of hybridisation buffer:

10× salts	125 μl
formamide	500 μl
50% dextran sulphate (Sigma)	250 μl
50 mg ml^{-1} yeast tRNA (Sigma)	25 μl
1 M DTT	10 μl
10 mg ml^{-1} poly (A) (Sigma)	62.5 μl
DEPC H_2O	27.5 μl

Dextran sulphate will go into solution readily when autoclaved. After a couple of months it should be discarded.

10× salts. Mix 3 M NaCl, 0.1 M Tris-HCl (pH 6.8), 0.1 M Na_2-phosphate (pH 6.8), and 50 mM Na_2-EDTA, in DEPC H_2O and autoclave.

P. Posthybridisation Treatment

Removal of excess probe is the objective of posthybridisation treatment and washes. One method, largely used to reduce excessive background with ^{35}S-labelled RNA probes (or in the detection of low abundance mRNAs), utilises lengthy washes in 50% formamide buffer solutions at elevated temperatures and RNase A treatment in high salt to remove non-hybridised single-stranded RNA probe. A second, less labour intensive, method has given acceptable levels of background with DIG-labelled transcripts.

Q. Posthybridisation Protocol

(1) *Method 1* (Raikhel et al., 1989)
(a) The rubber cement was carefully removed with forceps. The coverslip should not be moved. The slides were placed on a slide rack and submerged in a staining jar filled with 2 × SSC. The coverslips were floated off with gentle agitation (10–20 min). Gentle prying may be necessary for tight coverslips, but be forewarned that the sections are easily dislodged by this action.
(b) The coverslip-free slides were transferred to formamide wash buffer at 50°C for 4–5 h. The slides were then transferred to a solution of RNase A (20 μg ml^{-1}) in NTE solution at 37°C for 30 min. More or less background could be obtained by varying the length of treatment or concentration of RNase A. Care was taken to keep the salt concentration above 150 mM because RNase A could digest duplex RNA at low salt concentrations. The slides were washed in five successive solutions of NTE at 37°C for 15 min each. The slides were transferred to new formamide wash buffer at 50°C and left overnight. The slides were dehydrated in an ethanol series consisting of 30, 70, 95 and 100% ethanol and air-dried.

Solutions. Formamide wash buffer:

> 1 × salts (made from 10 × salts used in hybridisation)
> 50% formamide (analytical grade, Sigma)
> 10 mM DTT

NTE 500 mM NaCl, 10 mM Tris-HCl (pH 8.0), 1 mM EDTA, 10 mM DTT. Note that DEPC water is not required in any wash solution.

(2) *Method 2* (technical update 8812395/2M (1989) for the Genius (Non-radioactive DNA Labelling and Detection Kit, BMB))
The coverslip-free slides were washed twice in fresh 2 × SSC at room temperature for 1 h each. They were washed once in 1 × SSC for 1 h at room temperature and washed once in 0.5 × SSC for 30 min at room temperature.

Both posthybridisation washes worked well when used with DIG-labelled probes on pea roots. In all cases where DIG-labelling worked well, however, the abundance of target mRNA was at least moderate. DIG-labelling has been difficult to use on low abundance mRNAs (E. Y. Tanimoto, personal observation).

R. Immunodetection

(Technical update 8812395/2M (1989) for the Genius (Non-radioactive DNA Labelling and Detection Kit, BMB).

The detection of signal is determined by an indirect method. The DIG-labelled probe, by itself, does not generate any detectable signal. The colour detection is dependent on the specific binding of the antibody-linked alkaline phosphatase to the DIG moiety of UTP incorporated in the probe. This indirect signal level is dependent on the enzymatic amplification of coloured products by continuous alkaline phosphatase activity. Excess levels of non-specifically bound enzyme–antibody conjugate creates excess background. In addition, some tissues contain endogenous enzymes with similar reaction characteristics to the alkaline phosphatase used in the enzyme–antibody conjugates. This endogenous interference of the detection process can be controlled in some tissues. Levamisole, a potent inhibitor of mammalian alkaline phosphatases, does not work well on inhibiting endogenous plant alkaline phosphatases. We regret that we cannot recommend a suitable inhibitor. Sections that have not been treated with antibody conjugates should be used as a control to determine if endogenous alkaline phosphatases will be a problem.

S. Immunodetection and Staining Protocol

The following steps were done at room temperature:

(1) The slides were washed in buffer no. 1 for 5 min.
(2) Slides were incubated in 2% normal sheep serum (NSS) (CALBIOCHEM; La Jolla, CA, USA), 0.3% Triton X-100 in buffer no. 1 for 30 min.
(3) The anti-DIG-antibody conjugate was diluted 1:500 with buffer no. 1 that contained 1% NSS and 0.3% Triton X-100.
(4) Two hundred µl of diluted antibody conjugate was applied to the sections and incubated for 2 h in a humidified chamber. The slides were placed on styrofoam lids that floated levelly on H_2O. This apparatus can be placed inside plastic storage containers and fitted with a lid to maintain humidity.
(5) The slides were washed twice in buffer no. 1, 15 min each.
(6) Slides were washed once in buffer no. 3 for 2 min.
(7) Colour was developed in the dark by incubation of the slides with freshly made NBT/X-phosphate colour solution, 500 µl per slide. Additional background colour was reduced by the absence of light. Development time varied. High abundance mRNAs took 1–3 h to develop and moderate abundance mRNAs took longer. Longer development times were possible if the slides were kept in a humidified chamber. Inspection of colour development was made with a compound microscope under low power.
(8) The colour reaction was stopped in buffer no. 4 for 5 min.
(9) The slides were dehydrated through a graded ethanol series that contained 30%, 70%, 95%, and 100% ethanol, 2 min each. The slides were soaked for one minute in 1:1 (100% ethanol:xylene) and transferred quickly to 100% xylene for 5 min and the coverslips mounted with Permount® (Fisher). Appreciable loss of colour was observed if the slides were left in the 1:1 mixture for a longer time. Alternatively, one could go directly from 100% ethanol to xylene if residual water in the tissue was eliminated by two separate changes of 100% ethanol prior to the xylene step. No counter-staining of the sections was necessary to provide cytological detail for tissue recognition. The low background noise level provided adequate cellular detail.

Solutions
Buffer no. 1: 100 mM Tris-HCl, 150 mM NaCl (pH 7.5).

Buffer no. 3: 100 mM Tris-HCl, 100 mM NaCl, 50 mM $MgCl_2$ (pH 9.5).
Buffer no. 4: 10 mM Tris-HC1, 1 mM EDTA (pH 8.0).

III. CO-LOCALISATION PROTOCOL

A double labelling protocol was used for DIG-detection of pea histone H2A mRNA and autoradiographic detection of radioactively labelled S-phase cells.

We were presented with a problem of histologically demonstrating that pea histone H2A mRNA expression was correlated with DNA synthesis in the same cells. Additionally, we wished to investigate whether replication-independent expression of H2A mRNA could be found. We concluded from histological evidence and other means that pea histone H2A mRNA expression was replication dependent (Tanimoto, Rost, and Comai, 1992 and in preparation). Tissue preparation, solutions, and protocols used are described below.

A. Fixation, Embedding and Sectioning

Primary pea seedlings were axenically grown in the dark. One centimetre root tips were excised, placed in White's medium with 2% sucrose and immediately labelled for 1 h with methyl [^{3}H]thymidine, 1 μCi ml^{-1} (78.5 Ci mmol^{-1}, NEN/Dupont). The roots were rinsed with water and fixed in 4% paraformaldehyde in PBS (pH 7.4) for 12 h, dehydrated through a standard TBA series and embedded in Paraplast®. Median longitudinal sections (10 μm) were mounted on polylysine subbed slides.

B. Autoradiography

The slides were deparaffinised in xylene, dipped in 1:1 (xylene: 100% ethanol), then 100% ethanol, 5 min each, and air-dried. Slides were coated in diluted Kodak NTB-2 film emulsion (1:1 with H_2O) for autoradiography, dried and stored at 4°C in a sealed dark box for 1 week. The autoradiographs were developed for 5 min in Kodak D-19 developer, rinsed briefly in H_2O, and fixed for 10 min in Kodak general purpose fixer at 15°C. The autoradiographs were washed in DEPC-treated H_2O for 10 min All darkroom manipulations were done under a 15 W bulb and Kodak no. 2 red or Kodak GBX-2 filter.

C. Prehybridisation

The developed autoradiographs were dipped in 100 mM triethanolamine in DEPC-treated H_2O at room temperature. The slides were removed and acetic anhydride was added to make a 0.5% (v/v) solution and the slides were re-immersed for 10 min. The slides were rinsed twice in 2 × SSPE, 5 min each, and rinsed twice in DEPC H_2O, 2 min each. Prehybridisation consisted solely of this acetylation step in order to block non-specific binding of negatively charged nucleic acid probe *in situ* (Hayashi *et al.*, 1978). Contrary to conventional *in situ* hybridisation, no enhancement of signal by deproteinisation with HCl or proteinase K was done since it was determined that NTB-2 nuclear track film emulsion was made with gelatin.

The pea histone H2A probe was isolated from a root cDNA library by use of a heterologous tomato H2A cDNA probe (Koning et al., 1991). A 627 bp pea H2A cDNA was cloned into pGEM-1 as an EcoR1–Xba1 fragment (GenBank accession no. M64838). For in situ RNA hybridisations, the plasmids were linearised with EcoR1 or Xba1 digestion yielding the sense or antisense template for in vitro transcription from T7 or SP6 promoters. DIG-labelled RNA probes were synthesised from 1 µg of linearised plasmid using a RNA labelling kit from Boehringer Mannheim Biochemical. No alkaline hydrolysis of the transcript was done.

D. Hybridisation

Fifty µl of hybridisation solution (hyb-sol) containing 100 ng of DIG-RNA probe per slide was pipetted onto the slide over the wet developed film emulsion*. A 22 × 40 mm coverslip was gently lowered over the hyb-sol and rubber cement (Carter's) was applied in a ring around the edges of the coverslip. The slides were placed overnight at 60°C in a humid incubator (humidity provided by an open container of water).

E. Posthybridisation

The rubber cement was removed using forceps, care being taken not to peel off the coverslip. The slides were soaked with gentle agitation in 2 × SSC until the coverslips floated off and then the slides were washed twice in 2 × SSC for 1 h each; once in 1 × SSC for 1 h; and once in 0.5 × SSC for 1 h, all washings at room temperature.

F. Immunoliogical Detection

The slides were washed in buffer no. 1 for 5 min, then blocked by complete immersion in 2% normal sheep serum (NSS, Cal-Biochem, La Jolla, CA, USA), 0.3% Triton X-100 in buffer no. 1 for 30 min. Slides were taken out of the blocking solution, and 200 µl per slide of diluted antibody (diluted 1:500 in buffer no. 1 containing 1% NSS and 0.3% Triton X-100) was applied for 2 h. The slides were washed twice in buffer no. 1 for 15 min, then washed once in buffer no. 3 for 2 min. For detection, 500 µl of freshly made colour solution (nitroblue tetrazolium salt and X-phosphate) made in buffer no. 3 was applied to each slide and incubated in a humid chamber in the dark for 2–3 h. The colour reaction was stopped in buffer no. 4 for 5 min. The slides were dehydrated through an ethanol series, passed through xylene and coverslips were mounted with Permount®. All solutions were at room temperature and were made as previously described in this chapter.

---

*The probe concentration of 100 ng DIG-RNA probe per slide was estimated from results obtained during testing of probe activity. The long overnight detection time during probe testing, which was necessary for 10 pg of cDNA, indicated that the specific detectable activity of our DIG-transcripts was low. Published successful hybridisation results with DIG-labelled transcripts at 10 ng per slide (Bochenek and Hirsch, 1990) prompted us to try our hybridisation with higher concentrations of probe.

IV. RESULTS AND DISCUSSION

In situ RNA hybridisation with non-radioactive DIG-labelled antisense RNA provided a high degree of cell-to-cell resolution in the detectability of H2A mRNA. *In situ* hybridisation is easily accomplished for cells that have moderate to highly abundant mRNAs. Additional advantages include less radioactive exposure and less accumulation of radioactive waste. A disadvantage of both is found in relatively long detection times and high background problems caused by low signal-to-noise ratios. In addition, the non-radioactive probes which utilise indirect methods and some form of signal amplification are still less sensitive than the best radioactive probes (Kiyama *et al.*, 1990). A direct detection method for non-radioactive *in situ* hybridisation has been reported (Kiyama *et al.*, 1990) that utilises a method to couple alkaline phosphatase to modified nucleotides directly (Ruth *et al.*, 1985). Better signal-to-noise ratios can be acheived by lowering background caused by non-specific immunodetection. These probes have been used with equal degrees of sensitivity compared to the best radioactively labelled probes (Kiyama *et al.*, 1990).

Non-radioactive detection of H2A mRNA and autoradiographic detection of ^{3}H-labelled S phase cells provided less ambiguity than conventional double isotope-autoradiography techniques, which use two layers of film emulsion over the tissue to differentially detect low and high energy β-particles emitted by isotopes such as ^{3}H and ^{35}S.

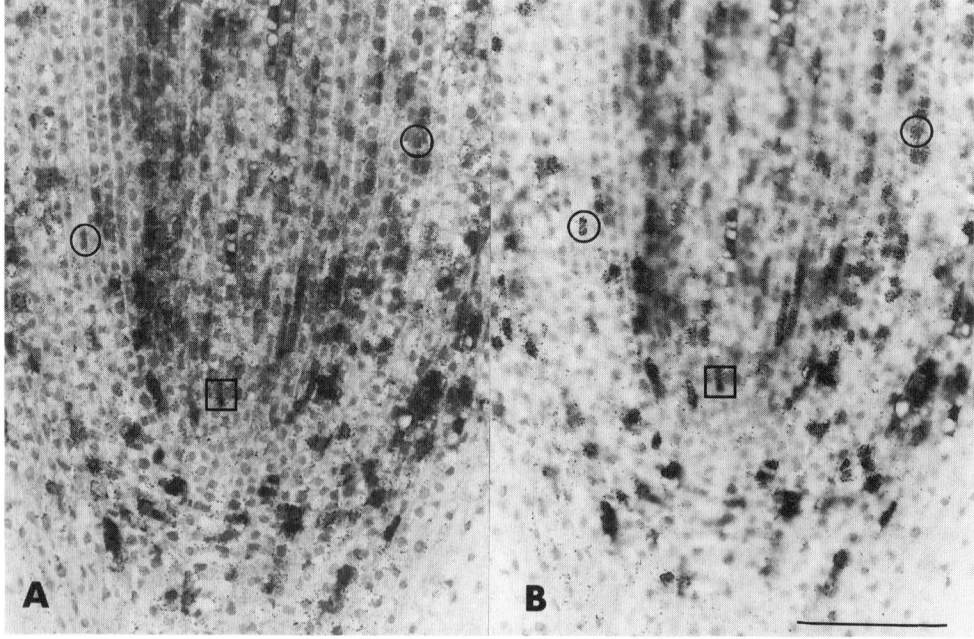

FIG. 7.1. (A) Longitudinal section of pea root showing DIG-labelled probe (dark coloured cells) with the focus at the level of the tissue. (B) The same section focused at the level of the photographic emulsion to show the silver grains. The circled cell has silver grains (cell incorporated ^{3}H-thymidine), but no dark stain (no H2A histone mRNA expression), this cell is probably in late S phase. The cell in the square has both silver grains and dark stain (this cell is in S phase and also shows H2A expression). Scale bar = 100 μm.

In devising this double label protocol, we found that we needed to proceed in a very unorthodox manner. Autoradiographic detection of ^{3}H-labelled S phase cells, which would normally have been done after *in situ* RNA hybridisation and colorimetric detection of the DIG label, was very problematic. We found that the substrates and/or products for colorimetric detection of alkaline phosphatase chemically exposed the film. If sections are hybridised prior to the application of film emulsion and colour detected after film development, no H2A mRNA could be detected on the slide. We were puzzled because our results appeared to be in contradiction to those of Kiyama *et al.* (1991), who found that autoradiographic detection of radioactively labelled probe could be done after using identical agents in the colorimetric detection of a DIG-labelled probe. We could find only one difference in their detection scheme – they used Ilford nuclear track emulsion and we used Kodak NTB II. We do not know if this caused the difference in chemographic effects.

In lieu of wholesale changes to our planned method of detection, we elected to do the autoradiography first, and hybridise and colour detect the antisense DIG-labelled transcript through the developed film emulsion (Fig. 7.1).

REFERENCES

Bochenek, B. and Hirsch, A. M. (1990). *Plant Mol. Biol. Rep.* **8**, 237–248.
Clavel, C., Binninger, I., Boutterin, M., Polette, M., Birembaut, P. (1991). *J. Virol. Meth.* **33**, 253–266.
Cox, K. H. and Goldberg, R. B. (1988). In "Plant Molecular Biology: A Practical Approach" (C. H. Shaw, ed.), pp. 1–35. IRL Press, Oxford.
Cox, K. H., DeLeon, D. V., Angerer, L. M. and Angerer, R. C. (1984). *Dev. Biol.* **101**, 485–502.
Cox, K. H., Angerer, L. M., Lee, J. J., Davidson, E. H., and Angerer, R. C. (1986). *J. Mol. Biol.* **188**, 159.
Hafen, E. and Levine, M. (1986). In "*Drosophila*: A Practical Approach" (D. B. Roberts, ed.), pp. 139–157. IRL Press, Oxford and Washington DC.
Hayashi, S., Gillam, I. C., Delaney, A. D. and Tener, G. M. (1978). *J. Histochem. Cytochem.* **26**, 677–679.
Johansen, D. A. (1940). "Plant Microtechnique", pp. 126–154. McGraw-Hill, New York and London.
Kiyama, H., Emson, P. C. and Tohyama, M. (1990). *Neurosci. Res.* **9**, 1–21.
Kiyama, H., McGowan, E.M. and Emson, P. C. (1991). *Mol. Brain Res.* **9**, 87–93.
Koning, A. J., Tanimoto, E. Y., Kiehne, K., Rost, T. and Comai, L. (1991). *The Plant Cell* **3**, 657–665.
Larsson, L. (1989). *Arch. Histol. Cytol.* **52** (Suppl.), 55–62.
Meyerowitz, E. M. (1987). *Plant Mol. Biol. Rep.* **5**, 242–250.
Negro, F., Berninger, M., Chiaberge, E., Gugliotta, P., Bussolati, G., Actis, G. C. and Bonino, F. (1985). *J. Med. Virol.* **15**, 373–382.
O'Brien, T. P. and McCully, M. E. (1981) "The Study of Plant Structure: Principles and Selected Methods." Termarcarphi Pty Ltd, Melbourne, Australia.
Raikhel, N. V., Bednarek, S. Y. and Lerner, D. (1989). In "Plant Molecular Biology Manual" (S. B. Gelvin and R. A. Schilperoot, eds), pp. 1–32. Kluwer Academic Publishers, Dordrecht, The Netherlands.
Ruth, J., Morgan, C. and Paska, A. (1985). *DNA* **4**, 93–96.
Scofield, M. A., Sun, L. and Pettinger, W. A. (1992). *BioTechniques* **12**, 20.
Singer, R. H., Lawrence, J. B. and Villnave, C. (1986). *BioTechniques* **4**, 230–250.
Smith, A. G., Hinchee, M. and Horsch, R. B. (1987). *Plant Mol. Biol. Rep.* **5**, 237–241.

Tanimoto, E. Y., Rost, T. L. and Comai, L. (1993). *In* "Molecular and Cell Biology of the Plant Cell Cycle" (D. Francis and J. C. Ormrod, eds), pp. 85-95. Kluwer Academic Publishers, Dordrecht, The Netherlands.

Tsukamoto, T., Kusakabe, M. and Saga, Y. (1991). *Int. J. Dev. Biol.* **35**, 25-32.

8 Immunolocalisation of Antigens in Plants with Light and Transmission Electron Microscopy

JUDITH A. JERNSTEDT[1], TODD J. JONES[2], and THOMAS L. ROST[3]

[1]*Department of Agronomy and Range Science and* [3]*Section of Botany, University of California, Davis, CA 95616, USA*

[2]*E. I. DuPont de Nemours and Co., DuPont Agricultural Products, Experimental Station, Wilmington, DE 19880-0402, USA*

I.	Introduction	160
II.	Background	161
	A. General considerations	161
	B. Tissue selection	161
	C. Fixation	162
	D. Embedding media	163
	E. Immunostaining	165
	F. Detection systems	166
	G. Production and storage of antibodies	170
	H. Photomicrography	172
III.	Protocols	172
	A. Protocol for cryosectioning and immunolocalisation of tubulin at the light microscope level	172
	B. Protocol for immunolocalisation of proteins in resin-embedded tissue for bright field and fluorescence microscopy	176
	C. Protocol for immunogold localisation of proteins using transmission electron microscopy	179

D. Recipes... 181
References.. 184

I. INTRODUCTION

Immunocytochemistry (ICC) is based on the conjugation of a labelled antibody to the corresponding antigen within a cell. The antigen–antibody–reporter molecule complex is then observed microscopically.

The diversity of antigens which have been localized in plant cells with varying degrees of sensitivity includes cytoskeletal elements (e.g. microtubules, microfilaments, microtubule associated proteins) (Doonan and Clayton, 1986; Cyr and Palevitz, 1989; Goodbody et al., 1989), storage proteins (e.g. lectins, wheat germ agglutinin, zeins and others) (Mishkind et al., 1982; Craig and Miller, 1984; Jones and Rost, in press), enzymes (Hattersley et al., 1977), and hormone carrier proteins (Jacobs and Gilbert, 1983).

There is a vast literature describing applications of ICC in biomedical research and diagnosis (Sternberger, 1986; Beltz and Burd, 1989; Bullock and Petrusz, 1989; Yadzi and Dardick, 1992). These fields have provided the driving force for development of new and more precise techniques, many of which have yet to be applied to plants. While the principles of ICC are the same for plant and animal cells, techniques developed for one group are not necessarily effective for another. The purpose of this chapter is to consider specifically those elements of ICC such as fixation, embedding media and staining which may require particular adaptation for plant tissues.

This chapter will not deal with the details of antibody production and/or purification. Both monoclonal and polyclonal antibodies against a variety of antigens are commercially available (Amersham, Boehringer Mannheim, Pierce, BioCell Research Laboratories and Jackson Immuno Research are a few examples of companies supplying these products), and custom antibody facilities will produce antibodies against antigens supplied to them by investigators. Readers interested in antibody production should consult any of a number of recent volumes devoted to antibody production and hybridoma technology (Harlow and Lane, 1988; Zola, 1990; Campbell, 1991).

The following steps are generally followed in ICC, regardless of the source of antibodies or tissues or the type of reporter molecule used for detection:

- Tissue preparation and microtechnique (fixation and embedding).
- Pretreatment of tissues.
- Antibody application.
- Postincubation washes.
- Detection of antigen–antibody–reporter molecule complex.

Each of these steps will be considered in a general way. The chapter will conclude by presenting three detailed protocols for immunolocalisation of specific antigens.

(1) Protocol for cryosectioning and immunolocalisation of tubulin at the light microscope level.
(2) Protocol for immunolocalisation of proteins in resin-embedded tissue for light and fluorescence microscopy.

(3) Protocol for immunogold localisation of proteins using transmission electron microscopy.

II. BACKGROUND

A. General Considerations

At the outset it must be recognised that ICC consists of a series of compromises. Depending on the abundance, location or conformation of the antigen, the cytology and anatomy of the tissue, the availability of plant material at appropriate stages, and/or the accessibility of specific instrumentation, investigators may choose a method which results in less than optimal preservation of structure, retention of antigens, or preservation of antigenicity. Because different investigators have different purposes and requirements for doing ICC, sensitivity, precision, reproducibility, rapidity or convenience may be the determining factor in selecting the appropriate method (Larsson, 1988). Thus, there is no single method that can be recommended for all situations.

A second caution is that ICC, like all histochemical and cytochemical techniques, is based on chemical interactions. As Larsson (1988) cogently points out, 'successful use of these methods and correct interpretations thereof will depend upon a thorough knowledge of the chemical interactions involved.' Immunocytochemical techniques are generally simple to perform, but correct interpretations of results obtained are often considerably more difficult. A major reason for this is that a given antibody will detect only a part of the antigen structure and may react with other molecules containing the same structure. These 'other molecules' may result in false-positive reactions which can only be determined as false if subsequent ICC uses multiple region-specific antibodies, recognising multiple epitopes on the antigen under investigation (Larsson, 1988).

An additional reality of ICC is that, with very few exceptions, the antigen of interest is no longer in its native state when antibodies are applied. This is a direct consequence of fixation, which is required to immobilise the antigens. Immobilisation is necessary for *in situ* localisation, which is the object of ICC in the first place. In contrast, native antigens are typically used in the production of antibodies, a fact which should be kept in mind when interpreting results, especially negative results. Antibodies for ICC can (and probably should) be made against chemically fixed antigens to avoid this.

B. Tissue Selection

ICC can be and has been performed on stems, roots, leaves, floral parts, fruits, seeds, embryos, gametophytes and on almost all tissues and cell types of the plant body. Specimens have ranged from single cells to embedded organs. The beginner is well advised to search out papers reporting results from organisms, organs and antigens corresponding as closely as possible to their particular system. In most cases, however, published protocols represent only a starting point for adapting methods to other tissues. Factors to take into account include taxonomic relationships (monocotyledon or dicotyledon); organ specificity (stems, roots or leaves); age specificity (mature or meristematic tissues); and cellular location (cytoplasmic or membrane-bound antigens).

Whatever the tissue, pieces should be as small as possible to allow rapid and thorough penetration of fixative, yet large enough to enable easy and efficient handling. Very small pieces usually fix rapidly and well, but are difficult to keep track of during subsequent solution changes. Very small pieces are also difficult to orientate precisely for sectioning in particular planes. A general rule of thumb is that tissue pieces should have at least one dimension that is <2 mm. In order to achieve this, it may be necessary to split roots, stems or other parts lengthwise several times. Another important consideration is the presence of an intact epidermis, which can severely limit fixative penetration. Splitting segments of an organ into halves or quarters can provide cut surfaces while maintaining topographic features to assist later in orientation.

C. Fixation

Most antigens are soluble in water or buffers and have to be chemically fixed or cross-linked in the tissue. Additionally, for intracellular antigens, the cell must be opened or permeabilised to facilitate antibody penetration. Procedures which permit antibody entrance also permit the antigen to leach or wash out. Thus, for most ICC, fixation, which is a means of immobilising the antigen, is required.

The choice of fixative depends to a certain degree on whether cells and tissues will be embedded and sectioned, or squashed on a slide or other substrate. In the former case, the question then becomes whether examination and detection of immunolabel will be with a light microscope or an electron microscope. An additional consideration is which of several usually contradictory criteria is of primary importance in a given application: retention of antigens; preservation of antigenicity; or permeabilisation of tissue for antibodies and reporter molecules (Larsson, 1988).

Fixatives fall into two general categories: denaturing (also termed precipitating or coagulating) and chemical (cross-linking) fixatives. Precipitating fixatives, such as methanol, ethanol and acetone, are gentle on antigens, but cellular preservation is poor. Additionally, not all proteins are rendered insoluble and some may wash out or move during incubation or wash steps.

Chemical fixatives, singly or in combination, are the most commonly used in ICC in both plant and animal tissues. Buffered paraformaldehyde (PFA) alone (1.5–8%; w/v) is often used for light microscopy, since it is relatively mild and penetrates moderately fast. A paraformaldehyde–glutaraldehyde (Glut) combination (e.g. 3% PFA + 0.2% Glut) is typically required for immunolocalisation at the electron microscope (EM) level, since addition of even small percentages of glutaraldehyde dramatically improves preservation of cellular structure. A potential limitation of PFA-Glut fixatives for light microscopic detection, however, is that glutaraldehyde can induce fluorescence in otherwise non-autofluorescent tissues (O'Brien and McCully, 1981). Acrolein is another aldehyde that penetrates and fixes tissues rapidly. It may substitute for glutaraldehyde as a primary fixative for light and electron microscopic ICC, although significant tissue shrinkage occurs during acrolein fixation. Acrolein is also highly toxic and extra safety precautions must be used for its storage and handling.

Addition of appropriate buffers to fixatives and postfixation rinses is critical for ICC success. Buffers serve to maintain pH within useful ranges and to create favourable osmolarities to minimise extraction of proteins and other solutes (Larsson, 1988). The most commonly used buffers for plant ICC are 50 mM potassium phosphate buffer (pH 6.8; Wick *et al.*, 1981), phosphate buffered saline (PBS, pH 7.4; Wick *et al.*, 1981), 0.1 M APES (pH 6.9; Falconer *et al.*, 1988); and 50 mM PIPES (Baskin *et al.*, 1992).

In some instances, additional compounds are added to the buffered fixative to promote penetration and rapid fixation. Brown *et al.* (1989) added 5% dimethyl sulphoxide (DMSO), while Wick and Duniec (1986) included detergents (Triton X-100, Nonidet P_{40} and saponin) in the fixative and subsequent rinse steps.

D. Embedding Media

Once tissue is fixed and subsequently rinsed, it must be infiltrated with some sort of supporting matrix if it is to be sectioned. An alternative to embedding and sectioning is the use of 'squashes', in which tissues are briefly incubated in cell wall-digesting enzymes and subsequently separated into individual cells or short files of cells (Wick *et al.*, 1981, 1989). This method has the dual advantage of being rapid (no lengthy infiltration period) and of promoting antibody penetration into cells through partially digested cell walls. Its major drawback is the loss of tissue integrity as cells are separated. Squashes are unsuitable for applications in which precise spatial orientations of cells and tissues are important and must be maintained.

The most common approach is to embed and section the tissue and to perform labelling on the sections. A variety of supporting matrix materials is available. The choice of matrix is influenced by requirements for certain degrees of resolution and cellular and subcellular preservation, ease of handling and sectioning, and the nature of the antigen. The four most common techniques for sectioned material are: cryosectioning, paraffin embedding, use of polyethylene glycol (PEG) or Steedman's polyester wax, and use of plastic resins. Each presents certain advantages and disadvantages which make it more or less suitable for a particular application.

1. Cryosectioning

Cryosectioning of frozen fixed tissue is the method of choice for beginning most ICC studies (Hogetsu and Oshima, 1986; Sakaguchi *et al.*, 1988; Brown *et al.*, 1989; Smith-Huerta and Jernstedt, 1989). With practice and a good instrument, sections down to 3–4 μm can be cut; sections in the range of 6–10 μm are easy to produce and usually quite adequate. The major drawback of cryosections lies in poor tissue preservation. On the other hand, antigenic preservation is excellent, as is antibody penetration. Furthermore, embedding in sucrose or Tissue-Tek™ O.C.T. compound, the supporting media of choice for plant tissues (Mishkind *et al.*, 1982; Saunders *et al.*, 1983; Smith-Huerta and Jernstedt, 1989), is easy and quick compared to using paraffin and plastic resins. However, another limitation of cryosections is the difficulty in attaching sections to slides. Cryosections have a disturbing tendency to detach during subsequent incubation and rinse stages, so slides must be treated very gently until a coverslip is finally mounted. Finally, formation of ice crystals in non-cryoprotected tissue can further decrease the quality of tissue preservation. Specimens are often cryoprotected by incubation in buffer plus 15–20% sucrose, glycerol or DMSO prior to freezing in supporting medium (e.g. O.C.T. compound) by liquid nitrogen (Mishkind *et al.*, 1982; Raikhel *et al.*, 1984).

2. Paraffin embedding

This is the method of classical botanical microtechnique and histochemistry (Johansen, 1940; Jensen, 1962) and offers several advantages for ICC. A major advantage of paraffin sections is that it is possible to obtain ribbons of serial sections. Also, tissue preservation in paraffin-embedded material is much better than in cryosections. Another advantage is the general availability of simple rotary microtomes and the relative simplicity of sectioning and subsequent handling of sections. The major disadvantages of paraffin embedding for ICC are the high temperatures experienced by tissues during infiltration and embedding (melting point 56–58°C), and the organic solvents (xylene or Hemo-De®, Fisher*brand*) required to remove the paraffin matrix prior to immunostaining. Thus, sensitive antigens may be harmed during processing, resulting in decreased antibody staining.

3. PEG and other 'soluble' waxes

The use of polyethylene glycol (PEG) and polyester waxes (e.g. Steedman's wax) for ICC has recently been advocated for certain antigens, including microtubules (Tiwari *et al.*, 1984; Wick *et al.*, 1985; Brown *et al.*, 1989). These waxes have lower melting points than paraffin, so preservation of antigens is improved over paraffin-embedded material. PEG is water soluble, so removal of matrix is a gentle process and damage to antigenicity is minimal. Disadvantages of PEG include difficulty with flattening curled sections, the slightly lower quality of tissue preservation compared to the harder matrix of paraffin, and the difficulty in obtaining PEG sections as thin as paraffin sections.

Steedman's wax (9 parts PEG 400 distearate + 1 part 1-hexadecanol; Brown *et al.*, 1989) is not water soluble, so sections can be flattened on water droplets. Steedman's wax is removed with ethanol, which may slightly increase the risk of antigen damage. As with paraffin wax, PEG and Steedman's wax are only suitable for light microscopy.

4. Plastic resins

Combined with appropriate fixatives, embedding in plastic resins provides the highest quality of tissue preservation. The rigid plastic matrix makes it possible to cut sections thin enough for transmission electron microscopy as well as high resolution light microscopy. Unfortunately, resin infiltration and polymerisation are harsh processes and some antigens do not survive these procedures. For example, the plant cytoskeleton, especially microtubules, is difficult to immunostain in resin sections. Two recent papers (Gubler, 1989; Baskin *et al.*, 1992) have reported success in staining plant microtubules in butyl-methylmethacrylate-embedded material, although similar results are not easily reproduced in other laboratories. Plastic resins commonly used for plant ICC are LR White, HistoResin®, glycol methacrylate (JB-4, Immunobed), and butyl-methylmethacrylate.

A critical step in immunostaining of resin sections is the pretreatment of sections with blocking proteins prior to application of primary antibody (Larsson, 1988; Baskin et al., 1992). Blocking with bovine serum albumin (BSA), powdered milk or other proteins aids in reducing non-specific binding and background staining (Larsson, 1988). Addition of glycine to blocking medium is useful for aldehyde-fixed material.

Recent work has shown that staining of sensitive antigens may be improved by low temperature embedding. Polymerisation reactions are often accompanied by significant temperature increases, the effects of which may be minimised by incubation in a chilled chamber (Baskin et al., 1992). Addition of dithiothreitol (DTT, 1-10 mM) to all solutions following fixation has also been shown to improve immunostaining in methacrylate sections. The basis for this effect is not known, although the antioxidant function of DTT may protect proteins from free radicals formed during UV polymerisation of methacrylate (Baskin et al., 1992). DTT has also been shown to suppress non-specific background fluorescence in PEG- and ester wax-embedded material (Brown et al., 1989).

An interesting variation on resin embedding for ICC is a procedure called reversible embedment cytochemistry (REC; Gorbsky and Borisy, 1985). Fixed tissues are infiltrated with fully polymerised polymethylmethacrylate (Plexiglas®, grade MC) dissolved in an organic solvent. Evaporation of the solvent leaves the tissue embedded in hard plastic. After sectioning by conventional methods the plastic is extracted with solvents (acetone or chloroform–acetone) and sections are immunostained. The major advantage of this method is that it avoids exposing antigens to potentially damaging polymerisation reactions (Gorbsky and Borisy, 1985). This technique has recently been used to visualise microtubules in plant root tissues (P. Lu, personal communication).

E. Immunostaining

Two types of ICC staining have been developed and applied to plant tissues. In the pre-embedding staining method, cells or tissues are directly exposed to the primary antibody, either before or after fixation. The specimen may then be observed directly, as with whole cells, or it may be sectioned. Pre-embedding immunostaining has not found wide application in plant biology because of the problem of antibody penetration into cells due to the presence of the cell wall (Knox, 1982). This difficulty has been partially overcome by the use of cell wall-digesting enzymes (cellulases, pectinases) and permeabilisation with various solvents (e.g. methanol at $-10°C$; Wick et al., 1981).

Post-embedding immunostaining is the most widely applied method for plant ICC. Postembedding labelling involves direct application of antibodies to sections of fixed or unfixed tissues, which permits direct access of antibodies to the cellular sites of antigens (Knox, 1982; Larsson, 1988).

Many factors influence the success of both pre- and postembedding immunostaining, in particular the strength of the signal obtained. These interrelated factors include specific antibody concentration, antibody avidity, antigen presentation, antigen concentration, and sensitivity and precision of the detection system. Readers wishing a comprehensive discussion of these factors are referred to Larsson (1988) and references cited therein.

As with any cytochemical investigation, optimal antibody concentrations (dilutions), incubation times and incubation temperatures (4°C, 37°C or room temperature) must be determined empirically. As a general rule, dilutions of an antibody should be as high as possible, without sacrificing any specific staining. Too high a concentration of primary antibody is wasteful, increases non-specific staining, and can reduce specific staining, perhaps because of steric hindrance (Bellon and Druet, 1974; Larsson, 1988).

Another general observation about immunolabelling is that primary antibody dilutions may be increased considerably if incubation time is prolonged or done at elevated temperature (37–45°C). An alternate strategy is double applications with an intermediate wash; this is sometimes more sensitive than a single prolonged incubation (Scopsi and Larsson, 1985). However, there can be adverse effects of prolonged incubations, including tissue damage and, potentially, extraction of antigens (Larsson, 1988). Thus, priorities must be set concerning required sensitivity, specificity, economy and speed, and no single concentration and/or incubation time can be applied to all systems.

It is equally difficult to specify optimal temperatures for incubation. Short incubations (30–60 min) are often carried out at room temperature or at 37°C, while prolonged (12 h to overnight) incubations at 4°C have proven satisfactory for some investigations (Larsson, 1988; Citterio et al., 1992).

Similar considerations apply for secondary antibody staining and any subsequent amplification steps. It is often the case with commercial reagents (e.g. secondary antibodies and fluorochromes) that dilutions are suggested by the manufacturer. These can be taken as starting points for optimising concentrations, times and temperatures for a particular system.

F. Detection Systems

1. Direct and indirect methods

The two antibody detection systems used in ICC are the direct and indirect methods. In the direct method (Fig. 8.1A) the antibody is tagged with a fluorescent or other label and it binds directly to an antigen. This method is the easiest to use but has limitations in cases where the amount of antigen is low. In such instances the indirect method is preferable, because by this means signal strength can be amplified as much as four to five times (Larsson, 1988).

In the indirect method tissues are first treated with the primary antibody which binds to the antigen. This is followed by treatment with a labelled secondary antibody which is antigenic to the primary antibody (Fig. 8.1B). This method has the advantage that the secondary antibody may attach to the primary antibody at more than one site, thereby increasing the signal strength. In addition, one secondary antibody may be used as a reporter for several different primary antibodies as long as they are all from the same animal species.

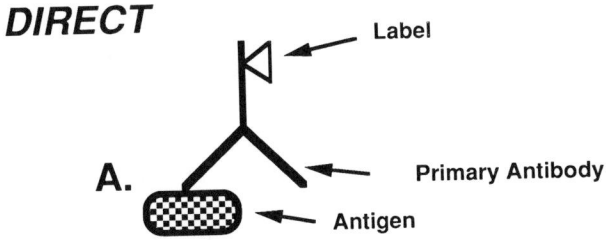
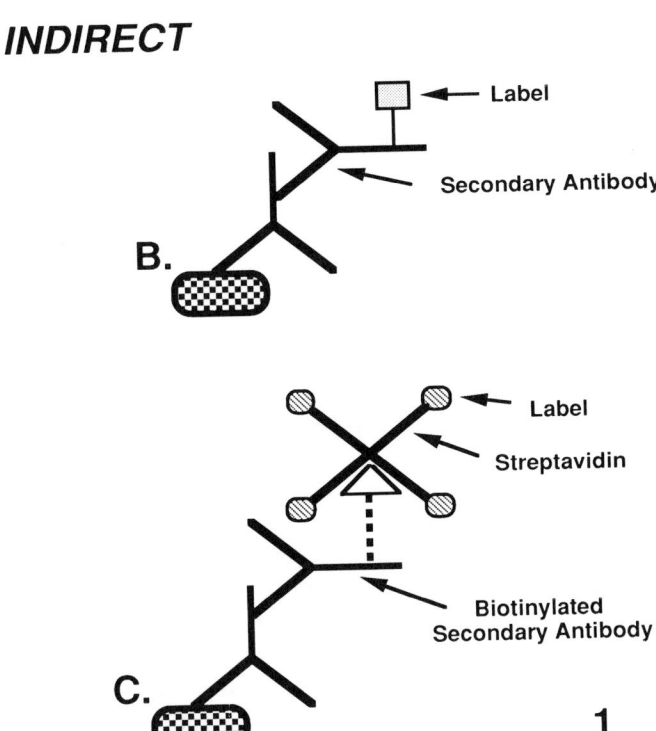

Fig. 8.1. Diagrams showing the direct (A) and indirect (B, C) methods of immunodetection. (A) The label is conjugated to the primary antibody. (B) A secondary antibody, made against the primary antibody, is tagged with a label. (C) The secondary antibody is tagged with biotin. Then the biotin is bound to streptavidin which contains the label. (Redrawn from Amersham technical bulletin "The biotin-streptavidin system." Amersham Corp., Arlington Heights, IL).

Another indirect method adds yet another step to further amplify the signal. In this method the secondary antibody is conjugated to biotin, a water-soluble vitamin. Biotinylated antibodies are then detected with labelled streptavidin (Fig. 8.1C), or less commonly, avidin. Both streptavidin, a bacterial protein, and avidin, a glycoprotein from egg yolk, have an extremely high affinity for biotin with association constants on the order of $K = 10^{15}$ M^{-1} and form essentially irreversible complexes upon binding (Coulton, 1990). Streptavidin is preferred to avidin because it is not a glycoprotein and has a more favourable p*I*, but both show low amounts of non-specific binding. This indirect method is very sensitive and has an added flexibility of allowing the use of different detection systems to be used with the same biotinylated secondary antibody. Amplification of signal can occur with this method because it is possible for more than one biotin to be conjugated to each antibody molecule, and many molecules of label can be attached to streptavidin. Further amplification can be achieved by utilising all four possible biotin binding sites present on the streptavidin molecule to form large streptavidin–biotin reporter molecule complexes.

2. *Types of label*

The choice of label to use will depend on whether detection will be done at the light or electron microscope level, and the availability of fluorescence microscopy. The following is a partial list of labels which are most commonly used for plant tissues:

- fluorescent labels;
- peroxidase;
- alkaline phosphatase;
- Protein A;
- ferritin:
- colloidal gold.

We will briefly discuss each of these. The reader is directed to several recent books for details (Sternberger, 1986; Larsson, 1988; Hayat, 1989; Bozzola and Russell, 1992).

(a) Fluorescent labels. Several fluorochromes are available conjugated to antibodies, streptavidin and protein A. The most important ones are fluorescein isothiocyanate (FITC), Texas Red™ (Molecular Probes, Inc.), BODIPY® (Molecular Probes, Inc.), and tetramethylrhodamine isothiocyanate (TRITC) (Hemmilä, 1991). O'Brien and McCully (1981) provide a good review of the general application of fluorochromes in plant microtechnique. One major problem of interpreting fluorescent labels in light microscope ICC is that of autofluorescence. Chloroplasts, for example, show red autofluorescence, and Glut-fixed material shows strong autofluorescence at several wavelengths. In some instances very low concentrations of Glut will work, but generally speaking it should be avoided when using fluorescent labels, at least when you are experimenting with a new tissue (O'Brien and McCully, 1981). Another problem is that fluorescent stains tend to fade very rapidly under excitation light. Some antifade materials have been developed (e.g. Mowiol + *n*-propyl gallate), which can lengthen the usable signal for several minutes (see protocols; Wick *et al.*, 1985).

(b) Peroxidase. This enzyme label makes an insoluble reaction product which can be made electron dense with osmium postfixation for TEM, and at the light microscope level makes a brownish colour product. Treated tissues are reacted with 3,3′-diaminobenzidine in the presence of H_2O_2 to form an insoluble precipitate. A variation of this method is called the peroxidase–antiperoxidase method (PAP). In this method a primary antibody is used against the antigen in question, followed by a

secondary antibody which is not labelled. This is followed by treatment with a PAP complex consisting of peroxidase conjugated to antibodies made from the same species as the primary antibody, but since the secondary antibody was polyclonal, it will also attach to the PAP complex (Vandesande, 1983; Sternberger, 1986). This method is now rarely used at the TEM level.

(c) Alkaline phosphatase. Alkaline and acid phosphatase have been popular histochemical targets in plant tissues for many years (Gomori, 1952). Alkaline phosphatase conjugated to secondary antibodies is now available from several supply companies. This method has been successfully used in both immunodetection studies for different antigens, and for *in situ* hybridisation studies for localisation of mRNAs (see Chapter 7, this volume). Several staining reactions have been used, but the nitroblue tetrazolium (NBT) method is probably the best for immunolocalisations (Larsson, 1988; Tanimoto *et al.*, 1993). Secondary antibodies with conjugated alkaline phosphatase are reacted with primary antibodies using standard methods. Slides are then developed in the dark with freshly made NBT/X-phosphate colour solution (nitroblue tetrazolium–5-bromo-4-chloro-3-indolyl phosphate). The incubation time is variable depending on the material, but is usually 1–4 h in the dark in a humid chamber (Tanimoto *et al.*, 1993). The most commonly used control is to omit the secondary antibody but still do the NBT reaction to test for endogenous alkaline phosphatases.

(d) Protein A. Protein A is produced in strains of *Staphylococcus aureus* and has the property of binding specifically to the Fc portion of some immunoglobulins (Bozzola and Russell, 1992) (Fig. 8.2). A variety of labels can be conjugated to Protein A, for example, biotin, colloidal gold or FITC. This method is mostly used at the TEM level along with colloidal gold labelling (Larsson, 1988).

(e) Ferritin. Ferritin is a storage form of iron in mammals. It is a molecule of molecular weight about 450 000 and is electron dense (Bozzola and Russell, 1992). The

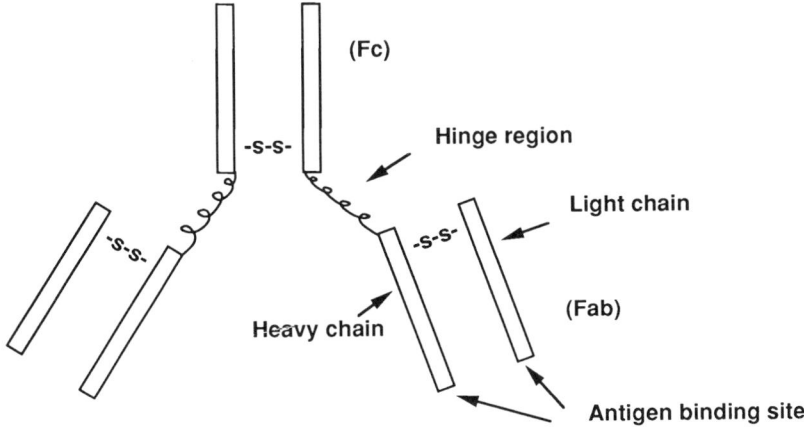

Fig. 8.2. Structure of immunoglobulin molecule. The antigen binding fragment (Fab) consists of a heavy and light chain connected by a disulphide bridge. The crystallisable fragment (Fc) is connected by a hinge region which can be digested away to release the Fab fragments.

ferritin technique had been extensively used in ICC as a label for TEM studies, but has fallen out of general use since the development of colloidal gold techniques. One major problem with its use is that ferritin is sometimes difficult to see in stained sections (Harris *et al.*, 1990).

(f) Colloidal gold. Heavy metals have been used as labels in ICC mostly at the TEM level. The most common is colloidal gold. Colloidal gold is easily made and can be conjugated to Protein A or to immunoglobulins (Bozzola and Russell, 1992). Hayat (1989) has edited a three volume text on all aspects of the colloidal gold technique, so we will only outline the general idea here (see also protocols in Section III).

Colloidal gold preparations can be purchased in different sizes starting from about 1 nm diameter. Under the electron microscope gold particles appear as small dots in the electron beam. As such they can be used in double-label preparations where two different size gold particles are used in conjunction with two different antibodies. Both direct and indirect methods can be used. Many different immunoglobulin and Protein A colloidal conjugates are commercially available. One problem with using colloidal gold conjugates is that colloidal gold does not irreversibly bind to an antibody and changes in pH may cause its release (Bozzola and Russell, 1992).

The signal strength of colloidal gold can be amplified by reacting cells with silver; this is called silver-enhancement (Merchenthaler *et al.*, 1989; Scopsi, 1989). In this method the gold particles act as nuclei around which silver accumulates, making each particle larger and denser. This method can make light microscope preparations easier to see and also has applications at the TEM level. The method involves treating slides or grids, after colloidal gold reaction, with a developer solution consisting of a colloid (usually gum arabic or polyethylene glycol), a reducing agent (usually hydroquinone) and a buffer (usually citrate buffer at pH 3.5) plus a silver source (usually silver lactate). The reducing agent reacts with the silver lactate to release silver ions which bind to the gold particles.

(g) Controls. Controls should always be used in ICC analyses. This is important because use of controls helps to eliminate the possibility of mistaking a false-positive signal as an antigenic binding site. Several different controls are commonly used. Preimmune serum from the same animal can be used in place of the primary antibody (Wick *et al.*, 1985; Bozzola and Russell, 1992). If the antigen is available, another control is to preadsorb the primary antibody with the antigen and then use the conjugate on the tissue in question. If all of the primary antibody has been bound, no reaction should occur in the tissue (Wick *et al.*, 1985). A simple control, to determine the level of non-specific binding of the primary or secondary antibody, is to delete either the primary or the secondary antibody from the reaction (Citterio *et al.*, 1989). Another control is to react the tissues with the reporter molecule alone, without the secondary antibody. Most workers will use more than one of these controls to be sure of the specificity of the labelling pattern.

G. Production and Storage of Antibodies

Antibodies, whether purchased from a commercial source or produced in your own laboratory, can be made by either of two methods (Sternberger, 1986). The most common method is to collect antibody-containing serum from an animal immunised against the antigen of interest. The serum will contain many different antibodies, but a proportion of them will specifically recognise the immunogen. Additionally, within the serum there are usually a number of antibodies that are specific for the antigen, but are directed against different epitopes on the antigen molecule. Each of these different antibody species is produced by plasma cells derived from a single B-cell clone. Therefore, this set of antigen-specific antibodies from serum is termed polyclonal. The other method of antibody production is to fuse an isolated antibody-producing plasma cell from an immunised mouse with a myeloma cell, a type of B-cell tumor. The resultant fusion hybrid cells are called hybridomas. These cells can be grown *in vitro*, where they continue to secrete highly specific antibodies that recognise only one epitope on the antigen molecule. Antibodies produced by hybridomas are therefore termed monoclonal.

For ICC purposes, both types of antibodies have advantages and disadvantages. Polyclonal antibodies, because they are a mixture of antibodies each capable of detecting a different epitope on an antigen molecule, tend to give stronger immunolabelling signals than do monoclonal antibodies. However, crude polyclonal preparations also contain many other, irrelevant antibodies (Stenram *et al.*, 1991; Vandenbosch, 1991). The contaminating antibodies may bind to tissue non-specifically, creating conditions of high background. Conversely, monoclonal antibodies are highly specific and non-specific background staining is usually not a problem. However, because monoclonal antibodies recognise only one epitope on the antigen, the signal strength is reduced compared to that from polyclonal antibodies. Furthermore, some epitopes may be damaged or masked during fixation and tissue preparation, further reducing the effectiveness of a particular monoclonal antibody. One way to avoid this problem is to pool a number of different monoclonal antibodies directed against the same antigen, but presumably to different epitopes. This combines the advantages of high specificity and low background with an increase in signal strength.

For all practical purposes, polyclonal antibodies are only useful for immunolabelling studies if the non-specific antibodies have been removed from the preparation. This is most effectively done by immunoaffinity purification. Antigen-specific antibodies are removed from the preparation by binding to pure antigen covalently bound to a solid support within a column. Non-specific antibodies are removed from the column by washing and the antigen-specific antibodies are then collected by elution from the column. Most commercially available polyclonal antibodies have already been immunoaffinity purified.

Antibody proteins are reasonably stable molecules and can be stored for years at or below $-20°C$. Antibody preparations can be conveniently frozen in the serum or supernatant in which they were collected or in phosphate buffered saline. Sodium azide should be added to the solution to prevent bacterial and fungal contamination (Harlow and Lane, 1988). (Care must be taken when handling sodium azide as it blocks the cytochrome electron transport system and is poisonous.) Antibodies can aggregate after repeated cycles of freezing and thawing, however, so the most prudent method of long-term storage is to freeze conveniently sized aliquots. Once thawed, working solutions are stable at $4°C$ for up to 6 months.

H. Photomicrography

Photography of immunochemical preparations tagged with fluorescent labels can be tricky. The following discussion focuses on some of the problems.

The choice of objectives is influenced by the numerical aperture (NA) of the lens, the depth of field required, and size of field of view needed. Brightness of image (intensity of signal) increases with increasing NA, but this must be balanced against the accompanying loss of depth of field (Roberts, 1986). Oil immersion objectives provide greater brightness as well, but this can be offset by the inconvenience of using immersion oil, especially when scanning slides at low to medium magnification.

Proper microscope alignment and centring of light source are essential for uniform illumination of the field of view. Non-uniform illumination may not be noticeable during observation, but it will become very apparent in photographs. Consult the instruction manual for procedures for microscope alignment and for centring of light source, and check these each time photographs are to be made.

For black and white negatives, Kodak T_{MAX} 400 and T_{MAX} 3200 have produced satisfactory images, using relatively short (15–30 s) exposures. With automatic cameras you will usually increase the ASA by four times to underexpose the photographs. Ektachrome 400 (daylight film) and Ektachrome 160 (tungsten-corrected colour film) have been used successfully for colour slides. Repeated exposures of the same field result in drastic reduction of brightness, due to bleaching of the fluorochrome. It is advisable to scan a field quickly at low magnification to locate cells for high magnification photographs, then to take the photographs before a more leisurely study of the cells. Continued observation is accompanied by further fading of the image.

III. PROTOCOLS

The protocols described are for three different ICC applications. These methods are known to work under the conditions described.

A. Protocol for Cryosectioning and Immunolocalisation of Tubulin at the Light Microscope Level

The procedure described below was optimised to enable visualisation of the microtubule cytoskeleton in cryosections of hyacinth roots and in shoot tissues of *Selaginella* (Fig. 8.3). It is a modification of published techniques (Wick *et al.*, 1981; Smith-Huerta and Jernstedt, 1989) and represents a starting point for adapting the methods to other tissues and species. Recipes for the necessary reagents are provided at the end of this section.

1. Fixations

Fix tissues for 1–2 h at room temperature in an 8% paraformaldehyde (PFA) solution made up in microtubule stabilising buffer (MTSB). It is very important that one does not over-fix. Two hours fixation is too long for some roots; $1\frac{1}{2}$ h fixation results in adequate cellular preservation and more distinct cortical microtubules.

8. IMMUNOLOCALISATION OF ANTIGENS IN PLANTS 173

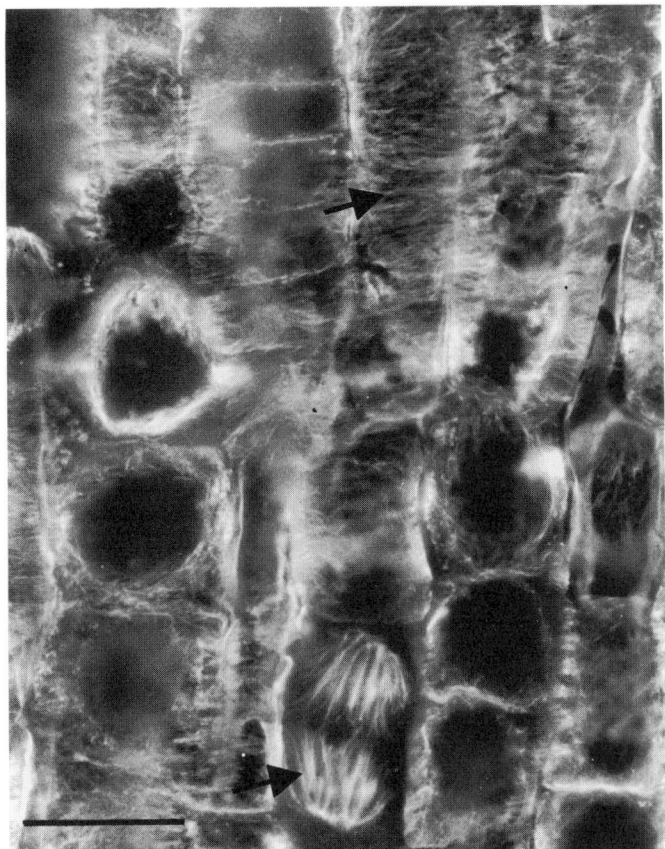

Fig. 8.3. Localisation of microtubules in longitudinal cryosections of hyacinth root tips using biotin–streptavidin–fluorescein detection system. Cortical microtubules are seen in the upper right (arrow) and spindle microtubules (arrow) are in the lower middle of the photograph. (Photograph by N. L. Smith-Huerta.) Scale bar = 25 μm.

2. Washing

Following fixation, samples are washed three or four times for 10–15 min each in plain MTSB. Adequate washing is necessary to remove residual PFA and also may reduce autofluorescence. Samples should be embedded immediately after washing, but can be stored indefinitely at −80°C once embedded.

3. Embedding

BEEM capsules (size 00) (Polysciences) or JB-4 embedding cups (Polysciences) are used as moulds. Place one or two drops of Tissue Tek® O.C.T. mounting medium (Scientific Products) or Histo-Prep® (Fisher) in the tip of a BEEM capsule or in the well of a JB-4 mould. Label capsules or moulds appropriately. (Warning: Some marking pens are not hexane- or liquid nitrogen-proof. Test your labels before combining samples!

An alternative labelling procedure is to write information on paper with pencil and embed the label in O.C.T. compound with the sample.) Blot excess MTSB from sample and place sample in BEEM capsule or the well of the JB-4 mould, orienting as required to obtain desired plane of section. Slowly and gently add O.C.T. compound to cover sample and fill capsule or well, by pouring down the sides. Samples may shift slightly as this viscous fluid is added, in which case they can be reorientated with a dissecting needle or fine forceps. For BEEM capsules, close lid and drop into a beaker or tube of hexane which has been chilled but not frozen solid in liquid nitrogen. For JB-4 moulds, gently lay a disk of filter paper inside embedding cup, making sure the paper is in contact with O.C.T. compound filling specimen well. Drop into chilled hexane and freeze until the O.C.T. compound becomes opaque (approximately 30 s). Do not over-freeze. If cracks are visible when blocks are removed from moulds, the time in chilled hexane was too long.

Blocks can be sectioned immediately after freezing or stored for up to several months at $-80°C$. Samples should not be allowed to thaw before sectioning. Transport blocks over liquid nitrogen or on dry ice to prevent thawing.

4. Cryosectioning

The main points to remember with this technique are (a) with practice anyone can do it, and (b) small temperature changes (from opening the cryostat chamber door or from fluctuations in ambient temperature) can make a big difference in how the sectioning goes. An additional point to remember is that difficulties in sectioning can arise when specimens are too cold or not cold enough. Often the only way to determine which is the case is to attempt sectioning at slightly higher (-16 to $-17°C$) or slightly lower (-20 to $-21°C$) temperatures to determine which works best for your specimens. Finally, allow 15–20 min for temperature equilibration of the specimen block inside the cryostat chamber.

The object of sectioning is to make sections as thin as possible and/or useful but thick enough that they do not fall apart during staining, washing, or mounting. It is difficult to obtain good cryostat sections below 5 μm. Sections as thick as 60–100 μm can be obtained and may be useful if your material contains very large, vacuolate cells.

Follow manufacturer's instructions provided with your particular model of cryostat. As with all microtomy, critical steps in the process include trimming the block (parallel block faces are necessary to obtain straight ribbons); using a sharp, clean knife; and experimenting with speed of sectioning and section thickness until you find the combination of conditions which result in uniform, intact sections.

Sections should form ribbons and collect on the knife edge. If this does not happen, the knife could be dull, the knife edge could be dirty, the temperature of the chamber could be too high or too low, the antiroll plate could be too close or too far from the knife edge, etc. If the latter is suspected, the antiroll plate should be advanced or retracted *slightly*, so that its edge is even with the knife edge.

Sections are most easily picked up by bringing a warm (room temperature) slide (coated [subbed] or not, as necessary) close to the ribbon. Without touching the knife, place slide near the knife edge. Sections will 'hop' onto the slide. Try to position sections away from edge of slide; this will make subsequent staining easier. After sections are picked up, slides may be stored for several days at room temperature or in the refrigerator. However, O.C.T. compound will eventually dry out, causing tissue distortion, so immunostaining should not be delayed too long.

The optimal temperature for sectioning must be determined for each species and sample; −15 or −16°C has worked well for bigger tissue pieces. A couple degrees higher seems better for meristematic tissues. If the temperature is too low for a given specimen, it will shatter rather than section. It takes from several hours to overnight to change the temperature of most cryostat chambers, chuck holders and knives, so plan ahead.

5. Immunostaining

The main points to remember for successful staining are:

- do not let sections dry out;
- treat the slides and sections gently, since sections are very weakly affixed to the slide;
- avoid contaminating later steps with solutions from earlier steps; and
- appropriate controls must always be done, especially when adapting this protocol to a new tissue or species — these controls are, simply, omission of primary and secondary antibodies and the fluorochrome, respectively.

(a) Remove slides with sections from storage and place extraction buffer on sections for 10-15 min to hydrate tissue and permeabilise membranes. Use about 100 µl per section. Depending on the material, this step may require up to one hour.

(b) Remove extraction buffer with pipette or with filter paper triangles; add approximately 10 µl primary antibody per section (or enough to cover section). Primary antibody is mouse monoclonal anti-alpha tubulin (Amersham), 1:100 dilution in PBS with 0.02% sodium azide. Place slides on supports in Petri dish containing moist filter paper, two slides per dish. Cover and incubate in primary antibody for 1 h in 37°C oven. Watch oven temperature and sections to see that they do not dry out during incubation.

(c) Add MTSB to sections (*c.* 200 µl) for 10 min to wash. It is not necessary to remove primary antibody prior to adding MTSB.

(d) Remove MTSB with filter paper triangles. Add secondary antibody (biotinylated sheep anti-mouse [Amersham]; 1:50 dilution in PBS plus 0.02% sodium azide). Use approximately 10 µl per section or just enough to cover each section with a small drop. Incubate for 1 h at 37°C.

(e) Wash again in MTSB for 10 min, using *c.* 200 µl solution.

(f) Remove MTSB with pipette and/or filter paper before adding fluorochrome. Stain with streptavidin-fluorescein [Amersham] (1:100 dilution in PBS + azide) for 15 min at room temperature in covered dishes. Use 10 µl solution or less per section. Cover closed dishes with aluminium foil to exclude all light.

(g) Rinse 5 min with MTSB in foil-covered dishes.

(h) Remove MTSB. Mount in antifade mounting medium containing Mowiol 40-88® (Aldrich) (sometimes spelled Moviol), using just enough to fill in under coverslip. Coverslips must be non-fluorescing (VanLab or Fisher). Use applicator stick to place several 'lines' of mounting medium on coverslip (centre and sides) and then slowly lower coverslip down onto slide. Slides should sit for 15–20 min before viewing with microscope. Always protect from light to preserve the fluorochrome.

(i) Store mounted slides horizontally in refrigerator inside plastic bag. These are semi-permanent preparations and will eventually dry out. However, slide life can be prolonged by ringing coverslips with melted wax. Critical observations and photographs should be made within a few weeks of staining to avoid problems with bacterial growth, drying of slides, shifting of coverslips, collapse of tissues, and fading of the fluorochrome.

6. Photomicrography

Select the appropriate filter set for fluorescein/FITC epifluorescence microscopy. This should include an excitation filter transmitting in the 450–490 nm region; a dichroic beam-splitting mirror, transmitting wavelengths above 510 nm; and a barrier filter, transmitting wavelengths greater than approximately 520 nm. Additional filters (e.g. a short-pass red filter, 560 nm) may also be added to the light path to reduce autofluorescence.

B. Protocol for Immunolocalisation of Proteins in Resin-embedded Tissue for Bright Field and Fluorescence Microscopy

The following protocol has been used to localise a variety of different proteins within plant tissues using both monoclonal and polyclonal antibodies (Fig. 8.4). Antigens have been detected in virtually every tissue of the plant body including leaves, roots, stems, embryos and endosperm. The utilisation of plastic-embedded tissue allows for precise cellular and subcellular localisation.

1. Fixation

Tissue suspected of containing the antigen of interest is fixed in a mixture of 3.0% EM-grade paraformaldehyde and 0.2% EM-grade glutaraldehyde in 0.1 M phosphate buffer, pH 7.0. It is essential that the paraformaldehyde solution be freshly made or, alternatively, be from frozen aliquots of a freshly made solution. The glutaraldehyde helps preserve structural detail, which is of particular importance if there is a chance that the tissue will also be observed with the electron microscope. Higher concentrations of glutaraldehyde, however, may have a detrimental effect on the antigenicity of the tissue, and should not be used if fluorescent labels will be applied (Mariac *et al.*, 1992). To facilitate fixation and resin infiltration, the tissue should be as small as reasonably possible. Fixation times will vary with the tissue, being several hours for small delicate tissues to overnight for dense difficult tissues such as seeds.

Following fixation, the tissue is rinsed three times with 0.1 M phosphate buffer, 15 min per rinse, and dehydrated through a graded alcohol series to 100% ethanol with changes every 15 min.

8. IMMUNOLOCALISATION OF ANTIGENS IN PLANTS

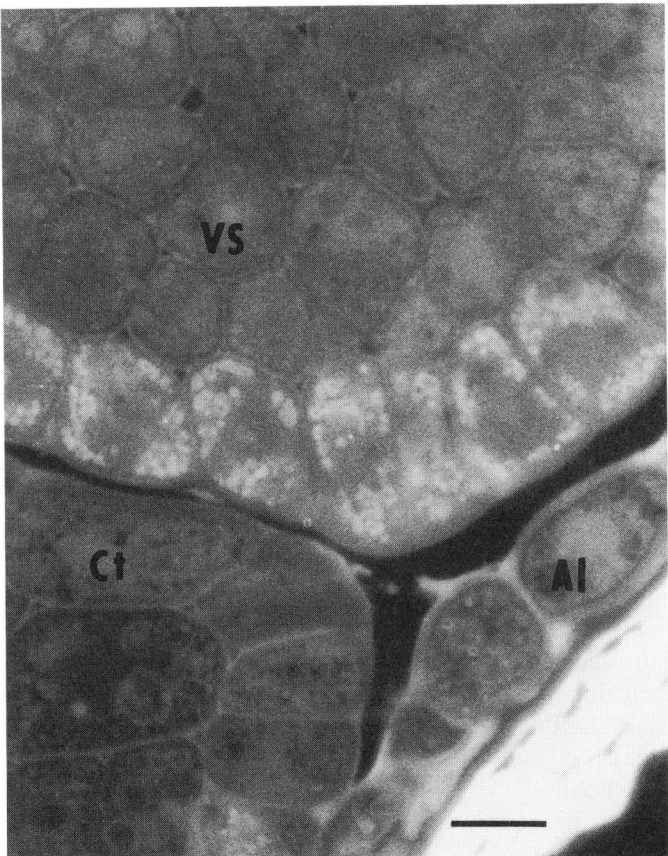

Fig. 8.4. Immunofluorescent localisation of rice lectin in a 2 μm thick section of rice embryo embedded in Historesin. The lectin is localised within discrete organelles (protein bodies) of the epidermal cells on the ventral scale (VS). The coleoptile (Ct) and the aleurone (Al) are unlabelled. Scale bar = 10 μm.

2. Embedding

After making two changes of 100% ethanol, resin may be introduced into the tissue. For light microscopy, we prefer Historesin (Reichert-Jung) for its ease of handling, consistent staining characteristics and the tenacity with which the sections stick to microscope slides. If there is a chance that the material may also be observed with transmission electron microscopy, the tissue should be embedded in LR White resin (London Resin Company). Infiltration should follow standard protocols with a stepwise introduction of resin from 50% resin (in ethanol) to 100% over several hours. Generally, two further changes of resin over the next 24 h is sufficient to infiltrate even the most difficult tissues.

Historesin is polymerised overnight at room temperature under a vacuum. If using LR White, polymerise in gelatin capsules for 24 h at 50–55°C in a vacuum oven. LR White resin polymerisation is particularly sensitive to oxygen and, consequently, flat embedding of tissue in LR White is difficult unless great care is taken to exclude air.

3. Sectioning

Section at 2 μm thickness with glass or diamond knives and mount slides on APES- (3-aminopropyl-triethoxysilane, Aldrich) or Vectabond-P (Vector Labs) treated slides. Poly-L-lysine coated slides are also acceptable but do not work as well with LR White sections. We find it convenient to put eight sections, in two rows of four, per slide. Each slide will then contain a pair of sections for the experimental labelling and three additional pairs of sequential sections for negative and positive controls. Prior to immunolabelling, individual sections or section pairs may be isolated from neighbouring sections by drawing a well around them with a PAP pen (Electron Microscopy Science).

4. Immunolabelling

The following protocol describes a technique for indirectly detecting antigens with a biotin/fluorescent-labelled streptavidin detection system. The procedure also works well when using gold–streptavidin and silver enhancement. Most of the reagents used for immunolabelling are prepared in phosphate buffered saline (PBS) although Tris buffered saline may be substituted. Sodium azide (0.02%) is typically added to the saline solution to prevent bacterial or fungal contamination.

Non-specific protein binding sites must first be blocked before incubation of the sections with the primary antibody. The following blocking solution has worked well for aldehyde-fixed plant tissue embedded in resin: 3.0% bovine serum albumin (BSA, Fraction V is fine); 1.5% glycine (0.2 M); 0.1% Tween 20 (polyoxyethylene sorbitan monolaurate) in PBS.

Block each section for 30 min and rinse with PBS-T (PBS with 0.1% Tween 20) for 10 min.

After blocking, incubate each section with a 10 μl drop of primary antibody diluted with PBS-T plus 0.1% BSA for 60 min at 37°C. The appropriate dilution will have to be determined empirically; 1:20 is a good place to start for most antibodies. This incubation must be performed in a humid incubation chamber. Following the incubation, rinse the slides with PBS-T three times for 10 min each rinse.

Incubate each section with a 10 μl drop of diluted biotin–secondary antibody (the dilution will again have to be determined empirically) for 60 min at 37°C. This must also be performed in a humid incubation chamber. Following the incubation of the secondary antibody, rinse the slides with PBS-T three times for 10 min each rinse.

Detection of the biotin–secondary antibody is accomplished by incubating each section with a 10 μl drop of a properly diluted streptavidin–fluorochrome conjugate for 30 min at 37°C. Again, this must be done in a humid incubation chamber. The choice of fluorochrome will depend upon the tissue examined. BODIPY®–streptavidin works exceptionally well, with excitation and emission characteristics of fluorescein but with much reduced quenching. If tissue autofluorescence is a problem, Texas Red–streptavidin is a good choice. Both are available from Molecular Probes, Inc.

The slides are then rinsed with PBS-T twice, for 10 min each rinse, followed by one rinse with double-distilled water to eliminate the salt. Slides are then dried and a coverslip is mounted with immersion oil or, if the slides are to be permanent, with Fluoromount or Permount.

Negative controls include preimmune serum if the primary antibody is a polyclonal or, if the primary antibody is monoclonal, another monoclonal antibody, not directed against the antigen of interest, but from the same species. Preabsorption of the primary antibody with antigen is also an adequate negative control. It is also advantageous to incubate some control sections in just the streptavidin–fluorochrome conjugate to detect the extent, if any, of non-specific streptavidin binding to the section.

C. Protocol for Immunogold Localisation of Proteins using Transmission Electron Microscopy

Several modifications of standard electron microscopic protocols have made possible the widespread, effective use of immunogold labelling as a means of localising proteins at the ultrastructural level. For instance, substitution of paraformaldehyde for glutaraldehyde as the primary fixative and embedding the tissue in hydrophilic polar acrylic resins, such as LR White, instead of the more conventional hydrophobic epoxy

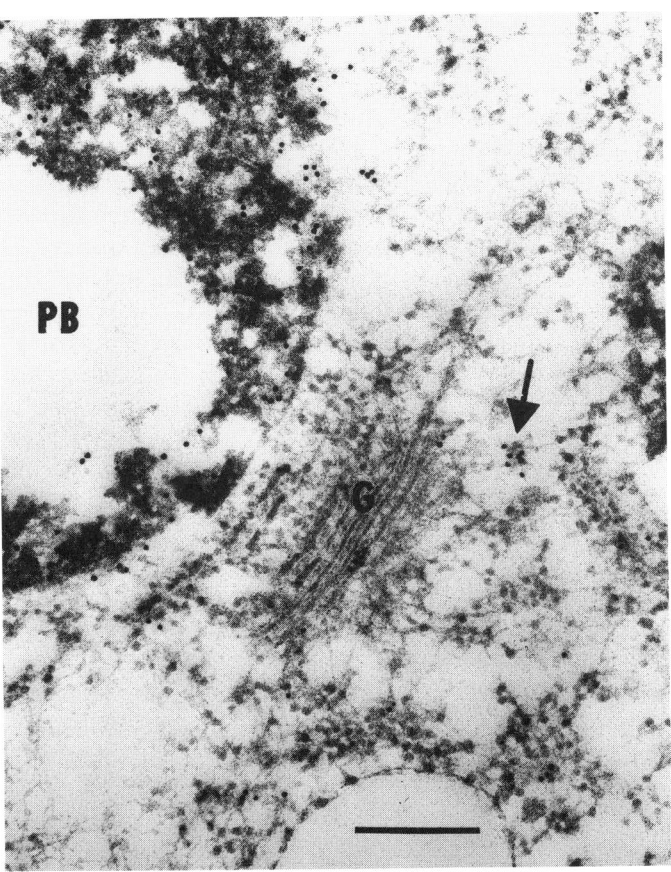

Fig. 8.5. Immunogold localisation of rice lectin within a radicle cell of a rice embryo. The lectin is preferentially localised to protein bodies (PB). Gold label is also evident over a Golgi body (G) and over what may be a transition vesicle (arrow) between the Golgi and one of the protein bodies. Scale bar = 0.25 μm.

resins, have helped increase the overall sensitivity of immunogold localisation (Craig and Miller, 1984). The following protocol has been used to effectively localise a variety of proteins within most organs of the plant body (Fig. 8.5).

1. Fixation

The conditions for primary fixation of tissues for electron microscopic immunolabelling are identical to those outlined in the previous protocol for light microscopy. Briefly, tissue is fixed in a mixture of 3.0% EM-grade paraformaldehyde and 0.2% EM-grade glutaraldehyde in 0.1 M phosphate buffer, pH 7.0, for 4–24 h depending on tissue. After thoroughly rinsing the tissue with buffer, postfixation with osmium tetroxide (1.0% for 2 h in the same buffer) is usually desirable for EM. However, osmium postfixation may have an adverse effect on antigenicity. This needs to be determined empirically for each antigen investigated. Following postfixation, the tissue is rinsed three times with buffer and dehydrated through a graded ethanol series to 100% ethanol with changes every 15 min.

2. Embedding

For electron microscopy, the tissue is infiltrated with LR White resin (hard grade, EM Sciences). Infiltration should follow standard protocols with a stepwise introduction of resin from 50% resin (in ethanol) to 100% over several hours. Generally, two further changes of pure resin over the next 24 h is sufficient to infiltrate even the most difficult tissues. Polymerisation is carried out in gelatin capsules for 24 h at 50–55°C in a vacuum oven. Due to the sensitivity of LR White resin polymerisation to oxygen, flat-embedding tissue in LR White is not recommended unless great care is taken to exclude air. Also, polymerisation of LR White with accelerator, using the conditions suggested by the London Resin Company, is not recommended due to the highly exothermic nature of the reaction and the adverse effects on tissue antigenicity.

3. Sectioning

Sections showing silver–gold interference colours (60–90 nm thick), are cut with diamond knives and collected on gold grids. Copper grids are avoided due to the potential reactivity of copper with some of the reagents used during immunolabelling. Grids coated with formvar-carbon are particularly useful for supporting LR White resin sections. For tissue postfixed with osmium tetroxide, it may be necessary to treat the sections with a 0.56 mM solution of sodium metaperiodate prior to immunolabelling to facilitate antigen detection (Harlow and Lane, 1988).

4. Immunolabelling

As with the previous immunolabelling protocol outlined for light microscopy, most of the reagents used for immunolabelling are prepared in phosphate buffered saline, although Tris buffered saline may be substituted. Sodium azide (0.02%) is typically added to the saline solution to prevent bacterial or fungal contamination. Most of the staining and rinsing steps are carried out in approximately 20 µl drops of reagents placed on small sheets of parafilm to keep the drops in nice beads. If the sections are collected on uncoated grids, the entire grid may be placed within the reagent drop. If collected on coated grids, the grid should be floated on the drop, section side down.

Again, non-specific protein binding sites must first be blocked before incubation of the sections with the primary antibody. Block the sections by placing the grids on drops of PBS containing 1.0% bovine serum albumin and 0.1% Tween-20 for 15 min. Rinse the grids in drops of PBS containing 0.1% Tween-20 (PBS-T).

Incubate each grid by floating on a 20 µl drop of primary antibody diluted in PBS containing 0.1% BSA and 0.1% Tween-20 (dilution will have to be determined empirically; 1:100 is a good place to start) for 90 min. After the incubation, gently rinse the grids with several drops of PBS-T from a pipette and then place the grids on three sequential drops of PBS-T for 10 min each drop.

The primary antibody is detected by incubating each grid on a 20 µl drop of diluted gold-conjugated secondary antibody directed against the primary antibody (dilution will again have to be determined empirically using the recommendations of the supplier to start; 1:20 is usually fine) for 30 min. Gold conjugate size should be within the 5–15 nm range. Rinse with PBS-T as indicated above for primary antibody incubation followed by a final rinse in distilled water.

Controls are essential for proper interpretation of immunolabelling results. Pre-immune serum, preabsorbed primary antibody or, if using monoclonals, a different antibody from the same animal as the primary (e.g. supernatant or ascites) are appropriate negative controls. It is also beneficial to always run positive controls using different known antibodies of the same class as the unknown experimental antibody.

5. Staining

Stain the sections with aqueous uranyl acetate (5–15 min) and lead citrate (1–3 min). LR White sections stain more readily than epoxy resins and therefore require shorter staining times.

D. Recipes

Microtubule stabilising buffer (MTSB)

The final solution should contain:

$$50 \text{ mM PIPES}$$
$$5 \text{ mM EGTA}$$
$$5 \text{ mM MgSO}_4 \cdot 7\text{H}_2\text{O}$$

These should be made up as separate stock solutions, which will keep for a few months in the refrigerator and longer in the freezer.

To make 100 ml final volume of MTSB, combine:

10 ml 10× PIPES stock
10 ml 10× EGTA stock
10 ml MgSO$_4$ stock
70 ml dI water

Use half to make fixative and half as wash buffer.

PIPES

To make a 10× stock:
(1) Dissolve 15.12 g PIPES (powder) in 100 ml deionised (dI) water. This will be very acidic (approximately pH 2.4) and milky.
(2) Adjust to pH 6.8 with KOH (at least 1 N KOH, or use KOH pellets). Solution will begin to clear around pH 5.5.
(3) Use 10 ml of this 10× stock for a final volume of 100 ml MTSB.

EGTA

To make a 10× stock:

(1) Dissolve 0.38 g EGTA in 12 ml dI water.
(2) Stir and slowly add KOH pellets (usually 2 or 3) until solution becomes clear.
(3) Adjust volume to 20 ml.
(4) Check pH; it should be between 6.8 and 8.5.
(5) Use 10 ml of this 10× stock for a final volume of 100 ml MTSB solution.

$MgSO_4 \cdot 7H_2O$

To make 10× stock:

(1) Dissolve 1.23 g $MgSO_4 \cdot 7H_2O$ in 100 ml dI water. (This goes into solution quite easily.)
(2) Use 10 ml of this 10× stock to make 100 ml final volume MTSB.

8% Paraformaldehyde (PFA) fixative

To make 50 ml in MTSB:

(1) Paraformaldehyde powder is extremely toxic. Avoid inhalation of powder or skin contact.
(2) Dissolve 4 g PFA in 50 ml freshly prepared MTSB.
(3) Heat to 60–70°C, stirring until PFA dissolves (several hours).
(4) If volume has been reduced by evaporation during heating, add dI water to bring up to 50 ml.
(5) This fixative has a shelf-life of about a month in the refrigerator.

Phosphate buffered saline (PBS)

(1) To 800 ml dI water, add 8 g NaCl, 0.2 g KCl, 1.15 g Na_2HPO_4 (anhydrous) and 0.2 g KH_2PO_4.
(2) Bring volume up to 1 litre.
(3) Adjust pH to 7.3.
(4) Refrigerate, or divide solution into 50 ml aliquots in plastic bottles and freeze. Keeps indefinitely when frozen.
(5) Prior to making antibody dilutions, thaw one 50 ml aliquot and add 10 mg

sodium azide (a mutagen) to make 0.02% azide–PBS. Use this for primary and secondary antibody dilutions and streptavidin–fluorescein dilutions.

Dilution and storage of antibodies and fluorochromes

(1) Antitubulin: keep in $-80°C$ freezer in $2\,\mu l$ aliquots in plastic microfuge tubes. Antibodies can be transferred to ordinary freezer prior to use. Carry on dry ice or liquid nitrogen. Just before use each day, dilute with $200\,\mu l$ PBS with 0.02% sodium azide (making a 1:100 dilution). Keep in refrigerator until ready to use. Antibodies should *not* be refrozen after dilution is made.
(2) Biotinylated sheep antimouse (secondary antibody): also keep in $2\,\mu l$ aliquots at $-80°C$. Dilute with $100\,\mu l$ PBS + azide to make 1:50 dilution.
(3) Fluorescein–streptavidin: also keep in $2\,\mu l$ aliquots at $-80°C$. Dilute with $200\,\mu l$ PBS + azide to make 1:100 dilution. Protect from light at all times.

Extraction buffer (EB)

$10\,\mu l$ Nonidet P_{40} (straight from bottle; can use Triton X-100 instead)
$1000\,\mu l$ MTSB
0.073 g mannitol

Antifade mounting medium

(1) In a 50 ml centrifuge tube add:
    6 g analytical grade glycerol
    2 g Mowiol 40-88 (Aldrich)
(2) Stir, then add 6 ml dI water.
(3) Leave at room temperature for 2 h, stirring occasionally.
(4) Add 12 ml 0.2 M Tris buffer, pH 9–10.
(5) Incubate in 50°C water bath for 10 min, stirring occasionally.
(6) Cool to room temperature.
(7) Add 0.5 g *n*-propyl gallate.
(8) Stir until no more *n*-propyl gallate dissolves.
(9) Clarify by centrifugation ($5000 \times g$ for 15 min).
(10) Place in 0.5 to 1 ml aliquots in *glass* vials (not plastic), stopper with a parafilm-covered cork, and freeze. Solution turns yellow in contact with air; do not use if yellowed.

To use, add a small amount (1 to 3 mg) sodium azide (NaN_3) to 0.5 to 1 ml thawed mounting medium to prevent bacterial growth in tube or on slides.

Chrome alum slide coating

(1) Use 600 ml tall beaker
(2) Dissolve 2.5 g gelatin in 500 ml distilled water at 30–35°C, stirring constantly.
(3) Add 0.25 g chromium potassium sulphate (chrome alum); stir until dissolved.
(4) Dip slides and dry in 35–45°C oven overnight. Store in dust-free slide box until use.

NOTE ADDED IN PROOF

There have been anecdotal reports that the diet (apparently the plant component) of animals producing the primary antibody can affect background staining and may result in false-positive reactions. Investigators should bear this possibility in mind if problems are encountered with excessive background staining.

REFERENCES

Baskin, T. I., Busby, C. H., Fowke, L. C., Sammut, M. and Gubler, F. (1992). *Planta* **187**, 405–413.
Bellon, B. and Druet, P. (1974). *Ann. Immunol. (Inst. Pasteur)* **125C**, 871–884.
Beltz, B. S. and Burd, G. D. (1989). "Immunocytochemical techniques: Principles and practice". Blackwell Scientific Publications, Cambridge, MA.
Bozzola, J. J. and Russell, L. D. (1992). "Electron Microscopy: Principles and Techniques for Biologists", pp. 234–250. Jones and Bartlett Publishers, Boston.
Brown, R. C., Lemmon, B. E., and Mullinax, J. B. (1989). *Botanica Acta* **102**, 54–61.
Bullock, G. R. and Petrusz, P. (eds) (1989). "Techniques in Immunocytochemistry", Vol. 4. Academic Press, London.
Campbell, A. M. (1991). "Laboratory Techniques in Biochemistry and Molecular Biology", Vol. 23. Elsevier, Amsterdam.
Citterio, S., Sgorbati, S., Levi, M., Colombo, B. M. and Sparvoli, E. (1992). *J. Cell Sci.* **102**, 71–78.
Coulton, G. (1990). In "*In situ* Hybridization: Application to Developmental Biology and Medicine". (N. Harris and D. G. Wilkinson, eds), pp. 1–32. Cambridge University Press, Cambridge.
Craig, S. and Miller, C. (1984). *Cell Biol. Int. Rep.* **6**, 879–888.
Cyr, R. J. and Palevitz, B. A. (1989). *Planta* **177**, 245–260.
Doonan, J. H. and Clayton, L. (1986). In "Immunology in Plant Science" (T. L. Wang, ed.), Society for Experimental Biology Seminar Series 29, pp. 111–136. Cambridge University Press, Cambridge.
Falconer, M. M., Donaldson, G. and Seagull, R. W. (1988). *Protoplasma* **144**, 46–55.
Gomori, G. (1952). "Microscopic Histochemistry Principles and Practices". University of Chicago Press, Chicago, IL.
Goodbody, K. C., Hargreaves, A. J. and Lloyd, C. W. (1989). *J. Cell Sci.* **93**, 427–438.
Gorbsky, G. and Borisy, G. G. (1985). *Proc. Natl. Acad. Sci. USA* **82**, 6889–6893.
Gubler, F. (1989). *Cell Biol. Int. Rep.* **13**, 137–145.
Harlow, E. and Lane, D. (1988). "Antibodies: A Laboratory Manual". Cold Spring Harbor Laboratory, Cold Spring Harbor, NY.
Harris, N., Mulcrone, J. and Grindley, H. (1990). In "*In situ* Hybridization: Application to Developmental Biology and Medicine" (N. Harris and D. G. Wilkinson, eds), pp. 175–188. Cambridge University Press, Cambridge.
Hattersley, P., Watson, L. and Osmond, C. B. (1977). *Aust. J. Plant Physiol.* **4**, 523–539.
Hayat, M. A. (ed.) (1989). "Colloidal Gold: Principles, Methods, and Applications", Vol. 1. Academic Press, New York.
Hemmilä, I. A. (1991). "Applications of Fluorescence in Immunoassays". Wiley and Sons, New York.
Hogetsu, T. and Oshima, Y. (1986). *Plant Cell Physiol.* **27**, 939–945.
Jacobs, M. and Gilbert, S. F. (1983). *Science* **220**, 1297–1300.
Jensen, W. A. (1962). "Botanical Histochemistry: Principles and Practice". W. H. Freeman and Co., San Francisco, CA.
Johansen, D. A. (1940). "Plant Microtechnique". McGraw-Hill, New York.

Jones, T. and Rost, T. (in press). *In* "Biotechnology in Agriculture and Forestry: Somatic Embryogenesis" (Y. P. S. Bajaj, ed.). Springer-Verlag, Berlin.
Knox, R. B. (1982). *In* "Techniques in Immunocytochemistry", Vol. 1 (G. R. Bullock and P. Petrusz, eds), pp. 205-238. Academic Press, New York.
Larsson, L. -I. (1988). "Immunocytochemistry: Theory and Practice". CRC Press, Boca Raton, FL.
Mariac, C., Rougler, M., Gaude, T. and Dumas, C. (1992). *Protoplasma* 166, 223-227.
Merchenthaler, I., Gallyas, F. and Liposits, Z. (1989). *In* "Techniques in Immunocytochemistry", Vol. 4 (G. R. Bullock and P. Petrusz, eds), pp. 217-252. Academic Press, New York.
Mishkind, M., Raikhel, N. V., Palevitz, B. A. and Keegstra, K. (1982). *J. Cell Biol.* 92, 753-764.
O'Brien, T. P. and McCully, M. E. (1981). "The Study of Plant Structure: Principles and Selected Methods", Termarcarphi Pty. Ltd, Melbourne, Australia.
Raikhel, N. V., Mishkind, M. and Palevitz, B. A. (1984). *Protoplasma* 121, 25-33.
Roberts, K. (1986). *In* "Immunology in Plant Science" (T. L. Wang, ed.), Society for Experimental Biology Seminar Series 29, pp. 89-110. Cambridge University Press, Cambridge.
Sakaguchi, S., Hogetsu, T. and Hara, N. (1988). *Planta* 175, 403-411.
Saunders, M. J., Cordonnier, M., Palevitz, B. A. and Pratt, L. H. (1983). *Planta* 159, 545-553.
Scopsi, L. (1989). *In* "Colloidal Gold: Principles, Methods and Applications", Vol. 1 (M. A. Hayat, ed.), pp. 251-295. Academic Press, New York.
Scopsi, L. and Larsson, L. -I. (1985). *Histochemistry* 82, 321-329.
Smith-Huerta, N. L. and Jernstedt, J. A. (1989). *Protoplasma* 151, 1-10.
Stenram, U., Heneen, W. K. and Skerritt, J. H. (1991). *J. Exp. Bot.* 42, 1347-1355.
Sternberger, L. A. (1986). "Immunocytochemistry", 3rd edn. Wiley and Sons, New York.
Tanimoto, E. Y., Rost, T. L. and Comai, L. (1993). *In* "Cell and Molecular Biology of the Plant Cell Cycle" (J. C. Ormrod and D. Francis, eds), pp. 85-95. Kluwer, Dordrecht, The Netherlands.
Tiwari, S. C., Wick, S. M., Williamson, R. E. and Gunning, B. E. S. (1984). *J. Cell Biol.* 99, 63s-69s.
Vandenbosch, K. A. (1991). *In* "Electron Microscopy of Plant Cells", pp. 181-215. Academic Press, New York.
Vandesande, F. (1983). *In* 'Immunocytochemistry" (A. C. Cuello, ed.), pp. 101-119. Wiley and Sons, Chichester.
Wick, S. M. and Duniec, J. (1986). *Protoplasma* 133, 1-18.
Wick, S. M., Seagull, R. W., Osborn, M., Weber, K. and Gunning, B. E. S. (1981). *J. Cell Biol.* 89, 685-690.
Wick, S. M., Muto, S. and Duniec, J. (1985). *Protoplasma* 126, 198-206.
Wick, S. M., Cho, S. and Mundelius, A. R. (1989). *Cell Biol. Int. Rep.* 13, 95-106.
Yadzi, H. M. and Dardick, I. (1992). "Diagnostic Immunocytochemistry and Electron Microscopy". Igaku-Shoin, New York.
Zola, H. (1990). "Laboratory Methods in Immunology", Vols I, II. CRC Press, Boca Raton, FL.

9 Protoplast Fusion

R. MARCHANT, M. R. DAVEY and J. B. POWER

Plant Genetic Manipulation Group, Department of Life Science, University of Nottingham, Nottingham NG7 2RD, UK

I.	Introduction	187
II.	Techniques for large-scale protoplast fusion	190
	A. Chemically induced fusion	190
	B. Electrofusion	194
III.	Small-scale fusion techniques	196
	A. Small-scale chemical fusion	197
	B. Small-scale electrofusion (microfusion)	197
IV.	Production of heterokaryons	198
V.	Selection of heterokaryons and hybrid cells	198
	A. Chlorophyll-deficient mutants (albinos)	198
	B. Nitrate reductase deficiency	199
	C. Resistance markers	199
	D. Antimetabolites	199
	E. Light-sensitive mutants	199
	F. Transformed cell lines	200
	G. Differential growth and plant regeneration	200
	H. Double mutants	200
VI.	Manual isolation of heterokaryons	200
VII.	Flow cytometry	201
VIII.	Influence of protoplast fusion on the nuclear genome	201
IX.	Effects of protoplast fusion on the cytoplasmic genome	202
	References	203

I. INTRODUCTION

Under suitable culture conditions, plant protoplasts exhibit totipotency (the ability to regenerate whole plants), which makes them suitable for plant genetic manipulations.

The subject of protoplast isolation and regeneration has been discussed in considerable detail and there is now a large number of species in which plants can be regenerated from protoplasts (see Bajaj, 1989). The aim of this chapter is to outline the methods that have been developed to fuse protoplasts.

The spontaneous fusion of protoplasts was first observed in 1909 by Kuster. Since that time, protoplasts have been induced to fuse by both chemical and electrical treatments. Power *et al.* (1970) were the first workers to fuse protoplasts chemically using sodium nitrate as the fusogen, whilst the first published report of electric fields being employed to fuse plant protoplasts was that of Senda *et al.* (1979). Procedures for the electrical fusion of both animal and plant cells were subsequently refined by Zimmermann and Scheurich (1981). Electric fields have also been used to promote DNA uptake into isolated protoplasts (Davey *et al.*, 1989; Davey and Rech, 1992). This has resulted in transformed tissues from which transgenic plants can be regenerated in several crops, an approach which is particularly important in crop plants such as cereals which cannot be transformed by *Agrobacterium*. Electric fields also stimulate the division of plant protoplasts (Rech *et al.*, 1987), as well as plant regeneration from protoplast-derived cells (Ochatt *et al.*, 1988).

Protoplast fusion permits plant genera, species or varieties, which are reproductively isolated due to natural incompatibility mechanisms at the whole plant level, to be combined together at the single cell level (Pelletier and Chupeau, 1984). Fusion at the membrane level is not species specific. This is illustrated by the fact that, provided cell wall degradation is complete, protoplasts from any two plants can be fused. Interkingdom cell fusion is also feasible. Thus, plant protoplasts can be fused with animal cells (Salhani *et al.*, 1985). In addition to producing somatic hybrids, the fusion of protoplasts can also be employed to study the physical membrane properties of plant cells (Mehrle *et al.*, 1990).

Fusion results in the formation of heterokaryons, which contain nuclei, organelles and cytoplasm originating from the two parental protoplast types. Under optimum conditions, the heterokaryons resynthesise a cell wall and undergo mitosis, resulting in the formation of somatic hybrid cells. Nuclear fusion occurs at interphase by the formation of nuclear bridges, or at the first synchronised mitosis (Fowke, 1989). Plant regeneration may follow via somatic embryogenesis or organogenesis from heterokaryon-derived tissues, giving rise to somatic hybrid plants. Somatic hybridisation not only provides a method of producing hybrids between sexually incompatible species, but is also a method of genetically manipulating vegetatively propagated plants, sterile or subfertile species and plants with naturally long life-cycles (Gleba and Sytnik, 1984). It has been shown that pollen tetrad protoplasts are an ideal system for the delivery of a haploid chromosome complement into a fertile diploid background (Pirrie and Power, 1986), and gametosomatic hybrids can be produced by fusing gametic protoplasts with somatic protoplasts (Pental *et al.*, 1989). Protoplast fusion also enables interspecific or intergeneric transfer of extranuclear genetic elements such as mitochondria, chloroplasts and cytoplasm (Kumar and Cocking, 1987). Indeed, characteristics such as cytoplasmic male sterility (CMS) and herbicide resistance are coded by such extranuclear genes in the mitochondria and chloroplasts, respectively, and can be transferred by protoplast fusion. This provides an additional approach to plant genome manipulation.

Since protoplast fusion is a random process, it does not result in the formation of only heterokaryons. Instead, a mixture of heterokaryons, homokaryons and unfused parental protoplasts remain after all treatments. Consequently, it is necessary to utilise a strategy following fusion to select the heterokaryons from homokaryons, unfused parental protoplasts and/or multiple fusion bodies. Unfortunately, a universally applicable selection strategy has not been developed. The approach adopted will depend not only on the nature and type of protoplasts being fused, but also on the method of fusion, and the nature of the desired end-product.

Recent achievements using protoplast fusion have included the transfer of resistance to Tomato Spotted Wilt Virus (TSWV) from wild *Nicotiana* species into *N. tabacum* (Atanassov *et al.*, 1991) and the transfer of resistance to *Phytophthora infestans* (the causative agent of Potato Late Blight) and the nematode pest *Globodera pallida*, into the potato gene pool by interspecific somatic hybridisation (Mattheij *et al.*, 1992). Hagimori and Nagaoka (1991) have also reported the development of a clubroot-resistant *Brassica* through protoplast fusion between cauliflower and Japanese radish. Table 9.1 gives additional, recent examples of plant protoplast fusion studies in crop plants.

New techniques in molecular biology have facilitated the identification of somatic hybrids at the nuclear and cytoplasmic levels. For example, restriction fragment length polymorphism (RFLP) analyses have been employed to identify somatic hybrids of *Brassica* and to study chromosome elimination in such hybrids (Sundberg *et al.*, 1991).

The merits and limitations of each of the methods of fusion must be considered when deciding which approach is most appropriate for the two particular protoplast types. In some cases, chemical approaches may be unsuitable as they may reduce protoplast viability or adversely affect protoplast development during the subsequent stages of culture (Chand *et al.*, 1988). Chemical techniques may also result in a relatively low yield of fusion products. Electrical techniques may overcome some of these problems and prevent excessive protoplast clumping, but, in general, require relatively expensive and complex equipment. Additionally, several parameters, such as the amplitude and duration of the applied voltage, may have to be determined and optimised to ensure efficient fusion. Such problems are common to both large-scale and small-scale fusion.

TABLE 9.1. Recent examples of plant protoplast fusion and somatic hybridisation.

Species fused	Fusion method	Reference
Citrus sudachi + *C. aurantifolia*	Electrofusion	Saito *et al.* (1991)
Medicago sativa + *M. falcata*	PEG	Mendis *et al.* (1991)
Hordeum marinum + *H. vulgare*	PEG	Funatsuki *et al.* (1991)
Lactuca sativa + *L. virosa*	Electrofusion	Matsumoto (1991)
Brassica oleracea + *B. campestris*	PEG	Christey *et al.* (1991)
Citrus reticulata + *Citropsis gilletiana*	PEG	Grosser *et al.* (1990)
Brassica napus + *Moricandia arvensis*	PEG	Mathias *et al.* (1991)

Typically, 3.5×10^6 protoplasts of each parental type are required for large-scale chemical fusion, while a minimum of 4.0×10^4 protoplasts of each type are needed for large-scale electrofusion. Small-scale techniques enable fusion to be carried out where the limited yield of protoplasts prevents the use of more conventional large-scale techniques. However, the number of fusion products that are produced by small-scale fusion methods can be limited (sometimes less than 1.0×10^4). This prevents techniques such as flow cytometry from being employed to select heterokaryons. However, such problems are not encountered with small-scale electrofusion in which only two protoplasts are fused and the resulting heterokaryon is cultured.

In the following sections, the approaches used to fuse protoplasts are discussed in detail. Where appropriate, protocols are given for fusion, together with examples of their use in somatic hybridisation.

II. TECHNIQUES FOR LARGE-SCALE PROTOPLAST FUSION

A. Chemically Induced Fusion

Chemical fusogens include salt solutions such as sodium nitrate (Power *et al.*, 1970), calcium nitrate (Binding *et al.*, 1988), polyvinyl alcohol (Nagata, 1978), dextran sulphate (Kishinami and Widholm, 1987) and polycations, e.g. poly-L-lysine. However, many of these fusogens have a detrimental effect on protoplast viability. Consequently, the most widely used method of chemical fusion involves polyethylene glycol (PEG) (Saunders and Bates, 1987) and/or high pH and Ca^{2+} solutions (Kao and Saleem, 1986). All these fusogens serve to reduce the net negative surface charge on protoplasts. This overcomes the natural repulsion between individual protoplasts, allowing the plasma membranes to come into close contact (<1.0 nm). The exact mechanism by which fusion occurs is not clear, but it is believed to take place when lateral diffusion of proteins from adhering membranes results in lipid-rich zones. The latter become destabilised and fuse (Boni and Hui, 1987). It has also been suggested that PEG, which has a high affinity for water, may immobilise water molecules in the vicinity of the membranes (Arnold *et al.*, 1985). This decreases the hydrophobic stability of the lipid bilayers, permitting fusion to occur. Subsequent mixing of cytoplasms takes place, together with nuclear fusion.

The following protocols describe standard, well proven methods for the chemical fusion of leaf mesophyll protoplasts of *Petunia parodii* with those of an albino cell suspension of *P. hybrida*. Such procedures have been used to produce a range of somatic hybrid plants. Heterokaryons are easy to identify as they contain chloroplasts from the mesophyll-derived protoplast partner in a richly cytoplasmic background originating from the cell suspension protoplasts. The use of such visual marker systems permits optimisation of the fusion conditions, such as temperature and duration of exposure to the fusogen.

1. A general procedure for the fusion of protoplasts

(a) Protoplasts of *P. parodii* and *P. hybrida* are isolated as described by Power *et al.* (1989), and the preparations each suspended in 28 ml of MSP19M medium (Table 9.2) at a density of $2.0 \times 10^{-5}\,ml^{-1}$.

TABLE 9.2. Composition of MSP19M protoplast culture medium and CPW salts solutions (Power et al., 1989).

MSP19M	MS salts (Murashige and Skoog, 1962)
	3.0% (w/v) sucrose
	9.0% (w/v) mannitol
	2.0 mg l^{-1} α-naphthaleneacetic acid (NAA)
	0.5 mg l^{-1} 6-benzylaminopurine (BAP)
CPW salts solution	
27.2 mg l^{-1}	KH_2PO_4
101.0 mg l^{-1}	KNO_3
1480.0 mg l^{-1}	$CaCl_2 \cdot H_2O$
246.0 mg l^{-1}	$MgSO_4 \cdot 7H_2O$
0.16 mg l^{-1}	KI
0.025 mg l^{-1}	$CuSO_4 \cdot 5H_2O$

CPW8M and CPW13M are as above, but with 8% and 13% (w/v) mannitol respectively. The pH is adjusted to 5.8 in all cases.

(b) Four ml of each preparation are dispensed into separate centrifuge tubes, to act as viability controls. A further 8 ml of each preparation is dispensed into two separate centrifuge tubes to act as selfed controls to monitor viability and potential cross-feeding following culture and selection of hybrids. The fusion treatment consists of an additional three centrifuge tubes into which is dispensed 4 ml of each of the two protoplast preparations.

(c) All the tubes (except the two viability controls) are centrifuged at 120 × g for 5 min to pellet the protoplasts and to facilitate removal of the medium. The pelleted protoplasts are resuspended in 0.5 ml of MSP19M liquid medium.

(d) The fusogen is added as detailed in the fusion protocols.

(e) Following fusion, protoplasts are washed so as to remove the fusogen, and 16 ml of MSP19M medium are added per tube.

(f) Each of the three 16 ml fusion treatments are then divided into 8 ml aliquots. The latter are mixed with 8 ml vols of MSP19M agar-solidified medium, contained in six individual 9 cm Petri dishes. Similarly, 8 ml aliquots of each of the four controls are added to four Petri dishes, each containing 8 ml of agar-solidified MSP19M medium. The 16 ml of the remaining selfed controls (8 ml of *P. parodii* and 8 ml of *P. hybrida*) are combined into one tube and an 8 ml aliquot of this is added to 4 ml of agar-solidified MSP19M medium in a 9 cm Petri dish.

(g) All 12 Petri dishes are sealed with Nescofilm (Bando Chemical Ind Ltd, Kobe, Japan) and cultured at 25°C with 50 μmol m^{-2} s^{-1} continuous illumination provided by 'Daylight' fluorescent tubes.

2. Fusion using polyethylene glycol (PEG)

(a) As stages (a)–(c) of the general procedure.

(b) Two ml of PEG fusion solution (Table 9.3) are added to each of the tubes, except the viability controls. After 10 min, the PEG solution is diluted by addition, to each tube, of 0.5, 1.0, 2.0, 3.0, 4.0 ml aliquots of MSP19M medium at 5 min intervals.

(c) Following dilution, the protoplasts are pelleted by centrifugation at 100 × g for

TABLE 9.3. Composition of protoplast fusion solutions.

PEG	polyethylene glycol (mol. wt 6000) 4.0% (w/v) sucrose 0.01 M $CaCl_2.2H_2O$ Sterilise by autoclaving (121°C, 20 min); store in the dark at 4°C until required.
Purified PEG	30% (w/v) low carbonyl content PEG (mol. wt 1540) (Koch-Light Ltd, Haverhill, Suffolk, UK) in 75 mM HEPES buffer, pH 8.0.
High pH/Ca^{2+}	0.05 M glycine-NaOH buffer 9.0% (w/v) mannitol 1.1% (w/v) $CaCl_2.6H_2O$ pH 10.4 Sterilise by filtration. Use only freshly made solution.
High pH/Ca^{2+}	(for high pH/water fusion) As above, but with 10% (w/v) mannitol.

10 min. The supernatant containing PEG is removed and replaced with liquid MSP19M medium. The protoplasts are resuspended, centrifuged, and the medium is again replaced to ensure that all the fusogen has been removed.
(d) Stages (e)-(g) of the general procedure, are followed thereafter.

3. High pH/Ca^{2+} fusion

(a) As stages (a)-(c) of the general procedure.
(b) Four ml of high pH/Ca^{2+} fusion solution (Table 9.3) are added to each of the tubes, except the viability controls.
(c) The tubes are incubated at 30°C for 10 min.
(d) After incubation, the tubes are centrifuged at 50 × g for 4 min. The protoplasts are washed by replacing the supernatant with CPW13M solution (Table 9.3), resuspending the protoplasts and recentrifuging (100 × g, 10 min).
(e) Stages (e)-(g) of the general procedure, are followed thereafter.

4. PEG/high pH fusion

(a) Follow stages (a)-(c) of the general procedure.
(b) Two ml of PEG solution are added to each of the tubes, except the viability controls.
(c) After 10 min, 8 ml of high pH/Ca^{2+} solution are added to each tube.
(d) Following incubation for a further 10 min, stage (d) of the high pH/Ca^{2+} procedure is followed.
(e) Stages (e)-(g) of the general procedure, are followed thereafter.

5. High pH/water fusion

(a) Follow stages (a)-(c) of the general procedure.
(b) Eight ml of high pH/Ca^{2+} fusion solution are added to each of the tubes,

except the viability controls. The tubes are centrifuged at 60 × g for 3 min.
(c) The tubes are incubated at 30°C for 12–17 min.
(d) Two ml of sterile water are added to each of the tubes and mixed with the fusion solution, taking care not to disturb the protoplasts. The tubes are incubated for a further 10 min.
(e) Following incubation, the supernatant is removed and the protoplasts washed once by resuspending in CPW13M solution prior to centrifugation (100 × g, 10 min).
(f) Stages (e)–(g) of the general procedure, are followed thereafter.

The fusion techniques as detailed have been shown to work for *Petunia* (Power *et al.*, 1976). They are not equally applicable to all species. In spite of the toxicity of PEG to protoplasts of some plants, this fusogen has been employed to produce a range of somatic hybrids. When chemical fusogens are used successfully, the fusion frequencies are typically in the range of 5–50%. The other difficulty associated with the use of chemical fusogens is in preventing multiple fusions, since there is a tendency, especially with the use of PEG, for large aggregates of protoplasts to be produced. This results in multiple fusions between more than two protoplasts, which is undesirable.

6. Protoplast fusion using purified PEG

Commercial preparations of PEG contain contaminants, which may include added antioxidants and polymerisation inhibitors (Honda *et al.*, 1981) as well as oxidative decomposition products such as aldehydes, ketones and acids (Hamburger *et al.*, 1975). The cytotoxic effects of PEG have been attributed to the presence of these contaminants, together with induced dehydration of protoplasts during fusogen treatment, causing irreversible pore formation in plasma membranes (Hahn-Hagerdal *et al.*, 1986).

A method has been developed for fusing plant protoplasts using purified PEG with a low carbonyl content (Chand *et al.*, 1988). This approach has been successful for protoplasts of a number of plants, including *Nicotiana tabacum*, *Chrysanthemum morifolium*, *Medicago sativa* and *Oryza sativa*, giving a greater frequency of protoplast survival than can be obtained using unpurified PEG.

(a) Protoplasts of each partner are suspended separately in 9.0% (w/v) mannitol at a density of 1.0×10^5 ml^{-1}, and are mixed in equal vols.
(b) One ml aliquots of the protoplast suspension are dispensed into the 100 mm square wells of a 5 × 5 well flat-bottomed replidish (Sterilin Ltd, Teddington, UK). The protoplasts are allowed to settle for 10 min.
(c) One ml of filter sterilised, purified, low carbonyl PEG solution (Table 9.3) is added slowly to each well and the mixture incubated for 20 min.
(d) The PEG solution is eluted over a short period (2–4 min), using 2 ml aliquots of Ca^{2+}-free hypotonic solution (pH 5.8) containing 5.0% (w/v) mannitol and 0.2% (w/v) bovine serum albumin, until the protoplasts have regained their full turgidity (as determined microscopically).
(e) Further deplasmolysis is achieved by replacing the hypotonic solution with 9.0%

(w/v) mannitol (pH 5.8). The protoplasts are left in this solution for 30 min prior to transfer to culture medium.

B. Electrofusion

Electrofusion involves two distinct stages. In the first, protoplasts are placed in a medium with low conductivity and are subjected to an alternating current (AC) field. During dielectrophoresis in the AC field, particles can be either positively or negatively charged, or have no net charge (electrically neutral) and yet still migrate in the alternating electric field. This induces transient dipoles in the particles. As a result, even particles with no charge become charged and hence move to the nearest electrode.

In a non-uniform, alternating field of appropriate frequency (usually 0.5–1.5 MHz), plant protoplasts form transient dipoles and become differentially charged, with one side of the cell surface exhibiting a negative charge, while the other becomes positive. Such an effect occurs regardless of the charge, or lack of charge, on a protoplast. As the polarity of the AC field changes, so do the dipoles. The induced charges on the cell surface cause the cells to be drawn to zones of high electric field strength. Because the electric field strength is highest in the region adjacent to the electrodes, the cells migrate to whichever electrode is nearest when the AC field is applied. As the protoplasts migrate towards the electrodes they form chains; this is a result of negatively charged regions of protoplasts being attracted to the positively charged regions of other individuals. The chains of protoplasts which form perpendicular to the electrode are referred to as 'pearl chains'. The formation of 'pearl chains', in which protoplasts are in membrane-to-membrane contact, is an essential prerequisite for the subsequent stage of electrofusion. The latter involves the application of a single high voltage DC pulse of sufficient duration to induce membrane breakdown. The medium surrounding the cell is of low conductivity in comparison to the interior structures of the cell. As a consequence, most of the current moves through the cell. Since membrane breakdown takes place at the point of cell contact, the application of a DC pulse results in cell fusion, rather than cell lysis. The DC fusion pulse must be in the form of a square wave with a duration between 10 and 200 μs and a field strength of 500–2000 V cm^{-1}.

A model has been proposed for the molecular events which take place during fusion. Following membranes of adjacent protoplasts being brought into contact by dielectrophoresis, the application of a single DC pulse leads to breakdown of the plasma membranes (often referred to as pore formation) at the poles of the protoplasts, where they are in contact. Bridges form between the bilayers of the two membranes during the reannealing process which follows the fusion pulse. Formerly adjacent membranes become mixed, resulting in protoplast fusion.

Electrofusion has several advantages over chemically induced fusion. The yield of fusion products can be as high as 90% (Bates *et al.*, 1987), compared to 5–50% obtained using chemical fusogens (Gleba and Sytnik, 1984). Electrofusion also avoids the use of chemicals which may be deleterious to protoplast viability and division, or inhibit subsequent events, such as plant regeneration. The main advantage of electrofusion is that it allows a greater degree of control to be exercised over the fusion process than is possible when using chemical fusogens. The optimum parameters for the fusion of protoplasts can be defined readily by microscopic observation of the response of protoplasts during the stages of protoplast fusion. In order to maintain viability of the protoplasts undergoing fusion, it is essential to keep the amplitude and duration of the applied voltages to a minimum. The alignment of protoplasts into pearl chains is influenced by the field strength and frequency and also by the alignment time and density of protoplasts.

Each of these parameters can be investigated, with the number of aligned protoplasts being expressed as a percentage of the total protoplast population. The production of dimers, where only two protoplasts are in contact (avoiding multiple fusions), can also be determined and expressed as a percentage of the total number of protoplasts. Thus, the parameters for the production of pearl chains/dimers can be optimised. The source of protoplasts can also influence their ability to fuse. For example, leaf protoplasts generally fuse more readily than those isolated from cell suspensions, while the size of protoplasts also influences their fusion. In general, larger protoplasts can be fused more easily than smaller protoplasts (Tempelaar and Jones, 1985).

The composition of the fusion solution also affects the fusion frequency and the inclusion of ions in the solution may increase fusion. However, a solution with a high concentration of ions is undesirable. If the conductivity of the solution rises above 5.0×10^{-4} mho cm^{-1}, then the ions will carry the current, rather than the protoplasts themselves, preventing dielectrophoresis and fusion (Gaynor, 1986). Hence, the solution normally used for electrofusion consists of 6–13% (w/v) mannitol supplemented with $CaCl_2$ to 1.0 mM.

The length of the 'pearl chains' affects the fusion frequency. In general, the longer the chain length, the greater is the fusion frequency. The number of protoplasts present in the pearl chain can be increased by raising the field strength, by increasing the concentration of protoplasts, or by a combination of both parameters. Likewise, the number of protoplasts in the pearl chains can be reduced by lowering the field strength or protoplast concentration.

Pulse duration response curves can be obtained by plotting the percentage of aligned protoplasts that fuse against varying pulse durations (Tempelaar and Jones, 1985). The optimum pulse duration (i.e. the duration which gives the maximum percentage fusion) can be obtained from such a curve. Longer pulses at higher frequencies generally lead to more multiple fusions. Because the fusion of protoplasts occurs over a period of approximately 10 min following the application of the pulse, it is also possible to count the number of protoplasts involved in each fusion. It might be expected that with chains of up to 10 protoplasts, multiple fusions would be frequent. However, by choosing the correct fusion parameters it is possible to reduce the number of multiple fusion products to about 35%, with most products being the result of 1:1 fusions at a relatively high frequency. The variability in the fusion characteristics of protoplasts obtained from different varieties or species of plant has already been mentioned.

The following protocol describes a method for the electrofusion of leaf mesophyll protoplasts of *Rudbekia hirta* with protoplasts isolated from *R. laciniata*. This technique has been used to produce interspecific somatic hybrid plants (Al-Atabee *et al.*, 1990).

(a) Protoplasts are isolated from leaf mesophyll of *R. hirta* and callus of *R. lactiniata* as described by Al-Atabee and Power (1987). Following isolation,

the callus-derived protoplasts are stained with 25 μg ml^{-1} of fluorescein diacetate (FDA) to enable the fusion products to be identified at the end of the procedure.
(b) Parental protoplasts are suspended at a density of 2.0×10^4 protoplasts ml^{-1} in CPW8M solution (Table 9.2). The two protoplast types are mixed in the ratio 1:1 (v/v), and pelleted by centrifugation (80 × g, 5 min).
(c) Following centrifugation, the supernatant is removed and the protoplasts washed twice in filter sterilised electrofusion solution (11% [w/v] mannitol, 0.5 mM $CaCl_2.H_2O$).
(d) One ml aliquots of protoplast suspension are transferred to each of the nine central wells of a Sterilin 5 × 5 well replidish.
(e) The dish is placed on the stage of an inverted microscope, located within a laminar flow cabinet to maintain sterility. The electrofusion electrode is sterilised by immersion in 75% ethanol for 1 min, followed by drying in the sterile air stream.
(f) The electrode is placed in turn into the nine central wells of the replidish. The protoplasts are aligned with an AC field of 20 V and a frequency of 500 kHz. Protoplasts are fused by applying two DC pulses of 300 V and 2 ms duration, with a 2-s interval.
(g) Following fusion, 2.0 ml of CPW13M solution (Table 9.3) is added to each well. After 15 min, the protoplasts are transferred to 8 ml capacity, screw-capped centrifuge tubes. Once the protoplasts have settled (approximately 15 min), the mixture of electrofusion solution and CPW13M solution is carefully removed without disturbing the protoplasts.
(h) The protoplasts are resuspended in medium and cultured as described (Al-Atabee and Power, 1987).

The equipment used for electrofusion can also be employed to expose protoplasts to electrical treatments in order to stimulate the division of protoplast-derived cells (Rech *et al.*, 1987) and, in turn, to promote shoot formation from protoplast-derived tissues of several plants, including woody species such as *Prunus avium* × *pseudocerasus* (Ochatt *et al.*, 1988). The electrotreatment of protoplasts prior to chemical fusion has also been shown to promote division of heterokaryons and has facilitated the recovery of somatic hybrids between *Pyrus communis* var. *pyraster* and *P. avium* × *pseudocerasus* (Ochatt *et al.*, 1989). Indeed, hybrid tissues were not obtained when protoplasts were not electrostimulated prior to fusion. The same equipment used for electrofusion can, therefore, be utilised to investigate the electrostimulation of growth (cells and protoplasts) and to develop an effective, and possibly generally applicable, selection strategy for somatic hybrids.

III. SMALL-SCALE FUSION TECHNIQUES

Small-scale fusion techniques are often necessary when a limited number of protoplasts are available. As with large-scale fusion techniques, small-scale fusion (microfusion) can be promoted using either chemical or electrical stimuli.

9. PROTOPLAST FUSION

A. Small-scale Chemical Fusion

The following is a general protocol for the fusion of small numbers of protoplasts.

(a) The two protoplast types are each suspended at a density of 2.5×10^{-5} ml^{-1} in CPW13M solution (Table 9.3). Equal vols of the two protoplast suspensions are mixed in a centrifuge tube. The volume of each protoplast suspension should exceed 2.0 ml.

(b) Using a Pasteur pipette, a small drop (approx. 200 µl) of the protoplast suspension is placed on a sterile microscope coverslip in the bottom of a 5 cm Petri dish. The protoplasts are left to settle for 10–15 min.

(c) Two equal sized drops (each approx. 200 µl) of fusion solution (Table 9.3) are placed on opposite sides of the drop of protoplast mixture. The drops are allowed to coalesce.

(d) The protoplasts are left undisturbed for 25 min; one drop (200 µl) of washing solution (Table 9.3) is added very gradually to the protoplast–fusogen mixture, while removing some of the fusion solution from around the protoplasts. This process is repeated at 5-min intervals over a 20-min period. During this procedure, it is essential that disturbance of the protoplasts is minimal.

(e) Following removal of the PEG solution, the washing solution is replaced with the same volume of culture medium. At least three washes with culture medium are required.

(f) The protoplasts are cultured by flooding the coverslip with culture medium and scattering several sitting drops of medium onto the base of the Petri dish prior to sealing with Nescofilm.

B. Small-scale Electrofusion (Microfusion)

Small-scale fusion techniques have allowed workers to demonstrate that electrofusion does not reduce the regeneration capacity of higher plant protoplasts (Koop and Schweiger, 1985). Small-scale electrofusion, in which single pairs of protoplasts are fused, does not depend on the availability of selection/marker systems for the recovery of hybrid tissues and plants. Additionally, by using defined partners, fusion occurs under controlled conditions. The technique, which is described in detail by Koop and Spanberg (1989), involves placing two protoplasts of different parental types into a 500 nl droplet of 0.4 M mannitol, which is held below a film of mineral oil on a microscope slide on the stage of an inverted microscope. Two platinum electrodes, each 50 µm in diameter, are inserted into the droplet. Dielectrophoretic alignment of the protoplasts takes place (1 MHz, 66 V cm^{-1}) and is followed by a single negative DC pulse (50 µs, 66 V cm^{-1}) which induces fusion. Using this technique, it is possible to fuse up to 50 pairs of preselected protoplasts within 1 h. Following fusion, the heterokaryons are cultured in droplets of medium. Usually, large-scale protoplast fusion mediates not only the combination of the nuclear genomes, but also introduces organellar genomes originating from both parents. Consequently, this results in fusion products with an unpredictable genetic constitution. Microfusion has been used to overcome this problem by permitting the transfer of defined numbers of organelles into a recipient protoplast. For example, it has been demonstrated that, by microfusion, it is possible to transfer chloroplasts, isolated from *Nicotiana tabacum* and contained within a microplast, into an albino protoplast of the same species. Restriction analysis of total DNA has confirmed that the plastid population in green plants recovered after the transfer of only two chloroplasts into an albino protoplast had originated from the organelles in the donor subprotoplast (Eigel *et al.*, 1991).

IV. PRODUCTION OF HETEROKARYONS

The main difficulty associated with protoplast fusion is that in order to favour heterokaryon survival and hybrid or cybrid production, the parental protoplasts need to be fused in a 1:1 ratio. It may also be necessary to optimise the fusion parameters for two protoplast types. Frequently, these parameters may differ considerably. There are two main approaches to overcome such problems and to optimise heterokaryon production when using electrofusion.

The first approach is applicable only to electrofusion. It involves the chronological regulation of protoplasts being introduced into the fusion chamber. Protoplasts of one parent are introduced into the fusion chamber while varying the AC field until a single layer of protoplasts has formed on the electrodes. Protoplasts of the second parent are then introduced into the chamber. By varying the protoplast density, AC field strength and rate of introduction of protoplasts of the second parent into the chamber, it is possible to optimise the number of 1:1 pairings between the two types of protoplasts. Once this is achieved, the normal fusion procedure, as previously detailed, is performed.

The second approach exploits the fact that some protoplasts can be fused more readily than others. The two types of protoplast are combined in a ratio which maximises fusion between them. Protoplasts which have a low rate of fusion are mixed with protoplasts which fuse more readily in ratios from 5:1 to 20:1. In this way, most of the highly fusable protoplasts will be in contact only with the less fusable type, resulting in the generation of a high proportion of heterokaryons.

V. SELECTION OF HETEROKARYONS AND HYBRID CELLS

Following fusion treatment, the production of viable, binucleate heterokaryons resulting from the fusion of protoplasts of the two parents is usually less than 10% of the total protoplast population. A major difficulty of somatic hybridisation is to encourage heterokaryons to develop in preference to homokaryons (resulting from the fusion of similar protoplasts) and unfused parental protoplasts. Some approaches frequently used to select heterokaryons are detailed below.

A. Chlorophyll-deficient Mutants (Albinos)

This is probably the most developed of all selection techniques. The fusion of non-allelic albino mutant protoplasts results in fusion products which complement to synthesise chlorophyll. Heterokaryon-derived products can be identified as green cell colonies when the cultures are exposed to light and can be isolated manually using fine forceps. This approach has been employed to recover somatic hybrid tissues between *Medicago sativa* and *M. borealis* (Gilmour *et al.*, 1989).

B. Nitrate Reductase Deficiency

Protoplasts which lack nitrate reductase cannot utilise nitrate in the culture medium. When such protoplasts are fused with protoplasts that do not exhibit this form of autotrophy, hybrids can be selected as the deficiency will be overcome by complementation from the other fusion partner. It is necessary to use another selectable marker in the cell line that does not exhibit nitrate reductase deficiency. This may be another form of autotrophy, as was used in the intraspecific fusion of *Datura innoxia* protoplasts (Saxena and King, 1987), or a characteristic such as the inability for one of the parent protoplasts to undergo sustained cell division, as employed in the selection of heterokaryons between *N. tabacum* and *N. glutinosa* (Pirrie and Power, 1986).

C. Resistance Markers

Resistance to antibiotics, herbicides and amino acid analogues has been exploited to select heterokaryons. Hybrid cells resulting from the fusion of two resistant lines will display the resistance characteristics of both parents. Unfused protoplasts and homokaryons are eliminated when plated on a medium containing both of the compounds to which the parental protoplasts are sensitive. In this way, hybrids between *Daucus carota* and *D. capillifolius* have been selected following fusion of protoplasts of cell lines showing resistance to 5-methyltryptophan and azetidine-2-carboxylate (Kameya *et al.*, 1981).

D. Antimetabolites

Selection systems have been developed based on the use of biochemical inhibitors, which block the metabolic pathways of the parental protoplasts prior to fusion. Inactivated parental lines will not undergo cell division. However, following fusion, the hybrid cells enter sustained division as a result of metabolic complementation. Iodoacetate pretreatment has aided the recovery of somatic hybrids between *Nicotiana sylvestris* and *N. tabacum* (Medgyesy *et al.*, 1980) and between *N. plumbaginifolia* and *N. tabacum* (Sidorov *et al.*, 1981).

E. Light-sensitive Mutants

Heterokaryon-derived green cell colonies have been selected following the fusion of mesophyll protoplasts from a light-sensitive mutant of *N. plumbaginifolia* with wild-type mesophyll protoplasts of *N. gossei*, the latter being irradiated with gamma rays from a cobalt source prior to fusion. Regenerated plants had the morphology of *N. plumbaginifolia* and normal chlorophyll synthesis (Medgyesy *et al.*, 1985).

F. Transformed Cell Lines

Genetic transformation can be used to introduce dominant antibiotic resistance markers into plants. Resistance to kanamycin sulphate, combined with the use of the metabolic inhibitor iodoacetate (see Section V.D), has been employed to select somatic hybrids between *Lotus corniculatus* and *L. tenuis* (Aziz *et al.*, 1991).

G. Differential Growth and Plant Regeneration

In many protoplast fusion combinations, hybrid tissues are capable of plant regeneration, provided tissues derived from one of the parental protoplasts are capable of plant regeneration. This has enabled culture media to be developed that permit growth of hybrid cells, while preventing the growth of protoplasts of at least one of the parents. For example, fusion of protoplasts of a kanamycin-resistant line of *N. sylvestris*, incapable of plant regeneration, with non-regenerating protoplasts of *N. knightiana*, enabled the recovery of interspecific somatic hybrids (Maliga *et al.*, 1977).

The fact that there are a number of ways of regenerating plants from protoplast-derived tissues can also provide a means of selecting somatic hybrid material. Plant regeneration in *Rudbeckia hirta* occurs through shoot formation, while shoot production in *R. lactinata* is via rhizogenesis. Somatic hybrid plants of *R. lactinata* have been regenerated through rhizogenesis. Somatic hybrids were identified by the presence of pigmented roots, which is a characteristic feature of *R. hirta* (Al-Atabee *et al.*, 1990).

H. Double Mutants

Many fusion partners are wild-type in character. Consequently, they do not possess any selectable markers. However, this problem can be overcome by creating a parental line carrying both positive and negative selectable markers. Such a technique has been used to select hybrids between *Sinapsis turgida* and *Brassica oleracea*, using protoplasts from a double nitrate reductase deficient S-methyltryptophan-resistant mutant line of *S. turgida* (Toriyama *et al.*, 1987).

VI. MANUAL ISOLATION OF HETEROKARYONS

Probably the most precise, but laborious, method for the isolation of fusion products is to identify potentially hybrid cells visually and then to isolate them mechanically. Thus, when morphologically distinct or fluorescently labelled protoplasts undergo fusion, the fusion products are easily identified under the microscope. This technique is particularly effective when combined with large-scale electrofusion, which avoids the persistent protoplast agglutination frequently seen with many chemical fusion techniques.

Following the fusion of green leaf protoplasts with protoplasts from cultured cells which lack chlorophyll, heterokaryons can be identified soon after fusion by the fact that they contain chloroplasts in one half of their cytoplasm and colourless plastids in the other half. The pigmentation of protoplasts isolated from petals has also been used to identify fusion products.

Similarly, heterokaryons have been identified by red chlorophyll autofluorescence in combination with the yellow-green fluorescence of fluorescein diacetate (FDA) (Patnaik et al., 1982), or FDA labelling combined with the use of rhodamine isothiocyanate (Barsby et al., 1984).

Following identification, the heterokaryons are removed manually from the protoplast population using a micropipette connected to a simple syringe with a screw control (Gilmour et al., 1987). Although this technique is time consuming and requires some degree of skill, it has the advantage of enabling multiple fusion products to be avoided.

VII. FLOW CYTOMETRY

This technique, originally developed for cultured animal cells, involves the automated selection of fluorescently labelled protoplasts. The parental protoplasts are labelled, prior to fusion, with different fluorochromes, often fluorescein and rhodamine, which fluoresce at different wavelengths.

The dual fluorescence of heterokaryons allows them to be discriminated from homokaryons and non-fused parental material. One drawback is that the flow cytometer is unable to discriminate between adhering, non-fused protoplasts (which often arise with the use of chemical fusogens) and heterokaryons. Also, the machine requires populations of several thousand protoplasts to be effective and, as a consequence, cannot be used for the selection of products from small-scale fusions, or where protoplast yield is low. Passage through the machine may also damage fragile protoplasts.

The range of plants recovered from flow-sorted heterokaryons is limited. However, sorting has recently been extended to protoplasts from a wide range of species (Hammatt et al., 1990), with somatic hybrid plant production in the case of *Brassica napus* and *B. nigra* (Sjodin and Glimelius, 1989).

VIII. INFLUENCE OF PROTOPLAST FUSION ON THE NUCLEAR GENOME

The fusion of protoplasts from different species combines the cytoplasm, organelles and parental nuclear genomes. Studies of fusion products undergoing mitosis have shown chromosomes of both parental species arranged in a common metaphase plate, illustrating that true hybrid cells with fused hybrid nuclei can be obtained, as in fusions between *Datura innoxia* and *Atropa belladonna*. This shows that nuclear fusion, as with fusion at the membrane level, is not influenced by any specific characteristics of the parents (Krumbiegel and Schieder, 1979). However, somatic incompatibility may occur at the nuclear level. As cell fusion rarely occurs in natural systems, such incompatibility may be attributed as being a consequence of the normal ontogeny of the nucleus, rather than of any specific incompatibility mechanism.

It may be expected that the product of fusion between two protoplasts would show an additive increase in the chromosome complement. However, this is rarely the case. Hybrid cells usually possess an asymmetric combination of the parental genomes, with parts of one or both genomes being eliminated at some stage during culture (Imamura et al., 1987). The processes by which chromosomes are eliminated remains unclear, but several general principles have emerged. Most can be regarded as passive mechanisms, in which the genetic complement of one of the parents, without the active involvement of the other, is hampered, delayed or prevented from developing within the hybrid cell. This can occur for a number of reasons. Mixing of the cytoplasm may be incomplete, resulting in an increase in segregation when cell division takes place. Cytoplasmic gradients can also be created through incomplete cytoplasmic mixing. Such gradients may play a role in the rapid directional sorting of plastids in somatic hybrid cells.

As the uncoordinated sequencing of cellular processes can also stimulate chromosome elimination, synchronisation of the cell cycle must take place in order to produce a stable hybrid karyotype. If the disparity between the cell cycles of the parental protoplasts is great, then this is less likely to occur and will result in increased chromosome elimination.

As early as 1974, it was shown that following fusion between protoplasts of different genera, the chromosomes of each of the parental types became arranged in blocks rather than being mixed in a common metaphase plate (Kao *et al.*, 1974). Such a situation results in the uneven distribution of chromosomes, leading to further irregularity and subsequent chromosome loss during subsequent mitoses.

It has been demonstrated that the fusion of protoplasts isolated from rapidly dividing cell suspensions, with protoplasts derived from mesophyll cells, which have a relatively longer cycle time, results in fusion products with a cell-cycle time that is intermediate to that of the two parents. This was the case in heterokaryons between *Arabidopsis thaliana* and *Brassica campestris* (Gleba and Hoffman, 1978). Thus, rapidly cycling nuclei had stimulated their mesophyll counterparts to proceed to mitosis more rapidly.

Szabados and Dudits (1980) showed that fusion of protoplasts in interphase with those in metaphase resulted in structural changes in the interphase nuclei, leading to prophasing, chromatin fragmentation and pulverisation. Premature condensation of the chromosomes will give rise to mitotic figures where one set of chromosomes lags behind that of the other. However, it is unclear whether this asynchrony persists throughout the cell cycle.

IX. EFFECTS OF PROTOPLAST FUSION ON THE CYTOPLASMIC GENOME

Characteristics such as male sterility and herbicide resistance are known to be encoded by cytoplasmic genes. Male sterility is of importance to plant breeders, as it prevents undesirable self-fertilisation and enables out-crossing. The introduction of male sterility also facilitates the generation and release of F1 hybrid seed.

Following fusion, the chloroplasts of the two parental types usually segregate, so that regenerated hybrid plants contain only one parental type (Fluhr, 1983). The reasons for this are still not clear. Several factors during fusion, selection and regeneration can have an effect. For example, the number of chloroplasts (plastids) per protoplast, the replication potential of the chloroplast genomes in the cultured product and the number of cell divisions between fusion and complete sorting, may influence the pattern of parental chloroplast segregation (Galun, 1982; Birky, 1983;

Davey and Kumar, 1983). The parental chloroplast segregation patterns will also be influenced by inherent somatic incompatibility operating at the cytoplasmic level.

The selection pressure itself may place constraints on the chloroplast segregation pattern. For example, a selection system based on screening green colonies for hybrid production will not be able to select all the hybrid/cybrid events if the parents are a nuclear albino mutant and a cytoplasmic albino mutant.

Compared to the results of sexual crossing, in which a uniparental contribution of organelles is normally observed, the products of protoplast fusion will give rise to a much greater range of nuclear-cytoplasmic combinations. This generates considerable diversity, out of which agronomically important gene combinations may occur. Although plastid segregation is the more common phenomenon, genome recombination has also been shown to occur between plastids. Recombination is common between mitochondria. Such recombination will also increase the potential for genetic diversity arising from protoplast fusion and the subsequent novel nuclear cytoplasmic interactions.

The cytoplasmic genome can be manipulated using a technique referred to as 'cybridisation'. This follows the basic procedures for fusion, except that the nuclear genome of one of the parental protoplast types is inactivated prior to fusion, often by irradiation. Fusion generates products with a diversity of cytoplasmic variations combined with the intact recipient parental nuclear genome. Thus, cybridisation allows cytoplasmically coded traits to be specifically manipulated, while the majority of the characteristics of the plant, encoded by the nuclear genome, remain unchanged. Manipulation of the cytoplasm has been used to transfer cytoplasmic male sterility in a number of species, including monocotyledons such as rice (Kyozuka *et al.*, 1989), and also to produce plants with cytoplasmically encoded atrazine resistance (Binding *et al.*, 1982). The capacity to alter plants at these levels can, and will, have profound effects on agricultural management strategies, especially for those crops of major agronomic value.

Rapid developments in the field of protoplast technology, and the handling of protoplast systems in general (Ochatt and Power, 1992), will mean that an ever increasing number of species can be considered as legitimate targets for genetic manipulation based on the somatic cell/protoplast. However, cell fusion technology in the broadest sense and the products generated therefrom, should always be considered as an adjunct to conventional breeding approaches.

REFERENCES

Al-Atabee, J. S. and Power, J. B. (1987). *Plant Cell Rep.* **6**, 414–417.
Al-Atabee, J. S., Mulligan, B. J. and Power, J. B. (1990). *Plant Cell Rep.* **8**, 517–520.
Arnold, K., Herrmann, A., Pratsch, L. and Gawrisch, K. (1985). *Biochem. Biophys. Acta* **815**, 515–518.
Atanassov, A., Dimanov, D., Atanassov, I., Dragoeva, A., Vassileva, Z., Vitanov, V., Jankulova, M. and Djilianov, D. (1991). *Physiol. Plant.* **82**, A23.
Aziz, M. A., Chand, P. K., Davey, M. R. and Power, J. B. (1991). *J. Exp. Bot.* **41**, 471–479.
Bajaj, Y. P. S (ed.) (1989). "Biotechnology in Agriculture and Forestry", Vol. 8, Plant Protoplast and Genetic Engineering I. Springer-Verlag, Berlin, Heidelberg, New York and Tokyo.

Barsby, T. L., Shepard, J. F., Kemble, R. J. and Wong, R. (1984). *Plant Cell Rep.* **3**, 165–167.
Bates, G. W., Nea, L. J. and Hasenkampf, C. A. (1987). *In* "Cell Fusion" (A. E. Sowers, ed.), pp. 479–496. Plenum Press, New York.
Binding, H., Jain, S. M., Finger, J., Mordhorst, G., Nehls, R. H. and Gressel, J. (1982). *Theor. Appl. Genet.* **63**, 273–277.
Binding, H., Zuba, M., Rudnick, J. and Mordhorst, G. (1988). *J. Plant Physiol.* **133**, 409–413.
Birky, C. W. (1983). *Science* **222**, 468–475.
Boni, L. T. and Hui, S. W. (1987). *In* "Cell Fusion" (A. E. Sowers, ed.), pp. 301–330. Plenum Press, New York.
Chand, P. K., Davey, M. R., Power, J. B. and Cocking, E. C. (1988). *J. Plant Physiol.* **133**, 480–485.
Christey, M. C., Makaroff, C. A. and Earle, E. D. (1991). *Theor. Appl. Genet.* **83**, 201–208.
Constabel, F., Koblitz, H., Kirkpatrick, J. W. and Rambold, S. (1980). *Can. J. Bot.* **58**, 1032–1034.
Davey, M. R. and Kumar, A. (1983). *In* "Plant Protoplasts" (K. L. Giles, ed.), pp. 219–263. *Int. Rev. Cytol.* (Suppl). **16**. Academic Press, New York.
Davey, M. R. and Rech, E. L. (1992). *In* "Plant Tissue Culture and Its Applications" (B. V. Charlwood, R. Dixon and J. M. Torres, eds). Edward Arnold, in press.
Davey, M. R., Rech, E. L. and Mulligan, B. J. (1989). *Plant. Mol. Biol.* **13**, 273–285.
Eigel, L., Oelmuller, R. and Koop, H.-U. (1991). *Mol. Gen. Genet.* **227**, 446–451.
Fluhr, R. (1983). *In* "Protoplasts 1983. Lecture proceedings, 6th International Protoplast Symposium" (I. Potrykus, C. T. Harms, A. Hinnen, R. Hutter, P. J. King and R. D. Shillito, eds), *Experimentia Supplementum* **46**, 85–92. Birkhauser-Verlag, Basel.
Fowke, L. C. (1989). *In* "Biotechnology in Agriculture and Forestry", Vol. 8, Plant Protoplasts and Genetic Engineering I. (Y. P. S. Bajaj, ed.), pp. 289–303. Springer-Verlag, Berlin, Heidelberg, New York and Tokyo.
Funatsuki, H., Lazzeri, P. A. and Lörz, H. (1991). *Physiol. Plant.* **82**(1), A22.
Galun, E. (1982). *In* "Methods in Chloroplast Molecular Biology" (M. Edelman, R. B. Hallick and N-H. Chua, eds), pp. 139–148. Elsevier Medical Press, Amsterdam.
Gaynor, J. J. (1986). *In* "Handbook of Plant Cell Culture" (D. A. Evans, W. R. Sharp and P. V. Ammirato, eds), pp. 149–171. Macmillan, New York.
Gilmour, D. M., Davey, M. R. and Cocking, E. C. (1987). *Plant Sci.* **53**, 263–270.
Gilmour, D. M., Davey, M. R. and Cocking, E. C. (1989). *Plant Cell Rep.* **8**, 29–32.
Gleba, Y. Y. and Sytnik, K. M. (1984). "Protoplast Fusion." Springer-Verlag, Berlin.
Gleba, Y. Y. and Hoffman, F. (1978). *Mol. Gen. Genet.* **165**, 257–264.
Grosser, J. W., Gmitter, F. G., Tusa, N. and Chandler, J. L. (1990). *Plant Cell Rep.* **8**, 656–659.
Hagimori, M. and Nagaoka, M. (1991). *Physiol. Plant.* **82**, A25.
Hahn-Hagerdal, H., Hosono, K., Zachrisson, A. and Bornman, C. H. (1986). *Physiol. Plant.* **67**, 359–364.
Hamburger, R., Azaz, E. and Donbrow, M. (1975). *Pharm. Acta Helv.* **50**, 10–17.
Hammatt, N., Lister, A., Blackhall, N. W., Gartland, J., Ghose, T. L., Gilmour, D. M., Power, J. B., Davey, M. R. and Cocking, E. C. (1990). *Protoplasma* **194**, 34–44.
Harms, C. T. (1983). *Q. Rev. Biol.* **58**, 325–353.
Honda, K., Maeda, Y., Sasakawa, S., Ohno, H. and Tsuchida, E. (1981). *Biochem. Biophys. Res. Commun.* **101**, 165–171.
Imamura J., Saul, M. W. and Potrykus, I. (1987). *Theor. Appl. Genet.* **74**, 445–450.
Kameya, T., Horn, M. E. and Widholm, J. M. (1981). *Z. Pflanzenphysiol.* **104**, 459–466.
Kao, K. N. and Saleem, M. (1986). *J. Plant Physiol.* **122**, 217–225.
Kao, K. N., Constabel, F., Michayluk, M. R. and Gamborg, O. L. (1974). *Planta* **120**, 215–227.
Kishinami, I. and Widholm, F. (1987). *Plant Cell Physiol.* **28**, 211–218.
Koop, H-U. and Schweiger, H.-G. (1985). *Eur. J. Cell Biol.* **39**, 46–49.
Koop, H-U. and Spanberg, G. (1989). In "Electroporation and Electrofusion in Cell Biology" (E. Neumann, A. Sowers and S. Woolford, eds), pp. 355–366. Plenum Press, New York.
Krumbeigel, G. and Schieder, O. (1979). *Planta* **145**, 371–375.
Kumar, A. and Cocking, E. C. (1987). *Am. J. Bot.* **74**, 1289–1303.
Kuster, E. (1909). *Ber. Dtsch. Bot. Ges.* **27**, 589–598.

Kyozuka, J., Kaneda, T. and Shimamoto, K. (1989). *Bio/Technol.* **7**, 1171-1174.
Maliga, P., Lazar, G., Joo, F., Nagh, A. H. and Menczel, L. (1977). *Mol. Gen. Genet.* **157**, 291-296.
Mathias, R. J., O'Neill, C. M. and Murata, T. (1991). *Physiol. Plant.* **82**, A25
Matsumoto, E. (1991). *Plant Cell Rep.* **9**, 531-534.
Mattheij, W. M., Eijlander, R., Dekoning, J. R. A. and Louwes, K. M. (1992). *Theor. Appl. Genet.* **83**, 459-466.
Medgyesy, P., Menczel, L. and Maliga, P. (1980). *Mol. Gen. Genet.* **179**, 693-698.
Medgyesy, P., Golling, R. and Nagy, F. (1985). *Theor. Appl. Genet.* **70**, 590-594.
Mehrle, W., Naton, B. and Hampp, R. (1990). *Plant Cell Rep.* **8**, 687-691.
Mendis, M. H., Power, J. B. and Davey, M. R. (1991). *J. Exp. Bot.* **42**, 1565-1573.
Murashige, T. and Skoog, F. (1962) *Physiol. Plant* **15**, 473-479.
Nagata, T. (1978). *Naturwissenschaften* **65**, 263-264.
Ochatt, S. J. and Power, J. B. (1992). *In* "Comprehensive Biotechnology", Suppl. 2 (M. Moo-Young, G. S. Warren and M. W. Fowler, eds), pp. 99-127. Pergamon Press, New York.
Ochatt, S. J., Chand, P. K., Rech, E. L., Davey, M. R. and Power, J. B. (1988). *Plant Sci.* **54**, 165-169.
Ochatt, S. J., Patat-Ochatt, E. M., Rech, E. L., Davey, M. R. and Power, J. B. (1989). *Theor. Appl. Genet.* **78**, 35-41.
Patnaik, G., Cocking, E. C., Hamill, H. and Pental, D. (1982). *Plant Sci. Lett.* **24**, 105-110.
Pelletier, G. and Chupeau, Y. (1984). *Physiol. Veg.* **22**, 377-399.
Pental, D., Pradhal, A. K. and Mukhopadhyay, A. (1989). *Theor. Appl. Genet.* **78**, 547-552.
Pirrie, A. and Power, J. B. (1986). *Theor. Appl. Genet.* **72**, 48-52.
Power, J. B., Cummins, S. E. and Cocking, E. C. (1970). *Nature* **255**, 1016-1018.
Power, J. B., Frearson, E. M., Hayward, C., George, D., Evans, P. K., Berry, S. F. and Cocking, E. C. (1976). *Nature* **263**, 500-502.
Power, J. B., Davey, M. R., McLellan, M. and Wilson, D. (1989). "Laboratory Manual — Plant Tissue Culture". University of Nottingham, UK.
Rech, E. L., Ochatt, S. J., Chand, P. K., Power, J. B. and Davey, M. R. (1987). *Protoplasma* **141**, 169-176.
Saito, W., Ohgawara, T., Shimizu, J. and Ishii, S. (1991). *Plant Sci.* **77**, 125-130.
Salhani, N., Vienken, J., Zimmerman, U., Ward, M., Davey, M. R., Clothier, R. H., Balls, M., Cocking, E. C. and Lucy, J. A. (1985). *Protoplasma* **126**, 30-35.
Saunders, J. A. and Bates, G. W. (1987). *In* "Cell Fusion" (A. E. Sowers, ed.), pp. 497-520. Plenum Press, New York.
Saxena, P. K. and King, J. (1987). *Plant Cell, Tissue Organ Cult.* **9**, 61-71.
Senda, M., Takeda, J., Abe, S. and Nakamura, T. (1979). *Plant Cell Physiol.* **20**, 1441-1443.
Sidorov, V. A., Menczel, L., Nagy, F. and Maliga, P. (1981). *Planta* **152**, 341-345.
Sjodin, C. and Glimelius, K. (1989). *Theor. Appl. Genet.* **77**, 651-656.
Sundberg, E., Lagercrantz, U. and Glimelius, K. (1991). *Plant Sci.* **78**, 89-98.
Szabados, L. and Dudits, D. (1980). *Exp. Cell Res.* **127**, 442-446.
Tempelaar, M. J. and Jones, M. G. K. (1985). *Planta* **165**, 205-206.
Toriyama, K., Kameya, T. and Hinata, K. (1987). *Planta* **170**, 308-313.
Zimmerman, U. and Scheurich, P. (1981). *Planta* **151**, 26-32.

10 Import of In Vitro Synthesised Proteins into Intact Chloroplasts and Isolated Thylakoids from Higher Plants

COLIN ROBINSON

Department of Biological Sciences, University of Warwick, Coventry CV4 7AL, UK

I.	Introduction	207
II.	Isolation of Chloroplasts	209
	A. Growth conditions	209
	B. Chloroplast isolation	209
III.	The synthesis *in vitro* of nuclear-encoded chloroplast proteins	211
	A. Choice of translation system	211
	B. *In vitro* transcription of cloned DNA	212
	C. *In vitro* translation	213
IV.	Import of proteins into intact chloroplasts	213
	A. The basic import assay	213
	B. Use of exogenous ATP to drive import into chloroplasts	215
V.	Import of proteins into isolated thylakoids	215
	A. General comments	215
	B. Incubation conditions	217
	References	218

I. INTRODUCTION

The biogenesis of chloroplasts in higher plants and green algae involves protein traffic on a particularly large scale. This is primarily because the chloroplast is the most complex organelle known, in structural terms, containing three distinct types of membrane (the outer and inner envelope membranes and the thylakoid membrane) which enclose three soluble phases (the interenvelope membrane space, the stroma and the thylakoid lumen). Each of these subcompartments carries out a specific set of metabolic functions, and contains accordingly a characteristic complement of proteins.

The biogenesis of the chloroplast is further complicated by the fact that the resident proteins are synthesised by two distinct genetic systems. Approximately 20% of the proteins are synthesised within the organelle, whereas the remainder are imported after synthesis in the cytosol. It is thus clear that chloroplast development requires both the import of numerous proteins into the organelle, and the operation of efficient sorting mechanisms to ensure the delivery of these proteins to the correct destinations. Studies on a wide variety of chloroplast proteins have indicated that chloroplast-encoded proteins either remain in the stroma or are targeted into, or across the thylakoid membrane. Cytosolically synthesised proteins are targeted into all of the organellar subcompartments (reviewed by Smeekens *et al.*, 1990). With the exception of outer envelope membrane proteins, which are imported by a distinct mechanism (Salomon *et al.*, 1990; Li *et al.*, 1991), nuclear-encoded chloroplast proteins are invariably synthesised as larger precursors containing amino terminal presequences. There is good evidence from chimaeric protein studies that these presequences contain most, if not all, of the information specifying targeting into the chloroplast (Van den Broeck *et al.*,

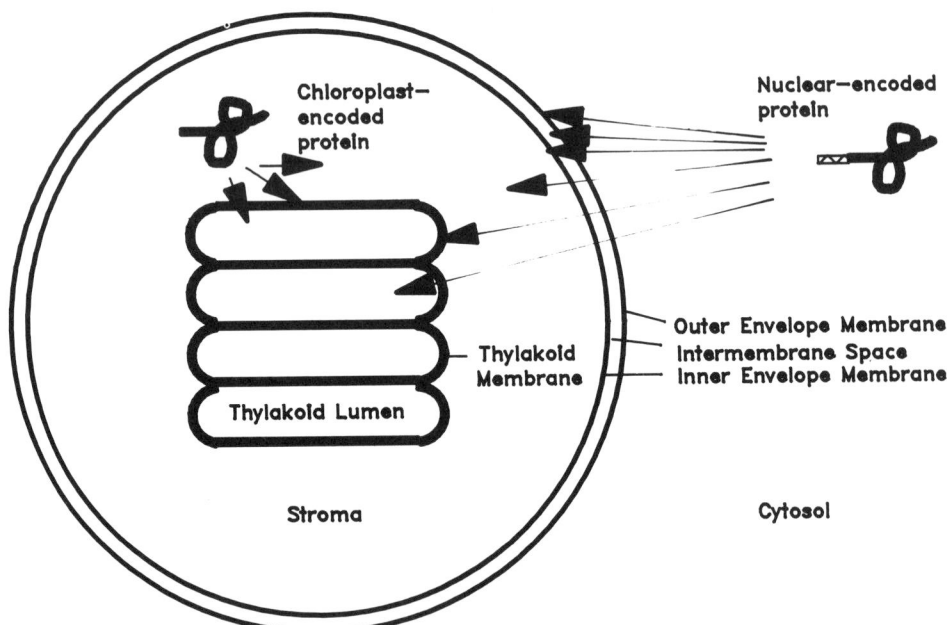

FIG. 10.1. Biogenesis of chloroplast proteins. Chloroplast proteins are synthesised by two distinct genetic systems. Approximately 20% are synthesised within the organelle; these include stromal, thylakoid membrane and thylakoid lumen proteins (there is no evidence to date for the insertion of chloroplast-encoded proteins into the envelope membranes). Nuclear-encoded proteins are synthesised with amino terminal presequences (with the exception of outer envelope membrane proteins) and targeted into all six of the chloroplast subcompartments.

1985). The origins and various destinations of chloroplast proteins are summarised in Fig. 10.1.

The main aims of this chapter are to describe protocols for the analysis *in vitro* of the biogenesis of chloroplast proteins. These techniques have been widely used in studies on the mechanisms involved in chloroplast protein transport and assembly, since these events normally occur too rapidly to be analysed *in vivo*. In addition, some of the basic import protocols are useful for other purposes. For example, a simple chloroplast import assay can unequivocally demonstrate whether a newly isolated cDNA clone encodes the precursor of a chloroplast protein. In this chapter I describe protocols for reconstituting the transport of proteins into both intact chloroplasts and isolated thylakoids. I also discuss methods for the fractionation of chloroplasts in order to pinpoint the precise location of an imported protein.

II. ISOLATION OF CHLOROPLASTS

We have found, perhaps not surprisingly, that efficient import assays depend totally on the rapid, careful isolation of chloroplasts from plant material grown under defined conditions. The choice of plant species is critical, since only a few have been shown to be suitable for these types of analysis. In our laboratory we use a dwarf pea variety (*Pisum sativum*, var. Feltham First, supplied by Sharpes seeds, Sleaford, Lincs.). Growth of seedlings is rapid and straightforward, and intact chloroplasts can be isolated with relative ease. Spinach can also be used for the isolation of import-competent chloroplasts (Chua and Schmidt, 1978). In general, monocotyledons such as wheat and barley appear to be a very poor source of chloroplasts for import assays, probably due to the difficulty in isolating intact organelles from these species. This is not usually a problem for studies on the import of monocotyledon proteins, since these proteins appear to be efficiently and faithfully imported by dicotyledon chloroplasts (Kirwin *et al.*, 1989).

A. Growth Conditions

Two factors are important for the isolation of import-competent chloroplasts from pea leaves. First, young expanding leaves must be used as starting material, since we have found that chloroplasts in mature leaves lose the ability to import proteins. Secondly, the seedlings must be grown under relatively low light intensities, otherwise accumulated starch grains tend to lyse the chloroplasts during the isolation procedure. We routinely grow pea plants from seed in compost (for example, Levingtons multipurpose) for 7–10 days under a 12-h photoperiod at 18–22°C. Illumination is provided by warm white light from fluorescent tubes (c. 40–50 microeinsteins $m^{-2} s^{-1}$).

B. Chloroplast Isolation

One simple rule for the isolation of import-competent chloroplasts is to work fast and keep the organelles at 0°C at all times. It is therefore important to be well prepared and for all of the relevant solutions and containers to be precooled. The grinding medium is placed at −20°C until an ice slurry forms, and the grinding container is also kept at −20°C prior to use. In our laboratory, a Polytron homogeniser (Northern Media Supplies, Hull) is routinely used for chloroplast isolation, but many simple domestic blenders can also give satisfactory results. The basic chloroplast isolation procedure is detailed below.

(1) The apical leaves of pea seedlings are removed and mixed with semi-frozen sucrose grinding medium (0.35 M sucrose, 25 mM HEPES-NaOH, pH 7.6, 2 mM EDTA) at a ratio of 20 g fresh weight leaves per 100 ml medium. Depending on the number of import assays to be carried out, we normally start with 40–100 g leaves.

(2) The leaves are homogenised with two to four 3-s bursts of the Polytron at 75% full speed. The leaves should be ground into small fragments, but excessive homogenisation should be avoided.

(3) The homogenate is strained immediately through eight layers of muslin into an ice-cold beaker, after which the contents are poured into 50 ml or 100 ml polycarbonate tubes.

(4) The tubes are placed in a swing-out rotor in a refrigerated centrifuge and accelerated rapidly to $2500 \times g$, held at this speed for 60 s, and then decelerated with the brake on.

(5) The tubes are removed and the supernatants discarded in one motion (the pellets are quite firm at this stage). The insides are then quickly wiped with tissue, and the pellets are resuspended in a small volume (4–8 ml) of sorbitol medium (50 mM HEPES-KOH, pH 8.4, 0.33 M sorbitol). We favour the use of medical cotton swabs to gently resuspend the chloroplasts, but a small paintbrush is an effective alternative.

It is very important to avoid resuspending chloroplasts by repeated pipetting, since they are large organelles which are easily lysed by shear forces. From this point, any pipetting steps should be carried out using wide-bore pipettes; when using small volumes, we widen the apertures of disposable plastic tips by cutting off the ends.

(6) The chloroplast suspension at this point consists of a mixture of broken and intact organelles. To greatly increase the proportion of intact organelles, the suspension is next centrifuged through a Percoll pad. This involves layering 4 ml of suspension slowly onto an equal volume of 40% Percoll in sorbitol medium in a 10 ml or 15 ml glass centrifuge tube. The tubes are then centrifuged at $2500 \times g$ for 7 min (brake off), after which intact chloroplasts are found in the pellet; lysed organelles fail to penetrate the Percoll pad. Figure 10.2 shows the results of a typical preparation. The pellet at this stage tends to be fairly sloppy, and it is best to remove most of the supernatant using a Pasteur pipette, and then quickly decant the remainder. The pellet is washed in 5 ml of sorbitol medium and then resuspended in 1 ml of sorbitol medium.

(7) At this point it is advisable to quickly assess the intactness of the chloroplast preparation. This is easily achieved by examination under phase-contrast microscopy; intact organelles appear bright green, often with a surrounding halo, whereas lysed chloroplasts appear darker and more opaque. The vast majority of the organelles (up to 95%) should be intact, although lower degrees of intactness will still probably produce decent results.

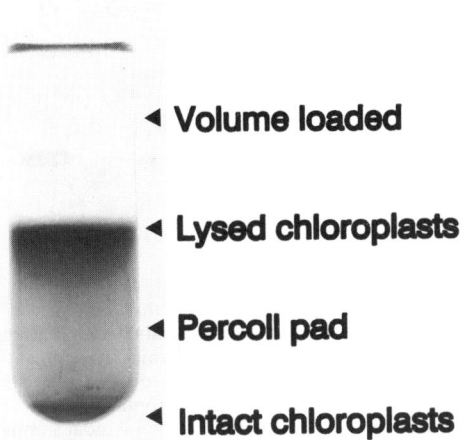

FIG. 10.2. Separation of intact and lysed chloroplasts. A mixture of broken and intact chloroplasts was layered on top of a Percoll pad and centrifuged. The pellet contains intact organelles whereas the broken chloroplasts fail to penetrate through the Percoll.

(8) The chlorophyll concentration of the suspension is determined as follows: a small volume (e.g. 20 µl) is removed into 80% aqueous acetone, and the absorbance is measured at 645 and 663 nm. The chlorophyll concentration is given by the dilution factor multiplied by:

$$[(20.2 \times A_{645}) + (8.02 \times A_{663})]$$

The concentration is adjusted to 1 mg ml^{-1} chlorophyll with sorbitol medium, and the chloroplasts are then ready for use in import assays. Ideally, they should be used immediately, but we have found that import-competence is retained for up to about 1 h if kept on ice.

III. THE SYNTHESIS *IN VITRO* OF NUCLEAR-ENCODED CHLOROPLAST PROTEINS

A. Choice of Translation System

Precursors of chloroplast proteins can be synthesised *in vitro* by programming a cell-free translation system with mRNA for the protein of interest. The two most commonly used translation systems are the wheatgerm lysate and the reticulocyte lysate, both of which are commercially available. In both cases, translation is carried out in the presence of a radiolabelled amino acid; it is necessary to generate a labelled protein because each of these systems synthesises very small quantities of protein.

If large numbers of translations are to be carried out, it may be more economical to prepare a large quantity of home-made lysate; this is reasonably straightforward in the case of the wheatgerm system (Anderson *et al.*, 1983). We have also found that the wheatgerm system is more suitable for chloroplast import studies; high concentrations of reticulocyte lysate cause lysis of isolated chloroplasts. The main

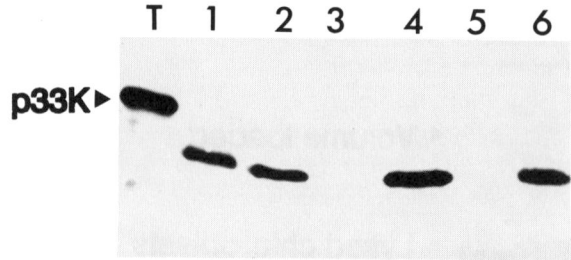

FIG. 10.3. Import of proteins into isolated chloroplasts and localisation of the imported protein. The precursor of the 33 kDa oxygen-evolving complex protein (p33K), a thylakoid lumen protein, was incubated with isolated intact chloroplasts. After incubation, a sample was washed and analysed directly (lane 1); this track shows the presence of mature-size protein (33K) and a small amount of precursor. Lane 2 shows that the precursor is digested by incubation of the chloroplasts with protease, whereas the mature protein is protected. A further sample was fractionated after import into stromal (lane 3) and thylakoidal (lane 4) samples; after fractionation of a parallel sample, the stromal and thylakoidal samples were incubated with protease (lanes 5 and 6).

As expected, the mature-size protein is found in the thylakoid fraction and is resistant to protease digestion, confirming that the protein is located in the lumenal space. Lane T: p33K translation product.

disadvantage with the wheatgerm system is that it is relatively poor at synthesising proteins above about 60 kDa in size.

mRNA encoding the protein of interest can be prepared either by extracting total mRNA from the appropriate plant tissue, or, if a full-length cDNA clone is available, by *in vitro* transcription. The latter is of course vastly preferable, since this leads to the synthesis of a single labelled product in the translation system. If total mRNA is used to programme the translation system, the protein of interest can usually only be visualised by immunoprecipitation from the translation products (Grossman *et al.*, 1982). Furthermore, it is difficult in practice to use total mRNA translation products for import studies, unless the protein of interest is an abundant translation product. Otherwise, very large translations are required, and the protein tends to be difficult to immunoprecipitate from the chloroplasts after the import reaction.

B. *In Vitro* Transcription of Cloned DNA

A number of protocols have been developed for the *in vitro* transcription of cloned genes, many of which use either SP6 or T7 RNA polymerase to generate large amounts of transcript (Melton *et al.*, 1984). The transcription protocol detailed below, which can be used with either SP6 or T7 RNA polymerase, is useful because the reaction conditions are compatible with those of the wheatgerm system. Thus, after transcription, the mix can be used directly in the translation reaction, thereby avoiding the necessity of phenol-extracting the RNA.

(1) Prepare clean (preferably CsCl-purified) DNA in water at $1\,\mu\mathrm{g\,ml}^{-1}$. Most protocols recommend that the DNA is linearised, but we have found that this is not generally necessary for most cDNA clones.

(2) Mix:

        2 μl DNA
        15 μl premix (40 mM Tris-Cl, pH 7.5,

6 mM MgCl$_2$, 2 mM spermidine, 10 mM
DTT, 0.5 mM ATP, CTP and UTP, 50 μl
GTP, 100 μg/ml^{-1} bovine serum albumin)
20 units RNasin (Promega)
0.25 units monomethyl cap
[m^{7}G(5′)ppp(5′)G]
1 μl of SP6 RNA polymerase (BRL, 15
units ml^{-1}) or T7 RNA polymerase
(BRL, 50 units ml^{-1})

Incubation is for 30 min at 37°C, after which 1 μl of 10 mM GTP is added; the incubation is then continued for a further 30 min. The transcription products can be stored at −80°C until required.

C. In Vitro Translation

In this laboratory we routinely use the wheatgerm lysate system, prepared by the method of Anderson et al. (1983). For optimal levels of translation, it is best to test varying amounts of transcription mix in translation reactions. We generally find that the optimum amount is about 1 μl of transcription product per 12.5 μl translation reaction. As a general rule, we would expect 1 μl of ^{35}S-methionine labelled translation product to give rise to a clearly visible band after electrophoresis, fluorography, and an overnight exposure to X-ray film.

After translation is completed, it is advisable to incubate the products in the appropriate import assay as rapidly as possible. We have, however, found that a variety of precursors are efficently imported after freezing at −80°C (although import competence is rapidly lost during subsequent rounds of freeze-thawing).

IV. IMPORT OF PROTEINS INTO INTACT CHLOROPLASTS

A. The Basic Import Assay

1. General comments

Provided that intact chloroplasts can be isolated, and the precursor can be prepared in a reasonably 'hot' form, it should be very easy to achieve reasonably efficient import. There are no real pitfalls to caution against. Different plastid protein precursors tend to be imported with somewhat differing efficiencies, but to my knowledge no authentic precursor has failed to be imported *in vitro*. If import assays are to be carried out for the first time, it would of course nevertheless make sense to first use a precursor which is known to be efficiently imported.

2. Incubation conditions

The following assay conditions, modified from those of Highfield and Ellis (1978) and Grossman *et al.* (1982), are routinely used for standard import reactions. Several incubations are set up, each containing:

55 μl intact chloroplasts (at 1 mg/ml^{-1} chlorophyll)
25 mM methionine (10 μl of 200 mM stock
   solution, made up in sorbitol medium)
3–8 μl translation mixture (depending
   on the efficiency of translation)

The methionine is included to prevent the labelled, unincorporated methionine in the translation mix from being incorporated into protein by the chloroplast translation machinery. If a different labelled amino acid is used in the translation reaction, an equivalent concentration of cold amino acid should be used in the import assay.

The above components are gently mixed and incubated at 25°C in an illuminated water bath, at an intensity of 300 microeinsteins m^{-2}s^{-1} for 20–40 min. The tubes should be shaken every 5 min or so to prevent the chloroplasts from settling out. After incubation, the mixtures are treated as described below.

3. Assessment of import efficiency

After incubation, one mixture is diluted by the addition of 5 ml of sorbitol medium, and the chloroplasts are pelleted by centrifugation at 4000 × g for 2 min. The (radioactive) supernatant is decanted off, the pellet is resuspended in a small volume (50–100 μl) of sorbitol medium, and the suspension is mixed with one volume of SDS-PAGE sample buffer and immediately boiled for 5 min. This sample contains imported proteins plus proteins bound to the chloroplast surface.

A second sample is transferred directly onto ice at the end of the import reaction, and mixed with 50 μg ml^{-1} proteinase K (1 mg ml^{-1} stock solution is made up in sorbitol medium). After incubation for 20 min on ice, the suspension is diluted and centrifuged as above, and the pellet is resuspended in the same volume of sorbitol medium, but in this case containing 2 mM phenyl methyl sulphonyl fluoride. At this point, the suspension should be mixed with an equal volume of sample buffer that is *already boiling*. This is important because proteinase K is a very potent protease which is difficult to inhibit.

In some experiments, particularly where the precursor under study is very susceptible to proteolysis, we have used thermolysin instead of proteinase K, at a final concentration of 0.2 mg ml^{-1}. The advantage is that this protease can be inhibited completely by the inclusion of 5 mM EDTA in the sorbitol medium in subsequent steps. With some precursors, however, incomplete digestion of unimported precursor is observed. Ideally, the above protocol should completely digest precursor molecules which are bound to the chloroplast surface, and thereby demonstrate both the efficiency of import and the sizes of the imported proteins.

4 Localisation of imported proteins

In order to determine the location(s) of the imported proteins, it is necessary to fractionate the chloroplasts after the import reaction. Most proteins are transported into either the stroma or the thylakoid network, in which case localisation is fairly straightforward. After import, chloroplasts are protease-treated, diluted with sorbitol medium and pelleted as described. The chloroplasts are then lysed in a small volume of 10 mM HEPES-KOH, pH 7.5, 5 mM $MgCl_2$, and centrifuged for 5 min in a microfuge to generate a stromal supernatant and a thylakoid pellet (envelope membranes also tend to pellet, but account for only a small fraction of total membrane).

Note: this procedure should be carried out as rapidly as possible because residual protease, bound to the chloroplast envelope, may otherwise digest imported proteins after lysis has occurred.

Further samples can be processed in the case of imported thylakoid proteins, in order to rigorously determine the ultimate location of the imported protein. For example, if the imported protein is believed to assemble into a larger complex, it may be desirable to isolate the complex in order to test whether assembly of the imported protein has indeed taken place. Alternatively, it is possible to test whether the protein is located in the thylakoid lumen by protease-treating the thylakoid fraction after lysis of the chloroplasts.

B. Use of Exogenous ATP to Drive Import into Chloroplasts

The import of proteins into the stroma requires ATP both in the stroma and at the chloroplast surface (Olsen *et al.*, 1989; Theg *et al.*, 1989). In the light-driven import assay described above, the external ATP is supplied by the translation mix, and the stromal ATP is generated by photophosphorylation. Light also plays a more direct role in the import of a number of thylakoid proteins, by generating a proton gradient required for the integration into, or the translocation across, the thylakoid membrane (Mould *et al.*, 1991; Cline *et al.*, 1992). In some circumstances, however, it may not be possible for the chloroplasts to generate ATP by photophosphorylation (for example, if inhibitors are used which affect the thylakoidal protonmotive force, or if it is necessary to conduct assays in the dark). In such cases, import must be driven by exogenously added ATP. The protocol is basically the same except that, before incubation with precursor, the chloroplasts are preincubated at 25°C for 15 min with 5 mM MgATP, in order to bring the stromal ATP concentration to the required level.

V. IMPORT OF PROTEINS INTO ISOLATED THYLAKOIDS

A. General Comments

During chloroplast biogenesis, a large number of proteins are transported into the stromal phase or the thylakoid membrane. A subset of these proteins, however, also undergo further translocation across the thylakoid membrane; known examples include plastocyanin and the proteins of the photosynthetic oxygen-evolving complex. The import of these proteins involves:

(1) Synthesis in the cytosol as larger precursors containing a presequence consisting of two targeting signals in tandem.
(2) The first, 'envelope transit' signal directs translocation into the stroma, after which it is removed by the stromal processing peptidase.
(3) The stromal intermediate form is transported across the thylakoid membrane,

after which complete maturation is carried out by a thylakoidal processing peptidase (Hageman *et al.*, 1986; James *et al.*, 1989; Ko and Cashmore, 1989).

Using the intact chloroplast import assay, it was found that the later stages of this two-stage pathway could not be analysed in detail, simply because of the speed with which localisation took place following transport across the envelope membranes. With this problem in mind, we have recently developed an *in vitro* assay for the import of proteins by isolated thylakoids; under these conditions, we observe efficient import of the 33, 23 and 16 kDa proteins of the oxygen-evolving complex (Mould and Robinson, 1991; Mould *et al.*, 1991; Klosgen *et al.*, 1992).

The two-phase import model described does not apply to the biogenesis of all thylakoidal proteins. Some integral membrane proteins, such as the light-harvesting chlorophyll binding protein of photosystem 2, are synthesised with only a stroma-targeting presequence. The information specifying integration into the thylakoid membrane is located in the mature protein sequence (Lamppa, 1988). An *in vitro* assay for the integration of this protein into isolated thylakoids has been developed by Cline (1986) and Fulson and Cline (1988).

In the assay for thylakoidal protein import, translocation is driven by light: each of these proteins requires a proton gradient across the thylakoid membrane for translocation to take place. Otherwise, however, the requirements of these proteins do differ. Whereas the 23 and 16 kDa proteins can be imported by simply illuminating a mixture of translation product and washed thylakoids, import of the 33 kDa protein also requires the presence of concentrated stromal extract (Mould *et al.*, 1991). We have recently found that this reflects the involvement of at least one stromal protein in the transport of the 33 kDa protein across the thylakoid membrane (submitted for publication). It is not yet known whether this factor is required for the import of other proteins into thylakoids, since so few proteins have been analysed. When testing for the import of novel proteins into isolated thylakoids, it would therefore be logical to conduct assays in both the presence and absence of stromal extract.

It is also important to note that different thylakoidal proteins can have other, significantly different, requirements for transport across the thylakoid membrane. For example, a proton gradient is completely unnecessary for the translocation of plastocyanin across the thylakoid membrane (Theg *et al.*, 1989), and ATP is probably

TABLE 10.1. Requirements for the import of proteins into isolated thylakoids.

Protein	Import requirements			
	pH	SPP cleavage	Stromal extract	ATP
23K, 16K	Yes	No	No	No
33K	Yes	ND	Yes	Yes
Plastocyanin	No	No	ND	Yes
CFo2	No	No	ND	ND

ND: not determined.
The table illustrates the requirements for the import into isolated thylakoids of the 33, 23 and 16 kDa proteins of the oxygen-evolving complex (33K, 23K, 16K), plastocyanin, and the CFo2 subunit of the ATP synthetase (C. Robinson and R. G. Herrmann, unpublished data). It is notable that most, if not all of these proteins can be imported by thylakoids as the full precursor form: cleavage by the stromal processing peptidase (SPP) to the intermediate forms is unnecessary.

required instead (our unpublished observations). Table 10.1 summarises the requirements for those proteins which have been studied to date, and includes some preliminary data, obtained in collaboration with the group of R.G. Herrmann on the import of the CFo 2 subunit of the ATP synthetase.

B. Incubation Conditions

The import assay is carried out as follows.

(1) Intact chloroplasts are isolated by the Percoll pad method detailed above. It is important to note that import has only been observed with pea thylakoids; the group of R.G. Herrmann (Munich) has obtained efficient import into pea thylakoids following the protocol given below, but observed no import whatsoever using spinach thylakoids (R.G. Herrmann, personal communication). We have not tested thylakoids from other plant species in this type of assay.
(2) The pelleted chloroplasts are lysed in 10 mM HEPES-KOH, pH 8.0, 5 mM $MgCl_2$ to give a chlorophyll concentration of 1 mg ml^{-1}. The sample is left on ice for 10 min to ensure complete lysis.
(3) The sample is spun in a microfuge for 5 min to generate a stromal supernatant and a thylakoid pellet. The stromal sample is kept on ice until required.
(4) The thylakoids are washed twice in 10 mM HEPES-KOH, pH 8.0, 5 mM $MgCl_2$ and then resuspended in either the same buffer or the stromal extract to 1 mg ml^{-1} chlorophyll.
(5) Import incubations (50 µl) contain thylakoids (30 µg chlorophyll), 10 mM

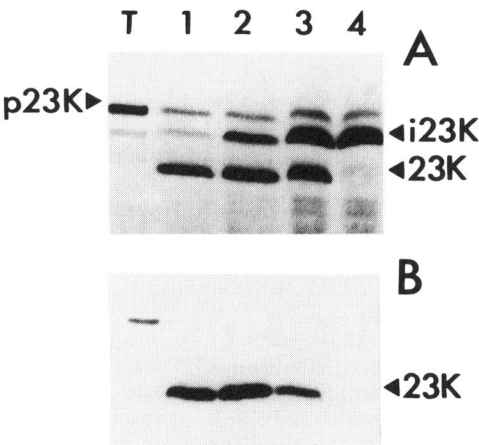

FIG. 10.4. Import of a lumenal protein into isolated thylakoids. The precursor of the 23 kDa oxygen-evolving protein (p23K) was incubated with pea thylakoids in the light in the absence of stromal extract (lane 1) or in the presence of dilute (lane 2) or concentrated (lane 3) stroma. After incubation, one half of each sample was analysed directly (upper panel) and the other half was treated with protease (lower panel). The figure shows that, with increasing levels of stromal extract present, more p23K is processed to the intermediate form (i23K) by stromal processing peptidase activity in the extract. In both the presence and absence of stroma, mature-size 23K is efficiently generated; this form is resistant to protease digestion, showing that import into the lumen has taken place. Lane T: p23K translation product.

MgCl$_2$, and 3–5 µl of translation mix. Incubation is for 30 min at 25°C under illumination (300 microeinsteins m^{-2} s^{-1}).

(6) After incubation, the sample is divided into two equal portions, one of which is incubated with protease (thermolysin at 0.5 mg ml^{-1} or proteinase K at 50 µg ml^{-1}) for 30 min on ice. Each sample is then diluted with 1 ml of 10 mM HEPES buffer, and the thylakoids are pelleted by centrifugation for 5 min in a microfuge. The pellets are resuspended in 50 µl of HEPES buffer and mixed with an equal volume of boiling SDS gel sample buffer.

Using this assay, we have obtained import of the 33, 23 and 16 kDa oxygen-evolving complex proteins with efficiencies of 40%, 60%, and 90%, respectively, of available precursor (Mould *et al.*, 1991; Klosgen *et al.*, 1992). Import of plastocyanin, however, is much less efficient (only about 5–10%). Figure 10.4 illustrates the results obtained with a typical import assay using the precursor of the 23 kDa protein (from wheat) as a substrate.

REFERENCES

Anderson, C. W., Straus, J. W. and Dudock, B. S. (1983). *Methods Enzymol.* **101**, 635–644.
Chua, N.-H. and Schmidt, G. W. (1978). *Proc. Natl. Acad. Sci. USA* **75**, 6110–6117.
Cline, K. (1986). *J. Biol. Chem.* **261**, 14 804–14 810.
Cline, K., Ettinger, W. and Theg, S. M. (1992). *J. Biol. Chem.* **267**, 2688–2696.
Fulson, D. R. and Cline, K. (1988). *Plant Physiol.* **88**, 1146–1151.
Grossman, A. R., Bartlett, S. G., Schmidt, G. W., Mullett, J. E. and Chua, N.-H. (1982). *J. Biol. Chem.* **257**, 1558–1563.
Hageman, J., Robinson, C., Smeekens, S. and Weisbeek, P. (1986). *Nature* **324**, 567–569.
Highfield, P. E. and Ellis, R. J. (1978). *Nature* **271**, 420–424.
James, H. E., Bartling, D., Musgrove, J. E., Kirwin, P. M., Herrmann, R. G. and Robinson, C. (1989). *J. Biol. Chem.* **264**, 19 573–19 576.
Kirwin, P. M., Meadows, J. W., Shackleton, J. B., Musgrove, J. E., Elderfield, P. D., Mould, R., Hay, N. A. and Robinson, C. (1989). *EMBO J.* **8**, 2251–2255.
Klosgen, R. B., Brock, I. W., Herrmann, R. G. and Robinson, C. (1992). *Plant Mol. Biol.* **18**, 1031–1034.
Ko, K. and Cashmore, A. R. (1989). *EMBO J.* **8**, 3187–3144.
Lamppa, G. K. (1988). *J. Biol. Chem.* **263**, 14996.
Li, H., Moore, T. and Keegstra, K. (1991). *Plant Cell.* **3**, 709–717.
Melton, D. A., Krieg, P., Rabagliciti, M. R., Maniatis, T., Zinn, K. and Green, M. R. (1984). *Nucl. Acids Res.* **12**, 7035–7056.
Mould, R. M. and Robinson, C. (1991). *J. Biol. Chem.* **266**, 12189–12193.
Mould, R. M., Shackleton, J. B. and Robinson, C. (1991). *J. Biol. Chem.* **266**, 17286–17289.
Olsen, L., Theg, S., Selman, B. and Keegstra, K. (1989). *J. Biol. Chem.* **264**, 6724–6729.
Salomon, M., Fischer, K., Flugge, U.-I. and Soll, J. (1990). *Proc. Natl. Acad. Sci. USA* **87**, 5778–5782.
Smeekens, S., Weisbeek, P. and Robinson, C. (1990). *Trends Biochem. Sci.* **15**, 73–76.
Theg, S., Bauerle, C., Olsen, L., Selman, B. and Keegstra, K. (1989). *J. Biol. Chem.* **264**, 6730–6736.
Van den Broeck, G., Timko, M. P., Kausch, A. P., Cashmore, A. R., Van Montagu, M. and Herrera-Estrella, L. (1985). *Nature* **313**, 358–363.

11 Seed Development

ANDREW C. CUMING

Department of Genetics, University of Leeds, Leeds LS2 9JT, UK

I.	Introduction	219
II.	Patterns of embryonic development	220
	A. Dicotyledonous plants	221
	B. Monocotyledonous plants	223
III.	Molecular analysis	225
IV.	Genetic analysis of embryogenesis	230
V.	Embryonic maturation	237
	References	247

I. INTRODUCTION

Sexual reproduction in higher plants characteristically culminates in the production of seeds. These remarkable structures contain an embryonic plant: the product of a developmental programme which, in a short space of time following fertilisation, can specify the formation of all the vegetative organs found in a mature plant. Subsequent growth of the plant elaborates these structures, but includes few additional changes in morphogenetic specification. Arguably, the only significant morphogenetic changes which occur, subsequent to seed formation, are those which follow on from the reprogramming of the shoot apical meristem in the transition from vegetative growth to the development of reproductive structures.

Accompanying the formation of the primordial organs of the embryo is the accumulation of nutrient reserves. The function of these reserves is to provide the embryo with a sufficient source of energy to support germinative growth, until the resulting seedling is competent to sustain itself through the assumption of an autotrophic habit. The synthesis and deposition of seed reserves generally occur later in the elaboration of the developmental programme, subsequent to the initial events of cellular determination, differentiation and proliferation which establish the identity of the embryonic organs. Typically, the bulk of the seed reserve substances are deposited either within the cotyledons (modified but recognisable, embryonic leaves) or within an accessory tissue, the endosperm, which like the embryo is also a product of sexual fusion. Whereas the initial morphogenetic events which establish the embryo — its polarity, symmetry and characteristic anatomy — appear largely to be under genetic control, both the deposition and the mobilisation of embryonic reserves are processes in which the genetic programme can be seen to interact with specific developmental and environmental cues. Both plant growth regulators ('plant hormones') and environmental factors can modify the timing and extent of reserve accumulation and utilisation.

The habit of life of plants is such that they are hostage to the vagaries of the environment. A variety of strategies have evolved to enable plants to withstand or evade adverse environmental conditions, and amongst these the production of resilient seeds ranks highly. Characteristically, the mature seed is a structure which by virtue of its mechanical properties (tough outer coat, low water content) is capable of withstanding extremes of temperature and desiccation which would be lethal to a vegetatively growing plant. Added to these properties are the capability for dispersal over wide areas, by a variety of agents, and the frequent occurrence of periods of dormancy, following maturation. The combined effects of these properties are to enable the propagules of a single plant to become distributed in both space and time over a wide area, thereby maximising the range over which progeny may successfully become established.

In considering the application of molecular biology techniques to the study of the unique attributes of seeds, this chapter must necessarily be selective both in its choice of those properties of seed biology which are discussed, and of the experimental approaches which have been adopted. In making this selection, I have chosen to highlight those aspects of the subject which are concerned with the biological nature of seeds, rather than those advances which are principally concerned with the manipulation of these properties for technological ends. I have further attempted to highlight not only those examples where molecular analysis has already yielded a wealth of information, but also those fields of study, as yet only investigated in pioneering fashion, in which the developing power of molecular analysis may be expected to combine synergistically with that of genetic analysis in expanding our understanding of the basic processes which underly plant development.

II. PATTERNS OF EMBRYONIC DEVELOPMENT

The development of the seed is initiated by the act of fertilisation. In the flowering plants, this event is characteristically a double fertilisation occurring within the female gametophyte. The processes of sexual development, fertilisation and subsequent embryonic development have been described following morphological studies in a large number of plant species. Whilst broadly the same developmental pathways are followed, there are, not surprisingly, a number of variations on the basic theme. For the purposes of this review, which will concentrate on the molecular analysis of embryogenesis, the brief description which follows is necessarily simplified, being based on the events occurring during the development of the embryos of model species.

11. SEED DEVELOPMENT

For dicotyledonous plants, the crucifer *Capsella bursa-pastoris* (Shepherd's purse) has long been recognised as a paradigmatic species, and its close relative, *Arabidopsis thaliana* (Thale cress), which has become a model species for molecular genetic analysis of plant development, follows an essentially identical embryogenetic course. For monocotyledonous plants I have taken *Zea mays* (Indian corn, maize) as an exemplar, again because of the wide utility of this species for molecular genetic analysis.

A. Dicotyledonous Plants

Haploid gametes are produced in the reproductive organs by meiosis. The female meiotic division initially gives rise to a tetrad of haploid cells, of which only one survives to become the female gamete — the megaspore. This in turn undergoes further mitotic divisions to produce the female gametophyte, or embryo sac. This comprises a single egg nucleus, together with two polar nuclei and additional accessory cells — antipodals and synergids. The whole is embedded deep within the surrounding maternal tissues, and for fertilisation to occur, contact has to be made with the contents of the pollen grains — the male gametes.

Pollen is produced in the anthers, each meiotic event producing a tetrad of microspores, each of which undergoes further mitotic division to produce a mature pollen grain containing two haploid sperm nuclei. Upon settling on a receptive stigmatic surface, the pollen grain germinates, producing a pollen tube which passes down the style towards the embryo sac. The sperm nuclei are transmitted within this tube and penetrate the embryo sac where zygotic fusion occurs.

The double fertilisation is achieved by the fusion of one of the sperm nuclei with the single haploid egg nucleus, to produce the zygote proper. The other sperm nucleus fuses with the polar nuclei to initiate a genotypically distinct lineage, which develops to become the endosperm tissue. Generally, the endosperm is a triploid tissue; the consequence of fusion of three haploid nuclei. However, in some species the numbers of polar nuclei vary, resulting in the production of endosperm with higher levels of ploidy.

The development of the zygote to form an embryo passes through a number of morphologically distinct stages, which in *Capsella* and *Arabidopsis* are the result of a relatively invariant series of cell divisions, enabling cell lineages to be traced with some clarity (Fig. 11.1). These are termed *preglobular (proembryo), globular, heart-stage* and *torpedo-stage*. In the formation of the crucifer embryo, cellular differentiation becomes apparent from the earliest stage as the first cell division is transverse and asymmetric, producing a small proembryonic apical cell and a larger basal cell which will give rise to the suspensor. The second division of each daughter cell differentiates the proembryo from the suspensor more clearly: the suspensor cell divisions continue in the same transverse plane, producing a file of cells, whereas the proembryonic cell's next two divisions occur perpendicular to one another, producing an eight-celled ball of embryonic tissue, the *octant* (Fig. 11.1, stage L). Control of the plane of cell division is of crucial importance in morphogenesis in higher plants. Because plant cells are bounded by a comparatively inflexible cell wall, morphogenesis cannot be modified by the relative movement of cells such as is characteristic of embryogenesis in animals (as for example at gastrulation). Consequently, many of the genetic controls which guide embryo development may be expected to function through directing the planes

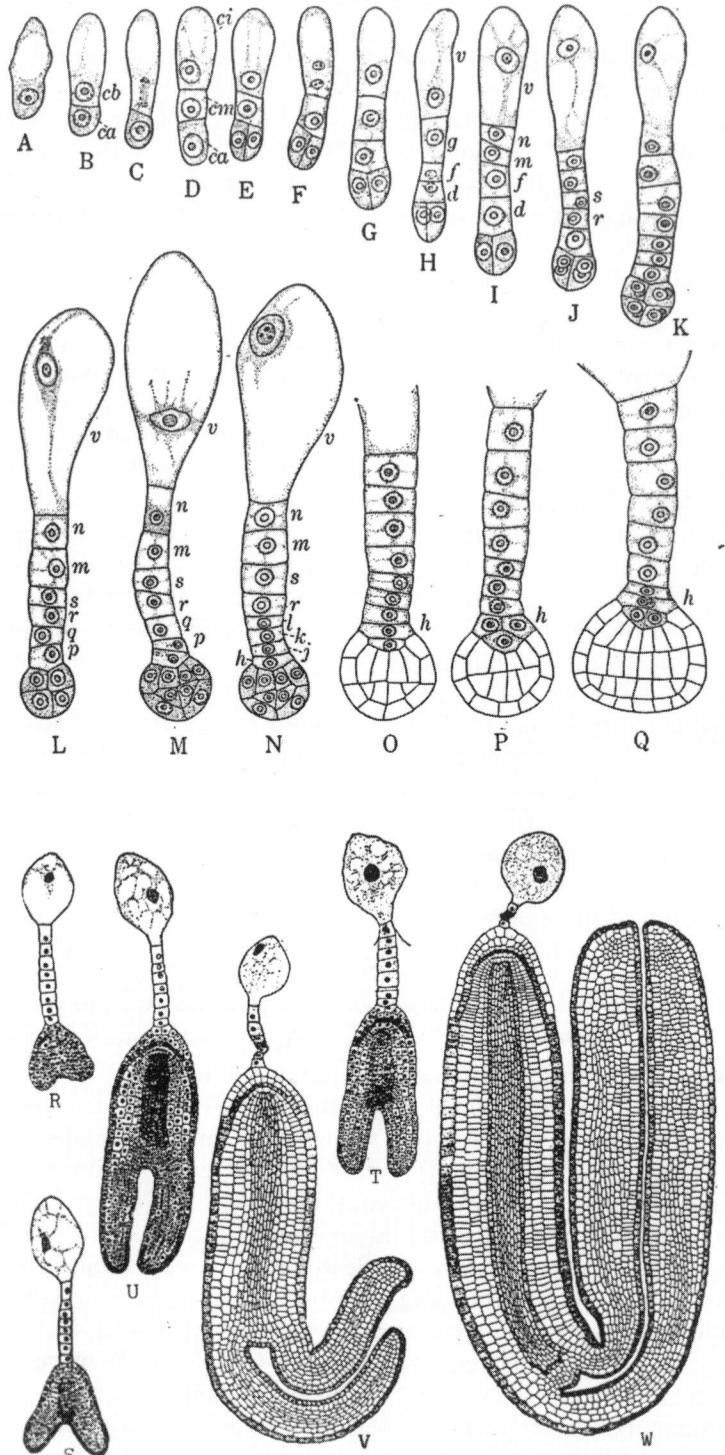

Fig. 11.1. Embryonic development in crucifers: *Capsella bursa-pastoris* and *Arabidopsis thaliana*. Reproduced from Maheshwari (1950).

of cell division and cell enlargement by appropriate temporal and spatial redistribution of cytoskeletal and cell wall components.

When the cells of the octant divide, further cellular differentiation becomes apparent as each cell undergoes a periclinal division (parallel to the surface). This establishes a distinct outer layer whose cells continue to divide in the anticlinal plane (perpendicular to the surface), thus forming a distinct epidermal layer (the *protoderm*: Fig. 11.1, stage M). The cells within the embryo continue to divide in the vertical and transverse planes, producing an ordered growth in volume of the globular embryo. The suspensor remains a relatively rudimentary tissue (in *Capsella* and *Arabidopsis*), with the function primarily of acting as a channel for nutrients to the developing embryo. However, the cell of the suspensor, the *hypophysis*, in contact with the globular proembryo, contributes to the continued development of the embryo proximal (denoted by '*h*' in Fig. 11.1, N–Q). This cell is the progenitor from which the root meristem and root cap will subsequently develop. Within the preglobular embryo, the progenitors of the 'adult' organ systems also become apparent at an early stage. The upper half of the proembryonic mass will form the shoot apex and the cotyledons, whilst the lower half will develop into the hypocotyl (the 'stem' connecting the cotyledons with the root).

As development proceeds, the symmetry of the embryo changes from a radial to a bilateral plan. The first intimations of this transition are marked by the development of 'shoulders' producing a heart-shaped embryo (Fig. 11.1, stage R) which gradually elongates through an increased proportion of transverse cell divisions (forming a torpedo-stage embryo; Fig. 11.1, stages T–W) with the concomitant differentiation occurring between proto-vascular tissue and parenchymatous cells, and with the appearance of distinct root and shoot apical meristems, together with an enlargement of the hypocotyl. The expansion of the shoulders of the heart-stage embryo produces the cotyledons of the mature embryo which ultimately occupy the bulk of the space available within the seed coat, and which become the primary site of reserve deposition within the seed.

In parallel with the development of the embryo the endosperm nuclei also proliferate, initially without accompanying cellularisation taking place. The subdivision of the syncytial endosperm by cell walls occurs as development proceeds further, but this tissue gradually degenerates as the embryo comes to take up the bulk of the space within the seed. (In *Capsella* and *Arabidopsis* the endosperm and the suspensor degenerate completely by maturity. However, this is by no means universal in dicotyledonous seeds, and the endosperm remains an important reserve storage tissue in a number of species.)

B. Monocotyledonous Plants

Monocotyledonous plants are principally differentiated from dicotyledonous plants by having only a single cotyledon within the seed. Thus their pattern of development is necessarily different. In *Zea*, as in graminaceous species generally, the mature embryo is characterised by a high degree of cellular differentiation, so that most of the structures recognised in the 'adult' plant are recognisably present, in miniature, within the seed. In particular a large number of leaf primordia can be identified, by contrast with the dicotyledonous crucifer embryos, where the shoot apical meristem remains small and inactive in initiating leaf primordia prior to germination. The stages of maize embryo development, illustrated in Fig. 11.2, have been termed *proembryonic*, *transition-stage* and *coleoptilar-stage* followed by a number of maturation stages numbered 1 through 6 (Abbé and Stein, 1954).

The initial, proembryonic stage of development is characterised by cellular proliferation to form an egg-shaped, globular embryo. Once again, the initial mitosis results in a transverse, asymmetric division to form a large basal cell and a small apical cell which subsequently develop into suspensor and proembryo, respectively. The subsequent cell divisions within the proembryo do not show the ordered pattern seen in the crucifer embryo. As the proembryo gradually elongates, so cellular differentiation becomes apparent with the lower cells, generally vacuolate and larger in size, forming the suspensor region, whilst the upper cells remain small and form the embryo proper. Elongation, particularly of the suspensor, continues until the proembryo takes on an appearance which has been vividly likened to an ice-cream cone — the transition stage.

During the coleoptilar stage, the embryo takes on the characteristic shape of the mature embryo, as the primordial organs develop. The cotyledon, or scutellum, begins to enlarge, whilst the root and shoot apical meristems form on one face of the enlarging embryo. As its name suggests, this stage of development is marked by the first appearance of the coleoptile — a sheath around the shoot which first appears to surround the shoot apical meristem. In the subsequent developmental stages the embryo, and particularly the scutellum, continues to enlarge, whilst the suspensor gradually degenerates. During these stages the leaf primordia emerge from the base of the shoot apical meristem, and elongate to envelop it. These leaf primordia are supplied with vascular tissue which also differentiates through these stages of development, so that the mature embryo is functionally a fully differentiated, miniature plant.

Endosperm development in the maize kernel is considerably more prominent than in crucifers, the mature kernel retaining endosperm as its principal site of reserve deposition. (In other cereals, the endosperm contributes an even greater proportion of the reserve potential of the mature seed than in maize, where the relatively massive scutellum acts as a substantial reserve storage organ in its own right.) Additionally, the endosperm undergoes some cellular differentiation, the most important being the setting apart of the outermost cell layer(s) as the aleurone layer — a tissue whose function is to provide enzymes for the mobilisation of the endosperm reserves upon germination.

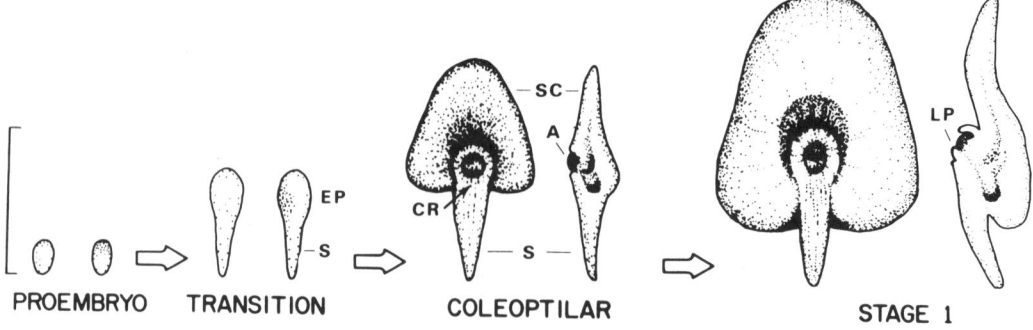

Fig. 11.2. Embryonic development in cereals: *Zea mays*. Reproduced from Sheridan and Clark (1987).

In both dicotyledonous and monocotyledonous seeds, the differentiation of the embryonic organs to produce an embryo capable of germination and growth occurs relatively rapidly. Certainly, the potential for germination is achieved long in advance of the final maturation of the seed. This can be demonstrated by the ability of immature embryos to germinate precociously if removed from within the developing seed and placed on a simple culture medium containing nutrients. Therefore the later stages of embryonic development are also marked by the imposition of dormancy upon the embryo. This occurs at about the time that reserve substances — lipids, carbohydrates and storage proteins — begin to accumulate massively in the storage organs (cotyledons and scutellum) and endosperm tissue of the developing seed. Finally, the developmental programme is brought to an end by the gradual desiccation of the seed which occurs prior to its being shed from the parent plant.

III. MOLECULAR ANALYSIS

Studies of the molecular biology of seed development have, until recently, focused principally on the later, maturation stages of embryonic and endosperm development and on the processes which occur subsequent to the initiation of germination. This concentration on the events following on from the principal morphogenetic events has been largely dictated by technical imperatives. During their early development, plant embryos are relatively inaccessible structures, buried within the maternal tissues and impossible to isolate in large numbers. The methods of molecular analysis available throughout the last decade have been based in biochemical practice, requiring that DNA and RNA be extracted in sufficient quantity for experimental manipulation — the generation of libraries of cloned DNA, the screening of these libraries and the analysis of the expression of individual genes by blot hybridisation. Only with the recent development of sequence amplification through the deployment of the polymerase chain reaction (PCR) (Saiki *et al.*, 1985) has it become feasible to generate substantial quantities of DNA from vanishingly small quantities of plant tissue. The molecular biochemical approach to the events occurring in seed development has therefore been of most value in the analysis of processes such as the deposition of storage reserves and their subsequent mobilisation during germination. This occurs relatively late in seed development, at times when large quantities of seed can readily be harvested for the facile extraction of proteins, RNA and DNA. Consequently, genes encoding the abundant storage proteins found in seed reserves were amongst the first plant genes to be isolated by molecular cloning, and this experimental approach has yielded a good understanding of the molecular controls which activate specific sets of genes in an organ-specific manner at the appropriate temporal stage of development.

For the analysis of events which occur during embryonic morphogenesis, and in the search for the controlling genetic switches which coordinate this complex process, two basic strategies have been followed. Continuing the theme of using molecular biochemical techniques to identify and isolate gene products associated with specific stages of embryogenesis, it has been necessary to develop experimental systems where large numbers of embryos can be prepared at synchronous stages of development. Such an end can be achieved by manipulating tissue cultures in order to induce the production in bulk of somatic embryos, *in vitro*. In the late 1950s the pioneering work of Steward and of Reinert demonstrated the developmental totipotency of individual plant cells by inducing dedifferentiated cells of carrot, maintained in cell suspension culture, to undergo a morphogenetic pathway indistinguishable from zygotic embryogenesis (Steward et al., 1958; Reinert, 1959). In brief, carrot cells cultured in the presence of the plant growth regulator 2,4-D (a synthetic auxin) proliferate in an essentially unorganised way, producing unorganised cell clusters. Withdrawal of 2,4-D causes these small, unorganised proembryonic cell masses to undergo somatic embryogenesis, passing through the characteristic preglobular, globular, heart and torpedo-stages of development. Under appropriate culture conditions, this pathway of somatic embryogenesis can be synchronously induced in the majority of proembryonic clusters present in the cell culture.

Using this experimental system, the laboratory of Sung has identified a number of gene products which act as molecular markers for the development of somatic embryos. The approach taken has been to identify proteins synthesised in cultures which have been induced to undergo embryogenesis, and which are not present in the uninduced, unorganised cell masses. This was carried out by harvesting cultures undergoing embryonic differentiation, which contained a mixture of somatic embryos at different stages of development (Choi et al., 1987). These embryos were homogenised to obtain a protein-rich lysate, which was used as an antigen for the immunisation of rabbits. The antiserum obtained from these animals therefore comprised a mixture of antibodies recognising proteins present in a wide range of somatic embryos. Of course, the bulk of the proteins present in a developing embryo will not be found exclusively within embryos, but will be proteins which all cells contain as a part of their normal range of biochemical and metabolic activities. Not surprisingly, the antiserum raised against a mixed population of proteins from developing embryos thus largely consisted of antibodies recognising these 'housekeeping' proteins. In order to identify antibodies specific for exclusively embryonic proteins, the whole serum was challenged with a mixture of proteins extracted from a cell suspension culture in which somatic embryogenesis had not been induced. By presenting this challenging antigen in an immobilised form — the undifferentiated cell extract was bound to a nitrocellulose membrane — it was possible to adsorb to the immobilised antigen all those antibodies corresponding to proteins synthesised in a non-embryo-specific manner. By carrying out repeated adsorptions to the challenging antigen, a residual antibody preparation was obtained which recognised only a few of the thousands of polypeptides present in the original embryonic lysate.

Antibodies can be used, not only directly as probes to identify specific antigens in mixed populations of polypeptides, but also as probes for the isolation of their cognate coding sequences from libraries of cloned DNA. By screening a cDNA library constructed in the phage expression vector λgtII (see Chapter 3) it was possible to isolate cDNA clones encoding two embryo-specific polypeptides, of 66 kDa and 59 kDa. Both these cDNA clones and the antibodies were then used in RNA blot hybridisation ('Northern blot') and immunoblot ('Western blot') analysis to detect the expression of the genes encoding these few polypeptides in somatic embryos as early as the formation of globular-stage embryos and their corresponding lack of expression in undifferentiated cells (Borkird et al., 1988). The expression of these genes could also be detected during the course of normal zygotic embryogenesis, during which their products accumulated to a high level in the torpedo-stage of development. Because these gene products failed to accumulate in cell lines which were only partially responsive to 2,4-D, and which were unable to proceed to the heart-stage of embryogenesis, it was inferred that these polypeptides were required for progression from the globular to the heart-stage.

A basically similar strategy has been employed by Bowles and her colleagues to identify genes expressed in an early embryo-specific manner in barley embryos. At first sight, this would seem inherently more difficult, since cereal tissue is notoriously refractory to manipulation in tissue culture. However, a high frequency of relatively synchronous somatic embryogenesis can be elicited, in some genotypes, by culturing developing anthers. The immature microspores produced in these organs are capable of undergoing an embryogenetic programme resulting in the production of haploid somatic embryos. By using comparative translation, *in vitro*, competitive immunoadsorption and differential cDNA hybridisation to identify embryo-specific markers, this group has identified a number of gene products whose expression is associated with embryo formation both in culture and in the course of zygotic development (Higgins and Bowles, 1990; Clark *et al.*, 1991; Smith *et al.*, 1992). Additionally, both the temporal and spatial specificity of these products have been analysed by a combination of Northern blot hybridisation and of hybridisation, *in situ*, to tissue sections (Fig. 11.3; see also Chapter 7). This further refinement should enable the development of experimental strategies to probe the mechanisms underlying the cell-type specificity of gene expression during embryonic organogenesis.

One of the consequences of initiating and maintaining plant cell cultures is the frequent generation of genetic aberrations. The observation that plants regenerated from tissue culture frequently differ from the plants which provided the source of the culture, both phenotypically and genotypically, is a common one and in some cases this 'somaclonal variation' is utilised by plant breeders as a source of additional genetic variation when handling overly uniform genetic material. The somatic mutations that arise in the course of tissue culture can frequently be maintained in culture, leading to the proliferation of mutant tissue culture lines, defective in a number of recognisably different ways. This property has been exploited to couple molecular biochemical analysis with a genetic approach, using mutant carrot cell suspension lines which have been found to exhibit aberrant behaviour upon induction of embryogenesis (Breton and Sung, 1982; Giuliano *et al.*, 1984; Widholm, 1984).

Biochemical studies of somatic embryogenesis indicated that the embryogenic capacity of cell cultures was correlated with their ability to secrete proteins into the liquid growth medium (Van Engelen and De Vries, 1992). Various lines of correlative evidence supported this conclusion. First, successful embryogenic induction required that the growth medium be 'conditioned': regular replacement of the growth medium with fresh medium prevented embryogenesis from proceeding following withdrawal of 2,4-D. Second, analysis of proteins secreted by cultured cells revealed discrete differences between non-embryogenic and embryogenic cells (Sterk *et al.*, 1991; Van Engelen *et al.*, 1991). Whereas non-embryogenic cells specifically secreted a glycoprotein (termed EP1, extracellular protein 1), embryogenic cell cultures could be identified by their secretion of a different spectrum of polypeptides, amongst which one (EP2) was identified as a lipid transfer protein.

Direct evidence for the involvement of secreted polypeptides in the formation of embryos comes from experiments based on direct supplementation of growth medium,

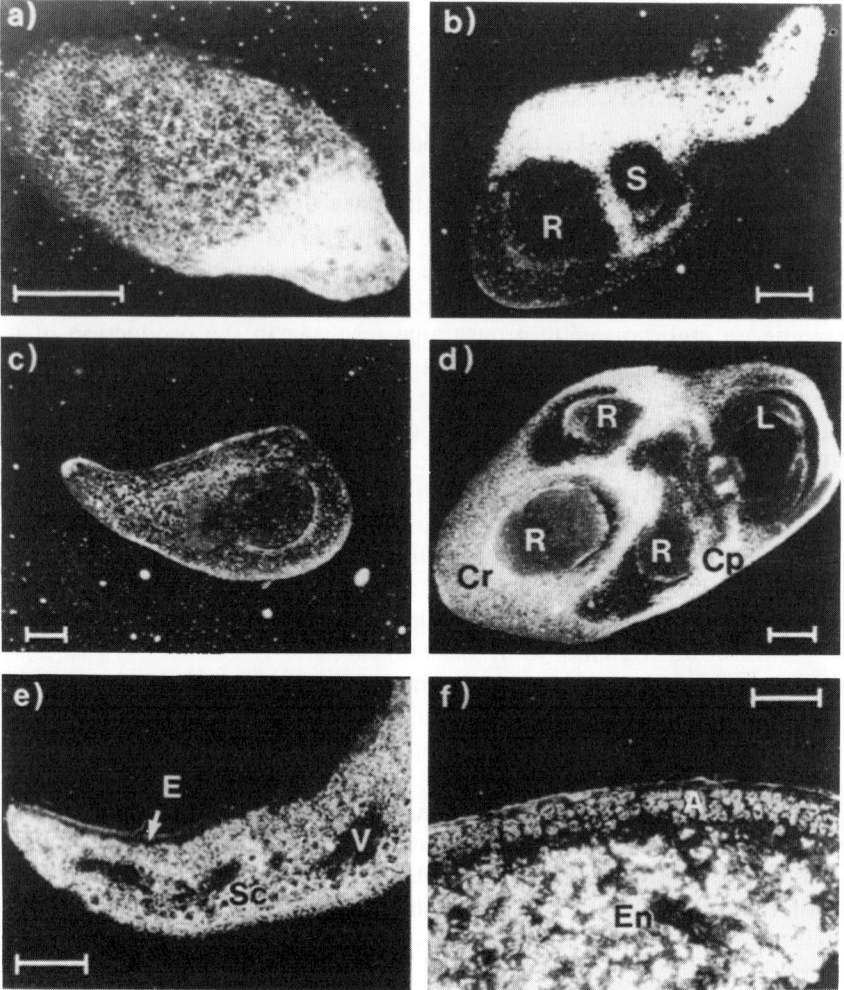

Fig. 11.3. Distribution of mRNA corresponding to a barley embryo-specific gene, revealed by hybridisation, *in situ*, with a cloned cDNA. Kindly provided by Professor D. J. Bowles, Leeds University.

and from studies of mutant cell lines blocked at specific stages of embryogenesis. Thus, in the first instance, the treatment of cell cultures with tunicamycin — an inhibitor of N-linked glycosylation — both resulted in the secretion by cells of an incorrectly glycosylated peroxidase, and simultaneously prevented somatic embryogenesis from proceeding from the preglobular to the globular stage. Supplementation of the medium with correctly glycosylated peroxidase permitted embryogenesis to proceed to the globular stage (De Vries *et al.*, 1988; Cordwener *et al.*, 1991). The significance of this may be merely mechanistic, rather than regulatory. The action of peroxidases generates free radicals, and within the cell wall these might act to cross-link cell wall components, thereby reducing the capacity of the cell to expand. Since proembryonic masses of carrot cells are distinguishable from clumps of non-embryogenic cells by being generally smaller and non-vacuolate, an enzymatic system which promoted cell wall 'tightening' would tend to maintain cells in the small, embryogenic condition. Cells lacking this property would 'escape' to become expanded and nonembryogenic.

Using a temperature-sensitive cell line (ts11: Giuliano *et al.*, 1984) in which development is normal at the permissive temperature of 24°C, but is arrested at the restrictive temperature of 32°C at the globular stage of development (the stage beyond which no further embryogenesis proceeds if normal cells are deprived of conditioned medium), the De Vries group was able to implicate a secreted protein as the target of the ts11 mutation (DeJong *et al.*, 1992). The mutant cells could be induced to develop normally at the restrictive temperature, if grown in conditioned medium derived from cultures of normal cells — the mutation could be complemented *in vitro*. This property provided a bioassay upon which the purification of the active factor could be achieved.

This factor in the wild-type conditioned medium could be identified as a protein, because pretreatment of the medium with proteases abolished its ability to complement the ts11 defect. Eventually, by using protein fractionation techniques, it proved possible to purify a protein from wild-type conditioned medium which, if added to the medium in which the mutant cells were growing, permitted them to complete their embryonic development.

This protein proved to be a 32 kD glycoprotein with acidic chitinase activity. Such proteins are commonly found to be secreted into the cell wall space (the apoplast) of mature plants, where they are believed to function primarily as antifungal agents. It was therefore surprising to find such an enzyme implicated in the embryonic morphogenetic pathway, particularly as plants are generally believed to lack chitin. It has been suggested that because chitinases are active against β-(1 $\Rightarrow$ 4) N-acetyl glucosamine (GlcNac) polymers, they may have an accessory role in modifying cell wall glycoproteins, many of which are known to contain GlcNac moieties in their sidechains. Alternatively they may be involved in loosening minor cell wall components. In either case, a consequence of their activity would be modification of the cell wall, perhaps crucially affecting the orientation in which cell wall components are laid down — cells of the ts11 line are arrested at the globular stage through their inability to form a proper protoderm, the outer layer of cells which results from a set of coordinated periclinal divisions. An aberrant chitinase is not the only consequence of the ts11 mutation. Mutant cells also show deficiencies in their ability to glycosylate a class of fucose-containing glycoproteins (LoSchiavo *et al.*, 1990), an activity which may also directly affect the morphogenetic fate of the cells.

Cell wall glycoproteins are increasingly being recognised as crucial components of the cell surface: monoclonal antibodies directed against their carbohydrate moieties can be used as morphogenetic markers for cellular differentiation, and it is likely that these glycoproteins play an important part in intercellular communication during plant development (Knox and Roberts, 1989; Stacey *et al.*, 1990; Knox, 1992). Similar properties of modifying the cell wall and thus its role in cell–cell contact can be adduced for the other extracellular proteins found to be implicated in embryogenesis: not only peroxidase but also the EP2 lipid transfer protein, which is speculatively linked with cuticle formation on the outer surface of the embryo, and which also could have a mechanical function in determining the rate of expansion of cell masses in culture, as cuticle components present an hydrophobic barrier to water uptake — the motive force for cell expansion through turgor pressure.

The development of mutant cell lines for the study of somatic embryogenesis has signally increased the range of experimental strategies available to the molecular biologist. This general line of approach, still firmly based on biochemical practices, has been successful in identifying molecular markers for the embryogenetic process, and in identifying components of the morphogenetic machinery.

Whilst this molecular biochemical approach is demonstrably a valuable means of identifying gene products whose synthesis and accumulation is characteristic of specific stages in embryo development, it suffers from the inherent limitation that the genes and gene products identified are likely, for the most part, to be those that are expressed as a consequence of regulation by the genetic programme controlling embryo development, and not themselves to be components of the genetic switching machinery. The identification of the regulated products should therefore be considered as only a first step towards the identification of the regulatory networks which coordinate embryogenesis. Progress towards this end, via the molecular biochemical route, will therefore necessitate the analysis of those regions of the regulated genes which control their expression (their promoters) and the subsequent identification of the various protein factors which interact with these *cis*-acting sequences to initiate and modulate their expression during development. At present, this experimental approach is both arduous and technically demanding.

IV. GENETIC ANALYSIS OF EMBRYOGENESIS

The fruit fly, *Drosophila melanogaster*, is an organism whose early development is perhaps better understood than that of any other complex eukaryote. The reason for this lies in the extensive history of this organism as a subject of genetic analysis. Techniques for generating and identifying mutants in a wide range of biological functions have been established through years of research, enabling the compilation of a comprehensive genetic map in which the relative disposition of genes can be determined on the chromosomes. This history of classical genetic study has now been supplemented by the development of molecular biological techniques which has enabled the cloning of genes responsible for defined biological functions through the application of such techniques as transposon tagging and chromosome walking from linked loci. Many mutants have been identified which disrupt the normal embryogenetic pathway in *Drosophila* (Nüsslein-Volhard and Wieschaus, 1980) and their cloning and structural analysis has yielded considerable insights into the genetic control of pattern formation.

It is the aim of plant developmental biologists to identify, isolate and analyse those genes which coordinate cellular differentiation and tissue specification during early embryogenesis, and consequently considerable efforts have been expended to identify mutations in genes which may have these functions. The first requirement for the pursuit of such a goal is the identification of suitable experimental systems, and to this end the foci of attention have been *Arabidopsis thaliana* as a representative of the dicotyledonous plants, and *Zea mays* as a representative monocotyledonous plant. Both species have long been subjects of genetic analysis, by classical means, and consequently have distinct adavantages over other species.

The properties of *Arabidopsis* which recommend it as a subject of study will be reviewed in Volume 11 of this series. For a brief outline, the reader is directed to previous reviews (Meyerowitz and Pruitt, 1985; Somerville, 1989). However, maize is in some ways less tractable than *Arabidopsis*, in that it posesses a much larger genome, and in addition is not yet routinely amenable to genetic transformation (although the efficiency with which transgenic maize may be regenerated has improved remarkably in the last two years). Nevertheless, it is a species which is extremely well characterised genetically, and in particular is the host to a number of resident transposable genetic elements which have been characterised in molecular detail, and which can be used to clone genes through the technique of transposon tagging.

To identify genes essential for the establishment of embryonic pattern, it is necessary to identify mutants which disrupt seed development (Meinke, 1991a). Great care must be exercised in developing a suitable screening procedure for such mutants, since mutations in functions which are responsible for embryonic development are likely to be phenotypically lethal. Certain general principles apply: mutagenisation (by chemical agents, radiation or by insertional inactivation of genes by transposons or exogenous DNA such as T-DNA), followed by self-fertilisation to reveal mutations in the homozygous state. However, mutations whose effects are lethal can only be maintained in the heterozygous state, and so a modified screening procedure must be operated. The normal procedure for screening for mutations which have lethal phenotypes is to screen the M2 generation by families (the seed within a single silique of *Arabidopsis* or an ear of maize are germinated in separate groups, rather than bulked for mass screening). This can be a laborious procedure. Fortunately, in the case of screening for embryonic lethals, such mutants become apparent during seed development, and frequently may easily be recognised by examination of the developing seed within the fruit which are formed on the M1 plant. A second caveat must also be observed. It must not be thought that any mutation which causes lethality in the developing embryo is necessarily due to an alteration in a gene whose principal function is in the control of embryogenesis. On the contrary, the majority of mutations which give rise to an embryonic lethal phenotype are likely to occur in genes which have an essential 'housekeeping' function in that they carry out processes required for the maintenance of normal cellular metabolism.

Over the past 15 years, a comprehensive study of embryonic-lethal mutants of *Arabidopsis* has been made, particularly by the laboratory of Meinke, whose recent review of this fascinating subject (Meinke, 1991b) is strongly recommended both as a model of clarity in its description of the experimental methodology attendant on the generation, screening and subsequent characterisation of embryonic-lethal mutants and as a valuable starting point for further reading. Several classes of embryonic mutation have been characterised. These range from housekeeping mutants, such as simple auxotrophs (Schneider *et al.*, 1989), through mutants which cause blocks in the embryogenetic process to occur at defined times in development, to those which cause a fundamental alteration in the pattern of embryonic development.

Mutants of this last class are of particular interest, since they may define genes with an overall regulatory role in the establishment of embryonic form: they may be considered analogous to the many mutants isolated which are disrupted in embryonic development in *Drosophila* and which have contributed so much to our understanding of the process of development in that organism. Recently a number of *Arabidopsis*

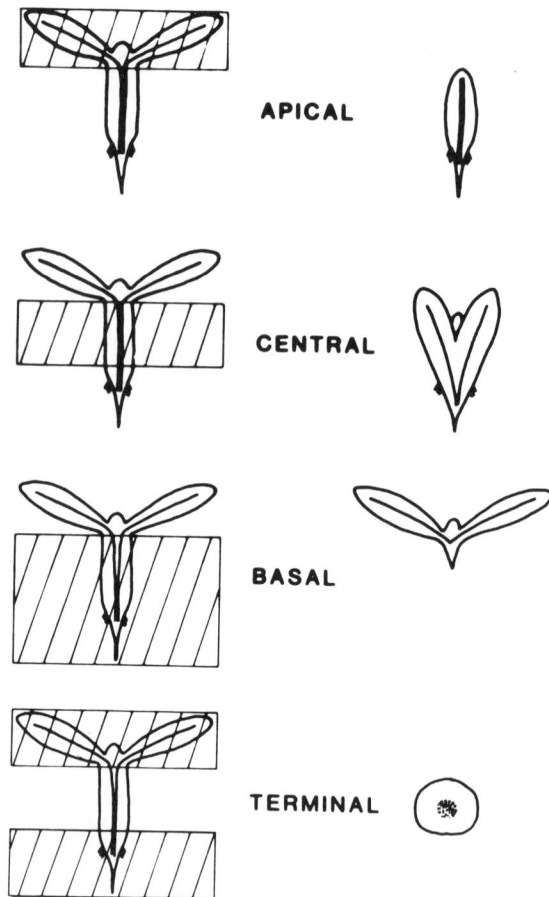

Fig. 11.4. Classes of embryonic-pattern mutants in *Arabidopsis thaliana*. (1) Lacking shoot apical meristem and cotyledons (e.g. *gurke*). (2) Lacking hypocotyl (e.g. *fackel*). (3) Lacking hypocotyl and root (e.g. *monopteros*). (4) Lacking shoot and root apical meristems (e.g. *gnom*). Reproduced from Mayer *et al.*, (1991).

mutants of this type have been isolated in the laboratory of Jürgens (Jürgens *et al.*, 1991; Mayer *et al.*, 1991), who reasoned that mutations causing disruption of processes which regulated the establishment of polarity and symmetry in the embryo might not be lethal during embryogenesis, but could be identified by screening in the germinated M2 seed for aberrant seedling morphology. Such a screen, following chemical mutagenesis, resulted in the isolation of three classes of mutant. The first class, illustrated in Fig. 11.4, comprised plants in which apical-basal polarity was perturbed. Thus mutants of the *gurke* type lacked shoot apices and cotyledons, whilst, by contrast, those of the *monopteros* type lacked root structures, due to deletion of both hypocotyl and root apex. In addition to these extreme polar mutants, *fackel* mutants lacked the hypocotyl alone, whilst *gnom* mutants lacked both the shoot and root poles, appearing to be derived from the hypocotyl lineage alone.

The second class of mutants comprised two types, both of which showed alterations in their pattern of radial development. Both *knolle* and *keule* showed aberrant structures due to the failure of the epidermal cell layer to differentiate properly. In the *knolle* mutants, this failure to differentiate could be observed to occur at a very early stage in the development of the globular embryo, the 16-cell stage at which the protodermal cells first become evident. In *keule* mutants the epidermal cells appeared abnormal, the first indications of this abnormality becoming apparent somewhat later. These mutants may thus prove valuable in dissecting the temporal regulation of pattern formation.

The third class of mutant included those in which the shape of the seedling was radically altered, not through the loss of any particular structure, but because of changes in cell shape in defined regions of the developing embryo. Thus *fass* mutants appeared short and fat due to a general compression of cell shape in all cell types, whereas in the *knopf* and *mickey* mutants (the latter named for the resemblance of the seedling for a famous Hollywood mouse!), cell shape changes appear principally in specific parts of the embryo: in the epidermal and vascular tissues in *knopf* mutants (where vascular differentiation appears lacking), and in the cotyledons in *mickey* mutants which are generally larger and anatomically less well defined than in the wild-type (Fig. 11.5).

Arabidopsis mutants with an apparently homoeotic effect have also been identified: the expected structure has been replaced with an alternative organ. Homoeotic mutations have been particularly fruitful as subjects of study by developmental biologists, since they occur in genes whose function is regulatory: they coordinate the action of the myriad structural genes with purely mechanistic functions in the specification of organ identity. The genes *agamous, pistillata* and *apetala*, in which mutations cause respecification of floral morphology in *Arabidopsis*, are examples for which molecular genetic analysis is leading to a good understanding of these developmental processes (Bowman *et al.*, 1989). Homoeotic mutations affecting embryonic pattern formation include the *doppelwurzel* mutant (Jürgens *et al.*, 1991) which, as its name suggests, has twin roots arising from either end of the embryo. In this case, instead of the embryo comprising clearly differentiated shoot apical and root apical meristems, the shoot apical half of the embryo has been deleted and replaced by a root apical structure. This mutation therefore defines a gene crucial in the early establishment of embryonic polarity and responsible for the early differentiation between shoot and root. Similarly, in the *toro* mutant the differentiation between the central shoot apex and the cotyledons has broken down, the cotyledons being transformed into additional shoots.

The characterisation of these mutants is, as yet, at an early stage and no molecular details have been revealed. However, the comprehensive genetic screening carried out in these studies of pattern formation has enabled an estimate to be made of the likely number of genes regulating pattern formation in *Arabidopsis*. On the basis of identifying *c.* 250 individual mutants it was reckoned that no more than 50 individual genes may be pivotal in regulating embryonic pattern formation. Significantly, microscopic analysis of the mutant embryos, in the course of their development, revealed that developmental abberrations were evident at very early stages in embryogenesis: prior to heart-stage for the lesions in the apical-basal axis, and in the globular stage for those mutants altered in radial patterning. This indicates that the principal elements in pattern formation – the generation of an apical-basal axis of polarity and the development of radial pattern – are established early in development,

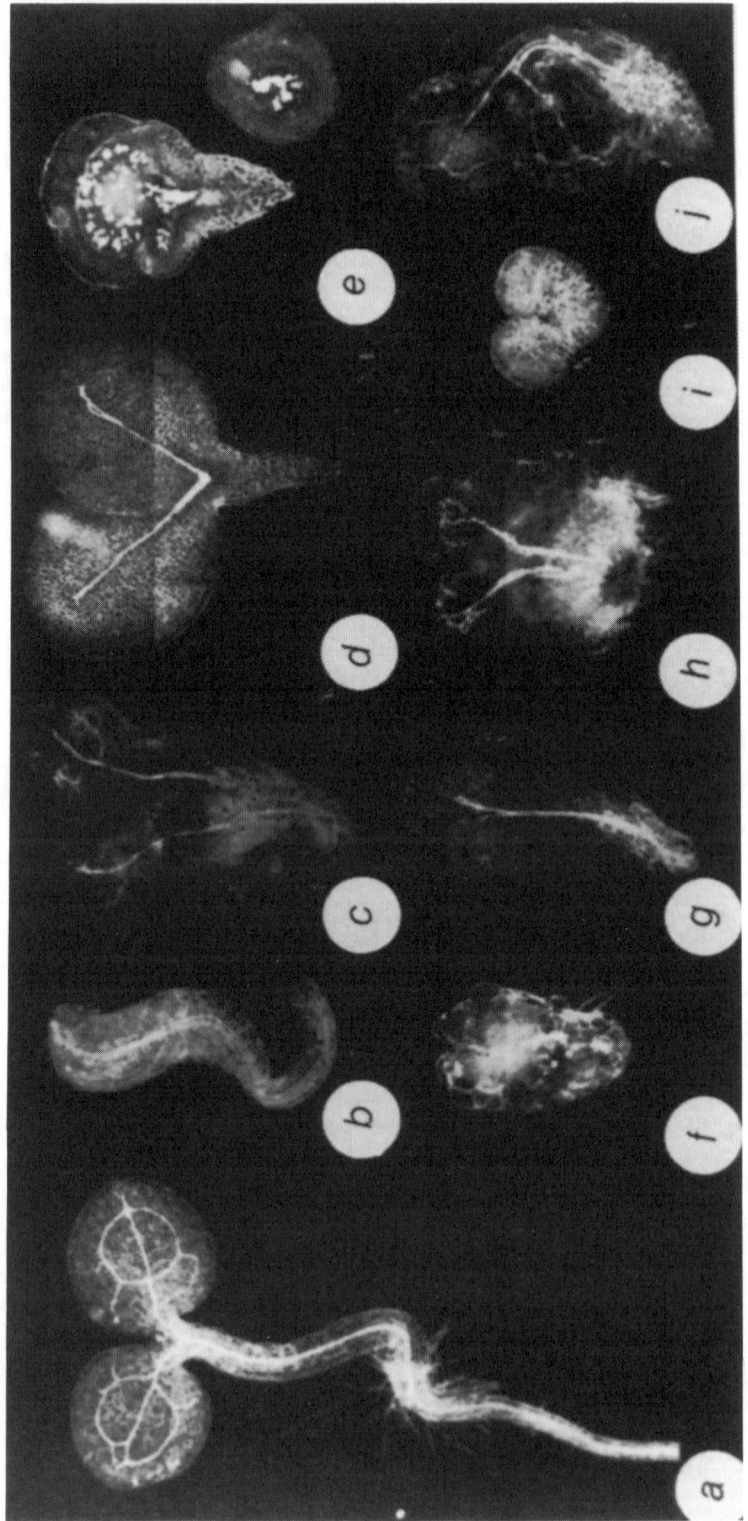

Fig. 11.5. Cellular patterning in mutant seedlings of *Arabidopsis thaliana*. Reproduced from Mayer *et al.* (1991).

as a consequence of the ordering of the planes of cell division, and that these elements of pattern are then carried through the subsequent growth of the embryo. Additionally, since the mutants in which the apical-basal patterning is disturbed still retain their characteristic radial patterning — divided into epidermal, cortical and vascular tissue — it is clear that the establishment of the apical-basal axis and radial patterning are separate and distinct processes (Jürgens, 1992).

Because these mutants were generated as a result of chemical mutagenesis, the isolation of the genes responsible must necessarily be effected by chromosome walking from linked loci. This is a relatively laborious procedure, necessitating a considerable investment in time to carry out the preliminary, essential linkage analysis.

The use of insertional mutagenesis by T-DNA following *Arabidopsis* seed transformation (Feldmann and Marks, 1987) has also enabled a wide range of T-DNA tagged embryonic lethal mutations to be identified (Errampalli *et al.*, 1991), and it is to be expected that a number of genes regulating embryo development will be cloned in this way in the near future (Meinke, 1991a, b). Screening of the collection of T-DNA tagged mutant lines is currently in progress in order to identify mutants of the embryonic-pattern type described above. A particularly powerful approach to the use of T-DNA tagging in the study of developmental gene expression has been adopted by Lindsey (Topping *et al.*, 1991; Lindsey and Topping, 1992; Lindsey *et al.*, 1993). This group's strategy is based on 'enhancer trapping'. Plants are transformed with a Ti plasmid in which the border sequence of the T-DNA lies immediately adjacent to a reporter gene encoding the bacterial enzyme β-glucuronidase (GUS), coupled with a truncated (and therefore barely active) promoter derived from the Cauliflower mosaic virus '35S' gene. This enzyme can be detected *in situ* in plant cells by histochemical staining with a chromogenic substrate. Insertions of this T-DNA into a plant gene, adjacent to its *cis*-acting regulatory elements, can result in the enhanced expression of the GUS gene under the control of these elements, often accompanied by the concomitant inactivation of the interrupted gene. This technique enables both the potential identification of developmentally significant mutations in plants homozygous for the T-DNA insertion, coupled with the ability to analyse the temporal and spatial pattern of expression of the interrupted gene by histochemical examination of plants heterozygous for the insertion. (These plants should be phenotypically wild-type as a result of the expression of the uninterrupted gene copy, yet synthesise the GUS enzyme as a result of the promoter activity of the interrupted gene.) Finally, the presence of the T-DNA acts as a tag for the cloning and analysis of the gene.

A genetic approach has also been used to identify genes coordinating embryogenesis in maize. Mutations affecting seed development in maize (*dek*: defective kernel) have been found to perturb the development of both embryo and endosperm (Neuffer and Sheridan, 1980; Sheridan and Clark, 1987). Amongst these mutants are those which show their effects coordinately in both embryo and endosperm, whilst others are restricted to the embryo, where their effects may be organ-specific in nature. Recently, a considerable advance has been made toward the isolation and characterisation of maize genes involved in embryogenesis through the identification of over 50 different embryo-specific (*emb*) mutants generated by the action of a transposon; Robertson's *Mutator* (*Mu*) (Robertson, 1978; Clark and Sheridan, 1991). The mutants obtained were varied in their phenotypes and a close examination of about half of them indicated that the mutations acted to block embryogenesis at distinct stages of their

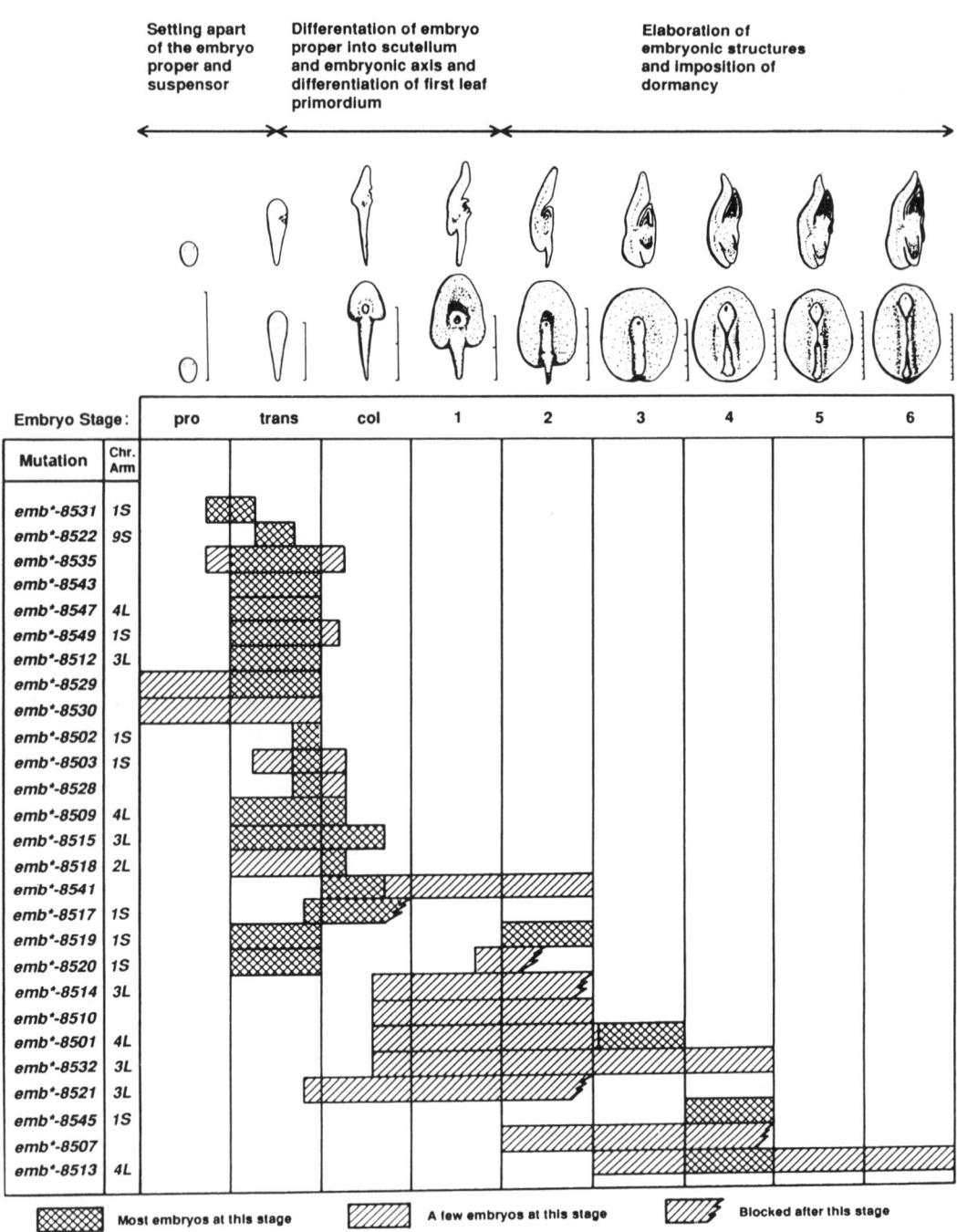

Fig. 11.6. Classes of *embryo-specific* mutants of *Zea mays*. Reproduced from Clark and Sheridan (1991).

development (see Fig. 11.6). The isolation of such mutants is of major importance, since the molecular cloning of the cognate genes should be facilitated through the use of the *Mu* element as a transposon tag. This element has successfully been used to isolate a number of maize genes, including the important regulator of seed maturation *Viviparous-1* (McCarty *et al.*, 1989a).

V EMBRYONIC MATURATION

As I have briefly outlined, the period during which embryonic pattern formation occurs occupies only a comparatively small part of the time which passes between fertilisation of the embryo sac and the dispersal of a mature seed. This is easily demonstrated by the ability of immature embryos, dissected from developing seeds, to germinate in culture media containing only simple nutrients. Such precocious germination does not usually occur within seeds which are developing normally on the parent plant, and indeed it is not in the interests of the plant for this to happen, since the maturation stages of development (which comprise that time elapsing after the acquisition of germination potential) are those in which the bulk of seed reserve substances accumulate. Thus a precociously germinating embryo would rapidly exhaust the reserves available to it.

Studies carried out in a wide variety of species indicate that the maintenance of the differentiated embryo in a quiescent (dormant) condition is linked to the action of the plant growth regulator, abscisic acid (ABA). Measurements of ABA, *in situ*, within developing seeds characteristically demonstrate a relatively high concentration of this substance during the period of embryonic maturation, the levels falling as the seed dehydrates and the low water content imposes a physical constraint on germination (King, 1976; Radley, 1976; Prevost and Le Page-Degivry, 1985; Barratt, 1986). Moreover, although immature embryos in culture will germinate precociously, this precocious germination is generally inhibited if the culture medium is supplemented with physiological concentrations of ABA (Eisenberg and Mascharenas, 1985; Quatrano, 1986; Dure *et al.*, 1981). As seed development progresses, it is frequently observed that there occurs a concomitant loss in sensitivity to ABA on the part of the maturing embryo (Morris *et al.*, 1989). At later stages of seed development it is likely that precocious germination is prevented by the physical block placed on the embryo by the lack of availability of water to provide intracellular turgor necessary for cell expansion: the concentration of ABA, exogenously supplied, which is required to prevent germination of a mature (dehydrated) embryo upon imbibition is frequently orders of magnitude greater than that necessary to prevent the precocious germination of an immature embryo removed from the developing seed (Walbot, 1978; Black, 1983; Xu *et al.*, 1990). The responses of immature embryos to ABA and water stress differ between species, and both these components may interact with other factors to prevent precocious germination.

Many studies of the gene products which accumulate in embryos in the presence of exogenous ABA or in response to the imposition of an osmotic stress reveal a pattern of gene expression which is essentially identical with that seen in quiescent embryos undergoing maturation within the seed (Crouch and Sussex, 1981; Quatrano *et al.*, 1983; Dure, 1985; Eisenberg and Mascharenas, 1985; Quatrano, 1986; Xu *et al.*, 1990).

Additionally, there is good genetic evidence to link the action of ABA with the prevention of precocious germination. In *Zea mays* there are a number of well-characterised mutants which exhibit the property of vivipary — the embryos germinate precociously from kernels within the developing ear (Mangelsdorf, 1930; Eyster, 1931). These *viviparous* (*vp*) mutants fall into two classes. Embryos which are homozygous for the *vp-1* mutation germinate precociously because they have a reduced sensitivity to ABA: they appear ABA-non-responsive (McDaniel *et al.*, 1977; Robichaud and Sussex, 1987). Mutations at the remaining *Vp* loci (*Vp-2* to *Vp-9*) cause lesions in the ABA biosynthetic pathway, so that embryos homozygous for the mutant alleles fail to synthesise and accumulate ABA (Robertson, 1955; Robertson *et al.*, 1978; Smith *et al.*, 1978; Neill *et al.*, 1986). Unlike *vp-1/vp-1* homozygotes, precocious germination by embryos of this latter class can therefore be inhibited by the application of exogenous ABA.

What of the genes which are expressed during embryonic maturation? Those which have been characterised fall into at least two distinctive classes. The first of these are proteins whose principal role is that of providing the seed with its primary storage reserves. These include lipids, carbohydrates (principally starch) and storage proteins which are deposited either in membrane-bound organelles within the cell, or in the case of storage lipids, as oil droplets bounded by a surface layer of proteins. The principal storage reserves occur either in the cotyledons of dicotyledonous plants, or in the endosperm of monocotyledonous plants (although this generalisation should be recognised as such: there are dicotyledonous plants in which the endosperm remains a significant site of reserve accumulation, whilst the scutellum of monocotyledonous embryos also contains storage compounds). The deposition of storage reserves during embryogenesis has been found to be influenced both by ABA and by the water status of the developing seed. Thus, whereas the storage globulins accumulating within the embryo may do so in response directly to the application of ABA in some species (Barratt, 1986; DeLisle and Crouch, 1989), in other species the imposition of water stress appears to be a more important factor in determining reserve accumulation (Dure *et al.*, 1981; Barratt *et al.*, 1989). Because of their importance as a food source the genetics, biochemistry and molecular biology of storage reserve deposition and catabolism have been extensively studied. I do not, therefore, propose to address these subjects further in this review. Instead I shall concentrate on the developmental programme which enables the accumulation of reserves — the period of embryonic dormancy and maturation, leading to dehydration of the mature seed.

In considering the history of the human species, perhaps the most significant cultural advance to have been made in the development of society has been the transition from a hunter-gatherer style of life to agriculturally based settlements. The regular cultivation of crops resulted in the accumulation of food reserves surplus to the immediate daily needs of the population, thus providing the resources for population growth and the evolution of domestic economies. Arguably the most important biological phenomenon underpinning this transition was exhibited by the mature seed: that of prolonged viability in the dry state. Selectively advantageous for the dispersal of the seed and the exploitation of the available habitats by the plant, this remarkable characteristic has been coincidentally exploited by the farmer both as a means of storing nutritionally valuable material to meet demands for future consumption and as a resource for the production of the following year's crops.

11. SEED DEVELOPMENT

Mature seeds achieve extraordinary levels of desiccation: moisture contents as low as 5% are frequently encountered, and are considered beneficial for the prolonged storage of viable seed (Priestley, 1986). Such levels of dehydration are lethal to most forms of life, being encountered only in specialised structures such as cysts, seeds and spores. Among higher plants such extremes of desiccation are lethal to vegetative tissues in all but a few striking examples (those plants whose vegetative tissues can withstand periods of desiccation and subsequently recover being aptly termed 'resurrection plants'). In addressing the mechanisms by which seeds may tolerate extreme desiccation, research has focused on the physical and metabolic changes which occur within cells, and the molecular and biochemical changes which occur as maturing seeds commence their preparation for dehydration.

The dry state is characterised by substantial physico-chemical changes within the cell. The withdrawal of free water — the principal solvent within the cell — causes an increase in the concentration of intracellular solutes. It also causes a change in the nature of hydrophobic–hydrophilic interfaces, which in the form of cellular membranes are essential for compartmentation of subcellular components and are the principal sites for bioenergetic conversions. Upon dehydration, the phospholipid bilayers undergo physical rearrangements, resulting in loss of integrity resulting in the opening of substantial 'pores' in the membranes. The existence of such pores is manifest upon the rehydration of dry seeds: during the first minutes of imbibition the water in which dry seeds are placed exhibits an increase in conductivity as ionic solutes leach out through these pores, prior to the reorganisation of the cellular membranes once more into semipermeable barriers (Senaratna and McKersie, 1983). Other potentially deleterious changes in the intracellular environment include damage which may result through the crystallisation of solutes, and the loss of 'bound' water from macromolecules. In addition to its role as a medium for free diffusion in which metabolic reactions may proceed, water is also essential in the maintenance of the structural integrity of macromolecules, particularly proteins, whose higher-order structure may depend on intramolecular hydrogen bonding occurring through water 'bridges' (Edsall and McKenzie, 1983). The stripping of such integral water molecules from proteins can cause their irreversible denaturation as illegitimate hydrophobic interactions result in aggregation.

The acquisition of desiccation tolerance occurs during embryonic maturation. This can be observed in experiments where immature embryos are excised from developing seeds, desiccated and subsequently rehydrated (Bartels *et al.*, 1988). In such an experiment it was observed that over a six-day period the germinability of barley embryos following rapid desiccation (and hence their viability) increased from zero at 12 days after pollination (dap) to 100% at 18 dap. It appears that ABA is implicated in this development of tolerance, just as it is in the maintenance of embryonic dormancy. Thus, when the 12-day-old embryos were cultured in the presence of 10^{-4} M ABA for three days, prior to dehydration, the subsequently rehydrated embryos exhibited a 90% germination rate. By contrast the viability of untreated embryos of the same chronological age (15 dap) remained low (20%). Such a role for ABA in initiating the acquisition of desiccation tolerance has been observed in a number of experimental systems (Anandarajah and McKersie, 1990; Iida *et al.*, 1992), as has a similar effect of the imposition of a level of osmotic stress sufficient to prevent precocious germination (Anandarajah and McKersie, 1990; Attree *et al.*, 1991).

Clearly, therefore, the biochemical events which occur during the period of embryonic maturation and which are potentiated by ABA and osmotic stress are crucial to the survival of the seed in the dry state. It is therefore highly pertinent to examine the changes which occur during this period of development.

Two principal aspects of seed maturation have attracted attention in relation to the enhancement of desiccation tolerance. These are the physical state adopted by the cytosol and its contents, and the genes expressed during the period of acquisition of tolerance. Tissues subjected to osmotic stress initially strive to retain water through achieving an osmotic balance with the medium surrounding them. Osmoregulation is frequently brought about through the accumulation of compatible solutes — compounds which exert an osmotic effect without being deleterious through altering the ionic balance of the cytosol. Among the compounds implicated in such osmoregulatory responses in higher plants are the amino acid proline, the quaternary amino compound betaine, carbohydrates such as sucrose and raffinose, and compounds such as inositol and sorbitol (McCue and Hansen, 1990). In other taxa, these compounds include glucosylglycerol, in cyanobacteria (Erdmann *et al.*, 1992) and trehalose, in a remarkable example of the acquisition of desiccation tolerance (or 'anhydrobiosis') exhibited by the dormant embryos of the brine shrimp (*Artemia salina*). These animals produce encysted embryos which may be stored dry for extended periods, until rehydrated by salt water, whereupon they hatch and commence their life-cycle. These have been found to accumulate a substantial quantity of trehalose — a carbohydrate associated with anhydrobiosis in a number of other organisms (Clegg, 1965) and which has been associated with the preservation of membrane integrity (Crowe *et al.*, 1984; Crowe and Crowe, 1986) — and a ribonucleoprotein complex which through its high affinity for water, and relative abundance retains a significant fraction of the residual water in the cysts (De Herdt *et al.*, 1981).

As desiccation of the developing seed proceeds, the osmotic deficit becomes too great to be overcome by the effects of solutes within the cell and water loss becomes significant. As dehydration becomes extreme, it has been proposed that a change in the physical state of the cytosol occurs, from being essentially liquid to vitreous in nature (Koster and Leopold, 1988): in effect the cytosol adopts the physical nature of a glass — a supercooled liquid — in which diffusion and molecular rearrangement is minimised. The evidence for the existence of such states within dry seeds has been drawn from physical studies (calorimetry and electron spin resonance) of maize, pea and soybean seed: embryos analysed in the desiccation tolerant stage of development showed the characteristic signs of the cytosol having undergone a vitreous transition, whereas embryos in the earlier, desiccation intolerant stage showed no signs of vitrification (Williams and Leopold, 1989; Bruni and Leopold, 1991).

That such changes in the state of a solution can occur is thought to be related to the nature and concentration of the solutes therein, and the principal agents implicated in the process have been sugars and a class of proteins generically termed 'LEA' proteins. Sugars frequently comprise a major fraction of the mass of dry embryos — in soybean seed and in dry wheat embryos sucrose and raffinose alone contribute a quarter of the total (D'Appolonia *et al.*, 1978; Koster and Leopold, 1988). In the leaves of the remarkable resurrection plant, *Craterostigma*, whose vegetative parts can withstand levels of dehydration normally only found in dry seeds, a novel 8-carbon sugar, octulose, forms some 50% of the dry matter (Bianchi *et al.*, 1991). This sugar is converted to sucrose upon desiccation. It is also significant that cells of this species can be propagated in culture, as undifferentiated callus tissue, under which conditions the cytoplasm can be induced to accumulate sucrose, by treatment with ABA, leading to the acquisition of desiccation tolerance (Bianchi *et al.*, 1992). This serves to reinforce the centrality of this plant growth regulator in the coordination of cellular processes in plants which lead to desiccation tolerance, as does the observation that leaves of *Craterostigma* also accumulate a subset of desiccation-specific polypeptides, which share extensive homology with the 'LEA' proteins found in developing seeds (Piatkowski *et al.*, 1990).

The class of proteins termed 'LEA' proteins are characteristically synthesised during the later stages of embryo development. The proteins themselves are relatively abundant, and their generic name derives from an acronym: Late Embryogenic Abundant proteins (Galau *et al.*, 1986, 1987). Within this class of protein a number of disparate molecular species exist, but which share the common feature that their synthesis is both temporally coordinated, and is initiated when immature embryos are treated with ABA. This has led to another frequently used designation of the genes encoding LEA proteins as esponsive to *AB*A (*Rab*) (Mundy and Chua 1988; Mundy, 1989; Skriver and Mundy, 1990). Interestingly, these genes are frequently also expressed in vegetative plant tissues, when these tissues are subjected to water stress — a property which has led to a subset of such proteins also being defined as '*dehydrins*' (Close *et al.*, 1989). This combination of properties has led to the hypothesis that such proteins form an integral part of the plant's drought acclimation response, and that in developing embryos the accumulation of large quantities of such proteins assists the acquisition of desiccation tolerance.

The means by which such proteins have been proposed to contribute to this property vary, depending on their molecular nature. One class of these proteins, first identified among the LEA proteins of cotton embryos, has been predicted (on the basis of their amino acid sequences) to adopt structures typified by the possession of a number of amphiphilic α-helices. These α-helices are composed of hydrophilic and hydrophobic residues disposed so that one face of the helix is predominantly hydrophobic in nature, whilst the opposite face is predominantly hydrophilic. It is postulated that such structures possess the ability simultaneously to interact with hydrophobic surfaces (e.g. biological membranes) and hydrophilic components of the cytosol, thereby enabling the stabilisation of cellular membranes in the dehydrated state (Dure *et al.*, 1989). Another LEA protein with the capacity to interact with other cellular components is an ABA-induced product, identified in maize embryos, which contains a structural domain characteristic of RNA-binding proteins. This might serve to protect parts of the translational machinery (Mortensen and Dreyfuss, 1989).

A second class of LEA protein is typified by perhaps the most abundant LEA protein found in wheat embryos: the wheat 'Em' protein (Grzelczak *et al.*, 1982). Proteins of this class are marked by a high degree of hydrophilicity, due to their being composed of a startlingly high proportion of amino acids with charged and uncharged polar side-groups (Litts *et al.*, 1987). Members of the Em family (distributed widely among both mono- and dicotyledonous plants: Baker *et al.*, 1988; Williams and Tsang, 1991; Delseny, 1991) are interesting in that their highly polar nature endows them with a surprisingly high capacity for binding water (McCubbin *et al.*, 1985). Moreover, although there is a high degree of sequence conservation between members of the Em family in relatively divergent species, this is not accompanied by the possession of common structural motifs: physical analysis indicates the Em polypeptides to adopt a random coil configuration with a high degree of flexibility (McCubbin et al., 1985). This has led to the hypothesis that proteins of this type are able to intercalate between other macromolecular species, and through their remarkable affinity for water, retain a shell of hydration about them, acting as a 'molecular sponge'. This would enable such polypeptides to stabilise other macromolecules by introducing 'shared water' through the formation of a protein–water–protein matrix (Lane, 1991).

Finally, a third class of ABA-induced protein for which a function in desiccation tolerance has been adduced comprises enzymes contributing to the synthesis of cryopreservative carbohydrates. Thus far, two such enzymes have been identified among the polypeptides accumulating in developing barley embryos. The best characterised of these was identified initially as the product encoded by a cDNA sequence, whose predicted sequence possessed homology with an aldose reductase previously found in rat eye lens. The significance of this homology lies in the function of this enzyme, which in mammalian cells catalyses the first step in the formation of sorbitol, a potent osmolyte. Subsequently, aldose reductase activity was identified in maturing barley embryos, and the gene encoding this enzyme was found to be both induced by ABA and the imposition of osmotic stress (Bartels et al., 1991). More recently, the analysis of another barley embryo cDNA sequence has revealed limited homology with bacterial dehydrogenases (glucose dehydrogenase in *Bacillus*, and ribitol dehydrogenase in *Klebsiella*). Although the degree of homology between these sequences was low (c. 30%), the degree of conservation of strategic residues (involved in the active sites of the bacterial enzymes) is strikingly high, fuelling speculation that this product represents a component of a carbohydrate biosynthetic pathway crucial to the development of tolerance (D. Bartels et al., unpublished findings).

It is a feature, then, of the molecular components implicated in desiccation tolerance, that their synthesis is initiated in response either to the dormancy-imposing plant growth regulator, ABA, or to the imposition of osmotic stress. It is consequently highly relevant to consider the molecular mechanisms utilised by developing embryos in the interpretation of these cues, and their subsequent responses. Little is known about the mode of action of plant growth regulators, particularly as regards their initial interactions with target cells. Despite many years of research, the unambiguous identification of plant hormone receptors has not been achieved (whilst there is increasingly strong circumstantial evidence for the identification of proteins which fulfil many of the criteria for auxin receptors, these have yet to be unequivocally identified as such). As a result, most studies have focused on the molecular details of hormone-regulated gene expression. In developing seeds, the actions of ABA and of osmotic stress have been demonstrated to elicit the expression of both storage proteins and of LEA proteins, and some components both of the transcriptional machinery and of the sensory response pathway have been identified.

Abscisic acid has long been implicated in responses to water stress, particularly in vegetative tissues where one of the most rapid physiological responses to water deficit – the closure of stomatal apertures – has been shown to be mediated by ABA. In this response, the subjection of tissues to an osmotic stress causes the rapid accumulation of ABA within the leaves. This ABA appears to act on the stomatal guard cells by triggering a release of intracellular calcium ions which are thought to

11. SEED DEVELOPMENT

act as 'second messengers' in a signal transduction chain ultimately causing the swelling of the guard cells and the closure of stomata (McAinsh *et al.*, 1990; Schroeder and Hagiwara, 1990).

This response system, in which a physical stress causes its effects through an increase in a plant growth regulator (Fig. 11.7) has been taken as a paradigm for the mode of action of osmotic stress in eliciting specific gene expression in developing embryos (Robertson *et al.*, 1989; Wilen *et al.*, 1990). The available evidence suggests that this conception may not be universal. Many of the proteins which are synthesised during embryonic maturation disappear from the mature embryo in the course of germination. However, the expression of the genes encoding them can be reinitiated if the young seedling is subsequently subjected either to dehydration, or to an application of exogenous ABA. In cases where seedlings have been caused to recapitulate the expression of genes encoding LEA proteins through the imposition of osmotic stress, concomitant measurements of endogenous ABA have sometimes revealed significant increases in the levels of the growth regulator (up to 10-fold: Gomez *et al.*, 1988). Whilst these observations support a model in which dehydration acts through an elevation of endogenous ABA to cause gene expression, they cannot universally be extended to cover the response of developing embryos.

Studies on the elicitation of storage protein synthesis in oilseed rape demonstrated that although levels of the mRNAs encoding the principal storage proteins, cruciferin and napin, were precociously elevated when immature embryos were cultured in the

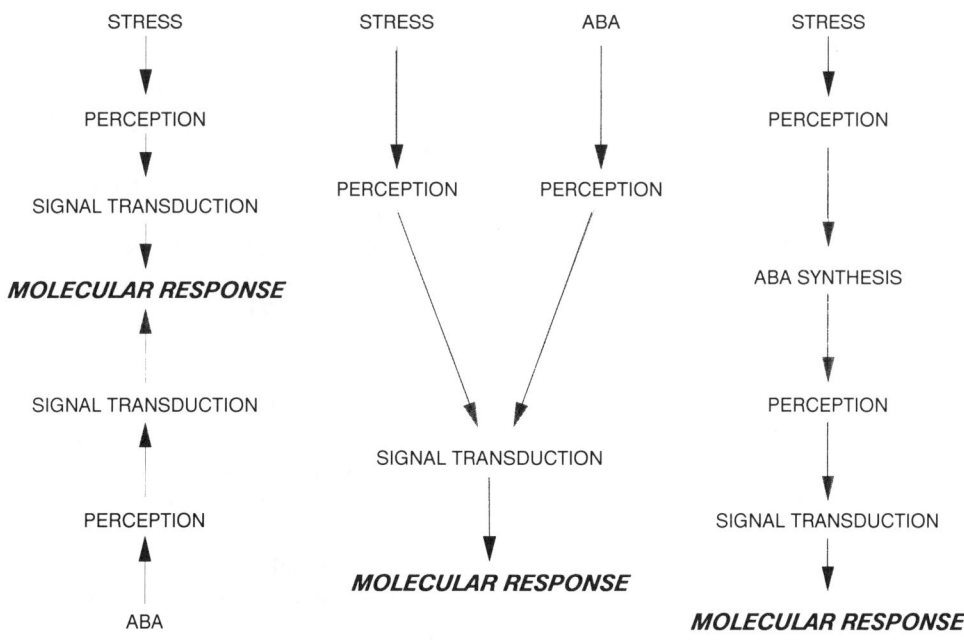

Fig. 11.7. Alternative pathways for the perception of and response to osmotic stress by plant cells.

presence of an osmotic agent, no increase occurred in the endogenous ABA concentration (Finkelstein and Crouch, 1986) and in developing pea pods the inhibition of ABA biosynthesis does not prevent the accumulation of LEA proteins in response to osmotic stress (Barratt *et al.*, 1989). Similarly, studies of the accumulation of 'Em' transcripts in immature wheat embryos revealed a similar pattern: the imposition of osmotic stress caused a rapid increase in mRNA, even in the presence of the herbicide norflurazon — an inhibitor of ABA biosynthesis. Interestingly, however, the levels of Em mRNA which accumulated in response to stress were much lower in embryos which had been exposed to norflurazon for extended periods, and in which the pre-existing ABA concentration had declined to undetectable levels (Morris *et al.*, 1990). These experiments suggest that whilst the stress–response pathway does not necessarily require an increase in endogenous ABA, there may be a necessity for the presence of this growth regulator in interpreting the perception of stress at the level of expression of individual genes.

The ABA-responsiveness of the cereal Em genes, and of a large number of other ABA-responsive genes including 'dehydrins' and '*rab*' genes resides, in part, in the nature of *cis*-acting sequence elements located in the promoters of these genes (Marcotte *et al.*, 1989). Thus, a large number of genes encoding LEA proteins have been found to share a common sequence motif lying within 200 base pairs of their 'TATA box' elements. Characteristically, this motif comprises an element known generically as a 'G-box': XACGTG, frequently followed by a G/C-rich motif. This motif has been demonstrated to contain the minimal information required to confer ABA-inducibility on a gene, in an elegant series of experiments in which the appropriate sequence was subcloned from a wheat 'Em' gene and placed upstream of a reporter gene comprising a truncated Cauliflower Mosaic Virus '35S' promoter (containing only a functional TATA box) fused with the bacterial gene encoding the enzyme β-glucuronidase (GUS). When this recombinant construct was used to transfect cereal protoplasts, GUS activity was found to be induced in response to treatment with ABA. Mutation of the G-box element revealed that alteration of two bases was sufficient to abolish this ABA-responsiveness.

Subsequently, the specificity of the ABA-responsive element (ABRE) has been demonstrated to be mediated through its interaction with a nuclear transcription factor — a member of the 'leucine zipper' class (Guiltinan *et al.*, 1990). Such factors consist of a DNA-binding domain and a leucine-rich α-helical domain through which they may form dimers with other proteins of the same class. This itself is significant, because the G-box sequence is by no means restricted to ABA-responsive genes, but has been observed to be present in the promoters of a number of plant genes whose regulation is ABA-independent (for example, a number of light-regulated photosynthetic genes: Schindler *et al.*, 1992). Some of the specificity in ABA responses may therefore reside in the available types of G-box-binding factors present in particular cell types, the nature of the dimeric complexes formed (heterodimers or homodimers), and interaction with other nuclear proteins causing their 'activation'.

Direct evidence for the necessity of the agency of additional nuclear factors in ABA and stress-responsive gene expression derives from studies of the maize viviparous mutant, *vp-1*. Phenotypically, *vp-1/vp-1* homozygotes appear ABA-non-responsive, in that immature embryos in culture will germinate in the presence of high concentrations of ABA (Neill *et al.*, 1987). Precocious germination of *vp-1/vp-1* embryos can only be arrested by the imposition of severe osmotic stress, which physically blocks germination (Butler and Cuming, 1993). (Germination requires cell expansion, which in turn is mediated through water uptake and consequent turgor-pressure. When embryos are incubated in the presence of an osmotic agent, there is net water loss, a reduction in turgor pressure and consequently no cell expansion.) Neither the imposition of stress, nor the application of ABA permit the expression of 'Em' gene expression in *vp-1/vp-1* maize mutants, conclusively demonstrating the centrality of the *VP-1* gene product in the signal transduction pathway leading from stress- and ABA-perception to Em gene expression, and the requirement for the stress-induction of gene expression in developing embryos to be conducted through an ABA-dependent pathway (McCarty *et al.*, 1989a; Butler and Cuming, 1993).

Direct evidence of the role of the *VP-1* gene product has been drawn from experiments which followed the cloning of this important regulatory gene by transposon tagging (McCarty *et al.*, 1989a). Whilst sequence analysis proved relatively uninformative about the possible nature of the *VP-1* gene product, experiments to test the function of *VP-1* proved more fruitful (McCarty *et al.*, 1991). A first indication that the *VP-1* protein might act as a transcriptional regulator followed experiments in which protoplasts derived from (wild-type) maize suspension culture cells were transfected with a recombinant plasmid in which the GUS reporter gene was under the control of the wheat Em gene. The ABA-regulated promoter conferred a moderate level of GUS gene expression when the protoplasts were stimulated by ABA, demonstrating the general ABA-responsiveness of the promoter fragment. However, when the protoplasts were co-transfected with both the Em-GUS construct and a second plasmid containing the *VP-1* coding sequence under the control of the powerful CaMV 35S promoter, then the level of ABA-stimulated and Em-driven GUS expression was greatly enhanced. This synergy between the overexpression of the *VP-1* gene and the effect of ABA on the Em promoter indicated that transcriptional activation of the ABA-regulated gene was occurring through the agency of the *VP-1* gene product.

This experiment did not by itself unambiguously identify the *VP-1* protein as a transcription factor. Direct evidence was obtained as a result of a sophisticated 'domain-switching' experiment. The design of this second experiment was suggested by the sequence of the *VP-1* gene. Although this possessed no obvious homologies with previously sequenced genes, it predicted that the *VP-1* protein comprised an *N*-terminal domain with a strong net negative charge and a *C*-terminal domain with an overall positive charge, the two regions of the polypeptide separated by a central, uncharged region. Regions with net negative charge have been found to occur in a number of proteins which function as transcription factors, and when it was observed that the amino acid sequence in this region of the *VP-1* protein had the potential to form a pair of amphipathic α-helices (also a feature of transcriptional activator sequences) a direct test of this function could be devised.

The experiment comprised the fusion of the *VP-1 N*-terminal acidic domain with the DNA-binding domain of the yeast transcription factor *GAL-4*, producing an 'activator construct'. The fusion gene was then cloned in a plasmid under the control of the CaMV 35S promoter to enable its transcription in plant cells. As a reporter for the activity of this construct, the chloramphenicol acetyl transferase (CAT) gene was cloned in a second plasmid, under the control of a TATA box and a sequence containing the *GAL-4* binding site. When the two plasmids were used to co-transfect carrot protoplasts, CAT activity was detected, indicating that transcription of the reporter gene had occurred. Control experiments demonstrated that this transcription was absolutely dependent on the presence of the *VP-1* sequence in the 'activator plasmid' and thereby supported the identification of the *VP-1* protein as an activator of transcription.

The *VP-1* gene therefore appears to be a stage-specific regulator of central importance in the programming of seed development, particularly as regards the operation of the switch between the embryogenetic developmental programme and the germinative programme. It appears to fulfil this role through controlling the expression of genes active during embryogenesis, presumably through interacting with the transcriptional machinery. The precise way in which this occurs is not yet known. There is (at the time of writing) no evidence to suggest that the *VP-1* protein is a DNA-binding protein, for example. It may interact at one step removed from the responsive DNA sequences in ABA- and stress-regulated genes by modifying the action of the DNA-binding proteins which are present in plant nuclei throughout development, causing them to activate transcription at only the appropriate time or place.

Additionally, it is likely that this gene is only one of several implicated in regulating embryo maturation and in linking the perception of stimuli such as ABA and osmotic stress with the expression of stage-specific genes. This conclusion is based on observations of the effects of different allelic variants of *VP-1*, exhibiting a range of differentially sensitive ABA-response phenotypes. These include mutants which fail to show vivipary, but which are blocked in the expression of *VP-1*-dependent genes (such as the *C1* gene which forms a part of the anthocyanin biosynthetic regulatory hierarchy in maize aleurone cells: McCarty *et al.*, 1989b; Hattori *et al.*, 1992). Conversely, a number of ABA- and osmotic stress-induced products have recently been identified which continue to accumulate in fully viviparous *vp-1/vp-1* embryos.

Thus although the Em polypeptide is not synthesised either in response to stress or to ABA in these embryos, mRNAs encoding a number of 28–30 kDa polypeptides have been found to be induced by ABA, in the otherwise 'insensitive' mutant (Pla *et al.*, 1991; Thomann *et al.*, 1992; Butler and Cuming, 1993). Similarly, these mRNAs, together with additional transcripts encoding a range of 'dehydrin'-like 20 kDa polypeptides, accumulate in response to osmotic stress (a treatment which, unlike the application of ABA, blocks precocious germination). These observations indicate that the signal transduction chains which link ABA and stress-perception are not simple linear communications, but exist as a branched network of interacting communicating factors within the cell. Within this network, the *VP-1* gene product would appear to have an important part to play, but considering its nature as a transcriptional activator it is most likely to be situated at a late point in the chain, with branch-points occurring at earlier positions. The identification of the components of the 'intracellular wiring diagram' remains one of the outstanding challenges in the study of seed development, and one in which the combination of genetic and molecular experimental techniques offers the greatest opportunities for success.

REFERENCES

Abbé, E. C. and Stein, O. L. (1954). *Am. J. Bot.* **41**, 285-293.
Anandarajah, K. and McKersie, B. D. (1990). *Plant Cell Rep.* **9**, 451-455.
Attree, S. M., Moore, D., Sawhney, V. K. and Fowke, L. C. (1991). *Ann. Bot.* **68**, 519-525.
Baker, J., Steele, J. and Dure, L. S. III (1988). *Plant Mol. Biol.* **11**, 277-291.
Barratt, D. H. P. (1986). *Ann. Bot.* **57**, 245-256.
Barratt, D. H. P., Whitford, P. N., Cook, S. K., Butcher, G. and Wang, T. L. (1989). *J. Exp. Bot.* **40**, 1009-1014.
Bartels, D., Singh, M. and Salamini, F. (1988). *Planta* **175**, 485-492.
Bartels, D., Engelhardt, K., Roncarati, R., Schneider, K., Rotter, M. and Salamini, F. (1991). *EMBO J.* **10**, 1037-1043.
Bianchi, G., Gamba, A., Murelli, C., Salamini, F. and Bartels, D. (1991). *Plant J.* **1**, 355-359.
Bianchi, G., Gamba, A., Murelli, C., Salamini, F. and Bartels, D. (1992). *Phytochemistry* **31**, 1917-1922.
Black, M. (1983). *In* "Abscisic Acid" (F. T. Addicott, ed.), pp. 331-363. Praeger, New York.
Borkird, C., Choi, J. H., Jin, Z-H., Franz, G., Hatzopoulos, P., Chorneau, R., Bonas, U., Pelegri, F. and Sung, Z. R. (1988). *Proc. Natl. Acad. Sci. USA* **85**, 6399-6403.
Bowman, J. L., Smyth, D. R. and Meyerowitz, E. M. (1989). *Plant Cell* **1**, 37-52.
Breton, A. M. and Sung, Z. R. (1982). *Dev. Biol.* **90**, 58-66.
Bruni, F. and Leopold, A. C. (1991). *Plant Physiol.* **96**, 660-663.
Butler, W. M. and Cuming, A. C. (1993). *Planta* **189**, 47-54.
Choi, J. H., Liu, L-S., Borkird, C. and Sung, Z. R. (1987). *Proc. Natl. Acad. Sci. USA* **84**, 1906-1910.
Clark, A. J., Higgins, P., Martin, H. and Bowles, D. J. (1991). *Eur. J. Biochem.* **199**, 115-121.
Clark, J. K. and Sheridan, W. F. (1991). *Plant Cell* **3**, 935-951.
Clegg, J. S. (1965). *Comp. Biochem. Physiol.* **14**, 135-143.
Close, T. J., Kortt, A. A. and Chandler, P. M. (1989). *Plant Mol. Biol.* **13**, 95-108.
Cordwener, J. H. G., Booij, H., van der Zandt, H., van Engelen, F. A., van Kammen, A. and De Vries S. C. (1991). *Planta* **184**, 478-486.
Crouch, M. L. and Sussex, I. M. (1981). *Planta* **153**, 64-74.
Crowe, J. H. and Crowe, L. M. (1986). *In* "Membranes, Metabolism and Dry Organisms" (A. C. Leopold, ed.). Comstock Publishing Assoc., Ithaca, NY.
Crowe, J. H., Crowe, L. M. and Chapman, D. (1984). *Science* **223**, 701-703.
D'Appolonia, B. L., Gilles, K. A., Osman, E. and Pomeranz, Y. (1978). *In* "Wheat Chemistry and Technology" (Y. Pomeranz, ed.). American Association of Cereal Chemists, St Paul, MN.
De Herdt, E., De Voeght, F., Clauwert, J., Kondo, M. and Slegers, H. (1981). *Biochem. J.* **194**, 9-17.
DeJong, A. J., Cordwener, J. H. G., LoSchiavo, F., Terzi, M., Vanderkerchove, J., van Kammen, A. and De Vries, S. C. (1992). *Plant Cell* **4**, 425-433.
DeLisle, A. J. and Crouch, M. L. (1989). *Plant Physiol.* **91**, 617-623.
Delseny, M. (1991). *J. Exp. Bot.* **42** (Suppl), P6.7.
De Vries, S. C., Booij, H., Janssens, R., Vogels, R., Saris, L., LoSchiavo, F., Terzi, M. and van Kammen, A. (1988). *Genes Dev.* **2**, 462-476.
Dure, L. S. III (1985). *In* "Oxford Surveys of Plant Molecular and Cell Biology", Vol. 2 (B. J. Miflin, ed.), pp. 179-197. Oxford University Press, Oxford.
Dure, L. S. III, Greenway, S. C. and Galau, G. A. (1981). *Biochemistry* **20**, 4162-4168.
Dure, L. S. III, Crouch, M., Harada, J., Ho, T-H. D., Mundy, J. C., Quatrano, R. S., Thomas, T. and Sung, Z. R. (1989). *Plant Mol. Biol.* **12**, 475-486.
Edsall, J. T. and McKenzie, H. A. (1983). *Adv. Biophys.* **16**, 53-183.
Eisenberg, A. J. and Mascharenas, J. P. (1985). *Planta* **166**, 505-514.
Erdmann, N., Fulda, S. and Hagemann, M. (1992). *J. Gen. Microbiol.* **138**, 363-368.
Errampalli, D., Patton, D., Castle, L., Mickelson, L., Hansen, K., Schnall, J., Feldman, K. and Meinke, D. W. (1991). *Plant Cell* **3**, 149-157.
Eyster, W. H. (1931). *Genetics* **16**, 574-590.

Feldmann, K. A. and Marks, M. D. (1987). *Mol. Gen. Genet.* **208**, 1-9.
Finkelstein, R. and Crouch, M. (1986). *Plant Physiol.* **81**, 907-912.
Galau, G. A., Hughes, D. W. and Dure, L. S. III (1986). *Plant Mol. Biol.* **7**, 155-170.
Galau, G. A., Bijaisoradat, N., and Hughes, D. W. (1987). *Dev. Biol.* **123**, 198-212.
Giuliano, G., LoSchiavo, F. and Terzi, M. (1984). *Theor. Appl. Genet.* **67**, 179-183.
Gomez, J., Sanchez-Martinez, D., Stiefel, V., Rigau, J., Puigdomenech, P. and Pagès, M. (1988). *Nature* **324**, 262-264.
Grzelczak, Z. F., Sattolo, M. H., Hanley-Bowdoin, L. K., Kennedy, T. D. and Lane, B. G. (1982). *Can. J. Biochem.* **60**, 389-397.
Guiltinan, M. J., Marcotte, W. R. Jr and Quatrano, R. S. (1990). *Science* **250**, 267-271.
Hattori, T., Vasil, V., Rosenkrans, L., Hannah L. C., McCarty, D. R. and Vasil, I. K. (1992). *Genes Dev.* **6**, 609-618.
Higgins, P. and Bowles, D. J. (1990). *Plant Sci.* **69**, 239-247.
Iida, Y., Watabe, K., Kamada, H. and Harada, H. (1992). *J. Plant Physiol.* **140**, 356-360.
Jürgens, G. (1992). *Science* **256**, 487-488.
Jürgens, G., Mayer, U., Torres Ruiz, R. A., Berleth, T. and Miséra, S. (1991). *Development* **91** (Suppl. 1), 27-38.
King, R. W. (1976). *Planta* **132**, 43-57.
Knox, J. P. (1992). *Protoplasma* **167**, 1-9.
Knox, J. P. and Roberts, K. (1989) *Protoplasma* **152**, 123-129.
Koster, K. L. and Leopold, A. C. (1988). *Plant Physiol.* **88**, 829-832.
Lane, B. G. (1991). *FASEB J.* **5**, 2893-2901.
Lindsey, K. and Topping, J. F. (1993). *J. Exp. Bot.* **44**, 359-374.
Lindsay, K., Wei, W., Clarke, M. C., McArdle, M. F., Rooke, L. M. and Topping, J. F. (1993). *Transgenic Res.* **2**, 33-47.
Litts, J. C., Colwell, G. W., Chakerian, R. L. and Quatrano, R. S. (1987). *Nucl. Acids Res.* **15**, 3607-3618.
LoSchiavo, F., Giuliano, G., de Vries, S. C., Genga, A., Bollini, R., Pitto, L., Cozzani, F., Nutironchi, V. and Terzi, M. (1990). *Mol. Gen. Genet.* **233**, 385-393.
Maheshwari, P. (1950). "An Introduction to the Embryology of Angiosperms". McGraw-Hill, New York.
Mangelsdorf, P. C. (1930). *Genetics* **15**, 462-494.
Marcotte, W. R. Jr, Russell, S. H. and Quatrano, R. S. (1989). *Plant Cell* **1**, 969-976.
Mayer, U., Torres Ruiz, R. A., Berleth, T., Miséra, S. and Jürgens, G. (1991). *Nature* **353**, 402-407.
McAinsh, M. R., Brownlee, C. and Hetherington, A. M. (1990). *Nature* **343**, 186-188.
McCarty, D. R., Carson, C. B., Stinard, P. S. and Robertson, D. S. (1989a) *Plant Cell* **1**, 523-532.
McCarty, D. R., Carson, C. B., Lazar, M. and Simonds, S. C. (1989b). *Dev. Genet.* **10**, 473-481.
McCarty, D. R., Hattori, T., Carson, C. B., Vasil, V. and Vasil, I. K. (1991). *Cell* **66**, 895-905.
McCubbin, W. D., Kay, C. M. and Lane, B. G. (1985). *Can. J. Biochem. Cell Biol.* **63**, 803-811.
McCue, K. F. and Hanson, A. D. (1990). *Trends Biotechnol.* **8**, 358-362.
McDaniel, S., Smith, J. D. and Price, H. J. (1977). *Maize Gen. Coop. Newslett.* **51**, 85-86.
Meinke D. W. (1991a). *Plant Cell* **3**, 857-866.
Meinke D. W. (1991b). *Dev. Genet.* **12**, 382-392.
Meyerowitz, E. M. and Pruitt, R. E. (1985). *Science* **229**, 1214-1218.
Morris, C. F., Moffatt, J. M., Sears, R. G. and Paulsen, G. M. (1989). *Plant Physiol.* **90**, 643-647.
Morris, P. C., Kumar, A., Bowles, D. J. and Cuming, A. C. (1990). *Eur. J. Biochem.* **190**, 625-630.
Mortensen, E. and Dreyfuss, G. (1989). *Nature* **337**, 312.
Mundy, J. C. (1989). *Plant Mol. Biol. Rep.* **7**, 276-283.
Mundy, J. C. and Chua, N.-H. (1988). *EMBO J.* **7**, 2279-2286.
Neill, S. J., Horgan, R. and Parry, A. D. (1986). *Planta* **169**, 87-96.
Neill, S. J., Horgan, R. and Rees, A. F. (1987). *Planta* **171**, 358-364.

Neuffer, M. G. and Sheridan, W. F. (1980). *Genetics* **95**, 929–944.
Nüsslein-Volhard, C. and Wieschaus, E. (1980). *Nature* **287**, 795–801.
Piatkowski, D., Schneider, K., Salamini, F. and Bartels, D. (1990). *Plant Physiol.* **94**, 1682–1688.
Pla, M., Gomez, J., Goday, A. and Pagès, M. (1991). *Mol. Gen. Genet.* **230**, 394–400.
Prevost, I. and Le Page-Degivry, M. Th. (1985). *J. Exp. Bot.* **36**, 1900–1905.
Priestley, D. A. (1986). "Seed Ageing." Comstock Publishing Assoc., Ithaca, NY.
Quatrano, R. S., Ballo, B. L., Williamson, J. D., Hamblin, M. T. and Mansfield, M. (1983). *In* UCLA Symp. Mol. Cell. Biol. N.S., Vol. 12: "Plant Molecular Biology" (R. B. Goldberg, ed.), pp. 343–353. Alan R. Liss, New York.
Quatrano, R. S. (1986). *In* "Oxford Surveys of Plant Molecular and Cell Biology", Vol. 3 (B. J. Miflin, ed.), pp. 467–477. Oxford University Press, Oxford.
Radley, M. (1976). *J. Exp. Bot.* **27**, 1009–1021.
Reinert, J. (1959). *Planta* **53**, 318–333.
Robertson, D. S. (1955). *Genetics* **40**, 745–760.
Robertson, D. S. (1978). *Mutation Res.* **51**, 21–28.
Robertson, D. S., Anderson, I. C. and Bachmann, M. D. (1978). *In* "Maize Breeding and Genetics" (D. B. Walden, ed.), pp. 461–494. John Wiley & Sons, New York.
Robertson, M., Walker-Simmons, M. K., Munro, D. and Hill, R. D. (1989). *Plant Physiol.* **91**, 415–420.
Robichaud, C. and Sussex, I. M. (1987). *J. Plant Physiol.* **130**, 181–188.
Saiki, R. K., Scharf, S., Faloona, F., Mullis, K. B., Horn, G. T., Erlich, H. A. and Arnheim, N. (1985). *Science* **230**, 1350–1354.
Schindler, U., Menkens, A. E., Beckmann, H., Ecker, J. R. and Cashmore, A. R. (1992). *EMBO J.* **11**, 1261–1273.
Schneider, T., Dinkins, R., Robinson, K., Shellhammer, J. and Meinke, D. W. (1989). *Dev. Biol.* **131**, 161–167.
Schroeder, J. I. and Hagiwara, S. (1990). *Proc. Natl. Acad. Sci. USA* **87**, 9305–9309.
Senaratna, T. and McKersie, R. D. (1983). *Plant Physiol.* **72**, 911–914.
Sheridan, W. F. and Clark, J. K. (1987). *Trends Genet.* **3**, 3–6.
Skriver, K. and Mundy, J. C. (1990). *Plant Cell* **2**, 503–512.
Smith, J. D., McDaniel, S. and Lively, S. (1978). *Maize Gen. Coop. Newslett.* **52**, 107–108.
Smith, L. M., Handley, J., Li, Y., Martin, H., Donovan, L. and Bowles, D. J. (1992). *Plant Mol. Biol.* **20**, 255–266.
Somerville, C. (1989). *Plant Cell* **1**, 1131–1135.
Stacey, N. J., Roberts, K. and Knox, J. P. (1990). *Planta* **180**, 285–292.
Sterk, P., Booij, H., Schellekens, G. A., van Kammen, A. and de Vries, S. C. (1991). *Plant Cell* **3**, 907–921.
Steward, F. C., Mapes, M. O. and Smith, J. (1958). *Am. J. Bot.* **45**, 693–703.
Thomann, E. B., Sollinger, J., White, C. and Rivin, C. J. (1992). *Plant Physiol.* **99**, 607–614.
Topping J. F., Wei, W. and Lindsey, K. (1991). *Development* **112**, 1009–1019.
Van Engelen, F. A. and de Vries, S. C. (1992). *Trends Genet.* **8**, 66–70.
Van Engelen, F. A., Sterk, P., Booij, H., Cordwener, J. H. G., Rook, W., van Kammen, A. and de Vries, S. C. (1991). *Plant Physiol.* **96**, 705–712.
Walbot, V. (1978). *In* "Dormancy and Developmental Arrest" (M. E. Clutter, ed.), pp. 113–116. Academic Press, New York.
Widholm, J. M. (1984). *Plant Mol. Biol. Rep.* **2**, 45–53.
Wilen, R. W., Mandel, R. M., Pharis, R. P., Holbrook, L. A. and Moloney, M. M. (1990). *Plant Physiol.* **94**, 875–881.
Williams, B. and Tsang, A. (1991). *Plant Mol. Biol.* **16**, 919–923.
Williams, R. J. and Leopold, A. C. (1989). *Plant Physiol.* **85**, 977–981.
Xu, N., Coulter, K. M. and Bewley, J. D. (1990). *Planta* **182**, 382–390.

12 Molecular and Genetic Analysis of Tomato Fruit Development and Ripening

JAMES J. GIOVANNONI

Department of Horticultural Sciences, Texas A & M University, College Station, TX 77843, USA

I.	Introduction	252
	A. Commonalities among species	252
	B. Climacteric versus non-climacteric ripening	252
	C. Ethylene: the gaseous ripening hormone	253
II.	Tomato as a model system for fruit development	254
	A. Tomato fruit ripening	254
	B. Ripening mutants	255
	C. Additional attributes of tomato as a model system	257
III.	Isolation of ripening-related genes	257
	A. Differential screening	257
	B. Ripening-related genes cloned on the basis of function	260
IV.	Determination of the physiological functions of ripening genes	263
	A. Antisense gene repression	263
	B. Co-suppression of gene expression	271
	C. Engineering of pathway blockage	272
	D. Complementation	272
V.	A model for induction and regulation of ripening	274
VI.	Strategy and progress toward cloning *rin*, *nor* and *Nr*	276
	A. Map-based cloning	276
	B. Isolation of markers tightly linked to a target locus	277
	C. Pooled sample analysis simplifies genetic mapping	280
VII.	Summary	281
	References	281

I. INTRODUCTION

Fruit development and ripening represent the manifestation of genetic programmes unique to the plant kingdom. As such, analysis of these processes has provided opportunities for both expansion of our collective grasp of fundamental plant biology and enhancement of agricultural productivity. In addition, researchers in these and related areas have made considerable contributions toward development of general plant molecular and genetic techniques including, most recently, antisense gene repression. The purpose of the following pages will be to present a summary of timely observations which have broadened our understanding of fruit development, particularly ripening, with emphasis on the contributions of these efforts toward expanding methodologies available for addressing additional questions of plant molecular biology.

For reasons which will be detailed shortly, much effort in recent decades has centred on fruit development, and especially ripening, of the cultivated tomato (*Lycopersicon esculentum*). Consequently, this biological system will be the logical focus of the following pages. Nevertheless, one must not lose sight of the potential benefits of expanding and supporting on-going research on fruit development and ripening of additional model species.

A. Commonalities Among Species

A stroll through the produce section of any market will convince most casual observers that fruit morphology, developmental patterns and ripening characteristics vary widely among plants. Nevertheless there are numerous unifying themes, and in many cases similar or identical biochemical processes, shared by maturing and ripening fruit of the most highly divergent species.

The evolved purposes of the fruit organ itself and completion of the ripening process are in the most general sense similar for many species. Regarding fleshy fruits such as tomato, peach and strawberry, the fruit is merely an expanded ovary initially supporting the developing seeds during the process of maturation. Ripening, the terminal phase of fruit development, ensues and can be loosely characterised as the summation of developmental changes selected to enhance the possibilities of seed dispersal and thus continued propagation of the species. Specific changes may occur in turgor or pathogen susceptibility which directly facilitate seed release from the fruit organ. In addition, changes often occur in texture, aroma, colour and flavour which render the fruit desirable and palatable for birds and other animals serving as seed dispersal vectors. In short, such animals provide the invaluable functions of seed liberation from the fruit organ, mobility from their site of origination and finally, in some instances, the biochemical pretreatment necessary for successful germination.

B. Climacteric Versus Non-climacteric Ripening

Historically, the ripening process of fruits from different species has been categorised into two general types differing in the occurrence or absence of the respiratory climacteric (Baile and Young, 1981). Following fruit expansion and seed maturation, climacteric fruit exhibit a dramatic increase in respiration concomitant with the onset of the ripening process. An additional marker for the initiation of climacteric fruit ripening is a marked and rapid increase in ethylene biosynthesis (McGlasson, 1985). The climacteric ethylene burst can be either preceded by a gradual increase in ethylene concentration in fruits such as banana, or display a sudden spike in ethylene evolution at the onset of ripening in other fruits including avocado and apple (Yang, 1985). The climacteric ethylene burst has been demonstrated to be autocatalytic in nature, thus responding positively to developmentally coordinated ethylene production.

In contrast to fruit which display climacteric ripening characteristics, non-climacteric fruit such as citrus, strawberry and grape fail to demonstrate increases in respiration or ethylene biosynthesis significantly above normally occurring preripening levels. It is important to note however, that although ethylene biosynthesis is not significantly induced in non-climacteric fruit, this hormone may still play an important regulatory role in the ripening process. Trewavas (1982) proposed that ethylene and other hormones may exert their influence, not only through altered concentration, but also via changes in a target tissue's ability to perceive steady state and/or relatively low levels of a given growth regulator. Several years following this initial hypothesis, Lincoln and Fischer (1988a) described an ethylene-inducible tomato fruit ripening-related gene (E17) which responds to ethylene, at least in part, via altered sensitivity to the hormone.

C. Ethylene: The Gaseous Ripening Hormone

Ethylene is an important plant growth regulator with wide ranging effects on plant development. Treatments of whole plants or plant tissues with ethylene or with inhibitors of ethylene synthesis or action have provided much insight into the role and effect of this gaseous hormone on plant processes. Exposure of seedlings to exogenous ethylene results in swelling of the stem and inhibition of stem elongation along with curvature of the hypocotyl (triple response), while treatment of mature plants fosters epinasty and root formation. In addition, senescence, abscission and fruit ripening have been accelerated with the application of external ethylene (Burg, 1962). Reciprocal experiments employing inhibitor compounds which block ethylene biosynthesis (AVG; aminoethoxyvinylglycine), or action (silver ions, CO_2, norbornadiene) confirm the role of ethylene in the regulation of these and other developmental processes (Beyer, 1981; Yang, 1985). Finally, ethylene biosynthesis is induced endogenously as a general stress response to a variety of environmental stimuli including mechanical injury, drought and pathogen infection (Abeles, 1973). Induced ethylene is known to function at least in part as a coordinator of the expression of stress response genes including proteinase inhibitors (Margossian *et al.*, 1988), cell wall hydrolases (Cass *et al.*, 1990; Percival *et al.*, 1991) and a variety of pathogenesis related (PR) proteins of both known and unknown function (Ecker and Davis, 1987 and references therein).

The biochemical pathway for the production of ethylene has been elucidated and shown to be responsible for most, if not all, ethylene biosynthesis in higher plants (Yang and Hoffman, 1984). The rate limiting process in this pathway is a tissue's capability to convert *s*-adenosyl methionine (SAM) to 1-aminocyclopropane-1-carboxylic acid (ACC) (Yang, 1980). The catalyst for this regulatory step is ACC synthase, an enzyme requiring pyrodoxal phosphate for maximal activity (Yu *et al.*, 1979). ACC is subsequently oxidised to ethylene, CO_2 and cyanide via the ethylene-forming enzyme, ACC oxidase, while the remainder of the modified amino acid skeleton resulting from ACC synthase activity is recycled back into methionine. The net result is a cyclic pathway regulated at the step catalysing ACC biosynthesis and otherwise capable of yielding high rates of ethylene production without depleting a given tissue of its methionine reserves. Finally, it is important to note that the action of a significant alternative pathway for ethylene biosynthesis in higher plants remains to be demonstrated.

II. TOMATO AS A MODEL SYSTEM FOR FRUIT DEVELOPMENT

A. Tomato Fruit Ripening

Tomato has long served as the system of choice for analysis of fruit development and ripening. One major factor contributing to this preference among fruit biologists is the pronounced ripening characteristics displayed by the cultivated tomato (*Lycopersicon esculentum*). The most noticeable phenotype of ripening tomato fruit is the dramatic change in colour occurring within the maturing plastids (Grierson, 1985). Ripening related chlorophyll degradation is coordinated with plastid accumulation of a variety of carotenoid pigments including β-carotene and the characteristic red carotenoid, lycopene (for a review of carotenoid biochemistry, see Goodwin, 1980). The net result is the conversion of chloroplasts to chromoplasts as the highly organised photosynthetic apparatus of the plastids is disrupted and carotenoids accumulate in crystalline form (Grierson *et al.*, 1987).

Genes encoding proteins responsible for numerous catalytic steps in the biochemical pathway for carotenoid biosynthesis have been cloned from prokaryotes (Bartley and Skolnik, 1989; Armstrong *et al.*, 1989, 1990; Misawa *et al.*, 1990) and fungi (Schmidhauser *et al.*, 1990). However, none of these sequences have proven useful in the isolation of plant genes with corresponding function (Pecker *et al.*, 1992). Nevertheless, tomato fruit phytoene desaturase has been cloned using a homologous probe isolated from the cyanobacteria *Synechococcus* (Pecker *et al.*, 1992). In addition, phytoene synthase, which catalyses the first committed step in carotenoid biosynthesis, has been cloned from ripening tomato fruit using differential hybridisation of a ripening fruit cDNA library with probes generated from green and ripening fruit (cDNA clone pTOM5; Slater *et al.*, 1985). As will be described in greater detail shortly, verification of gene function for pTOM5 has been demonstrated via antisense gene repression (Bird *et al.*, 1991). The isolation of genes encoding two of the primary catalytic events in the production of carotenoids represents the first step toward elucidating the genetic determinants of this process in higher plants.

In addition to obvious changes in colour, ripening tomato fruit demonstrate considerable changes in metabolism and ultrastructure of cell wall components. Several groups have reported the *de novo* synthesis of cell wall polymers during the ripening process of tomato (Lackey *et al.*, 1980; Mitcham *et al.*, 1989; Greve and Labavitch, 1991). However, most scientific effort has been directed toward the considerable cell wall disruption associated with ripening, especially that resulting from the activity of hydrolytic enzymes and their hypothesised relationships to the regulation of softening of ripening fruit (Fischer and Bennett, 1991 and references therein). Probably the most studied tomato fruit hydrolase activity is the ripening-associated disruption of middle lamella polyuronides (pectin) brought about by *endo*-polygalacturonase (Giovannoni *et al.*, 1992 and references therein). Experiments to be detailed later, employing antisense gene repression (Smith *et al.*, 1988; Sheehy *et al.*, 1988) and mutant complementation (Giovannoni *et al.*, 1989) have shown conclusively that polygalacturonase activity is responsible for ripening-related pectin depolymerisation. However, these same reports failed to demonstrate any significant correlation between pectin degradation and the substantial fruit softening previously attributed to polygalacturonase activity during ripening.

Finally, the most reproducible and quantitative characteristic of ripening tomato fruit is the evolution of large quantities of climacteric ethylene. Because of the importance of this hormone in the regulation and coordination of the ripening process, its own regulation and biosynthesis have been at the forefront of recent ripening research (for review, see Theologis, 1992). Recently, genes encoding ACC synthase (Sato and Theologis, 1989) and ACC oxidase (Slater *et al.*, 1985; Spanu *et al.*, 1991) have been cloned and repressed via antisense strategies to demonstrate their necessary roles in normal fruit development (Hamilton *et al.*, 1991; Oeller *et al.*, 1991). A similar strategy has implicated an ethylene-inducible, tomato fruit ripening-specific gene, E8 (Lincoln *et al.*, 1987), in the regulation of climacteric ethylene production (Penarrubia *et al.*, 1992).

B. Ripening Mutants

A variety of tomato mutants harbouring lesions in various aspects of the ripening process have been identified and described (for a review on tomato-ripening mutants, see Grierson *et al.*, 1987). Many of these mutations have been shown through linkage analysis to reside at unique loci (Rick, 1986), thus implying a complex regulatory mechanism for the control of fruit ripening. Most ripening mutants characterised to date influence normal colour development without significant effects on other ripening characters. Examples include the *greenflesh* (*gf*: Ramirez and Tomes, 1964) and *yellowflesh* (*r*: Darby, 1978) mutants which inhibit ripening-related chlorophyll degradation and lycopene accumulation, respectively.

Of particular interest to fruit ripening developmental biologists are several regulatory mutants exerting dramatic influence over numerous aspects of the normal ripening process. These include the *ripening-inhibitor* (*rin*: Robinson and Tomes, 1968), *non-ripening* (*nor*: Tigchelaar *et al.*, 1973), *Never-ripe* (*Nr*: Rick, 1956) and *alcobaca* (*alc*: Kopeliovitch *et al.*, 1980) mutations. These mutations result in the inhibition of ripening to varying degrees without obvious effects on growth and development of additional vegetative tissues.

The various ripening-impaired mutants have been used extensively as tools for analysis of the regulatory events coordinating the ripening process. In particular, the *rin* and *nor* mutants display the most extreme phenotypes in that their fruits fail to ripen (Tigchelaar *et al.*, 1978a). Thus, *rin* and *nor* represent recessive lesions at loci necessary for the initiation and completion of normal ripening. Although *rin* and *nor* fruit reach full size, they remain green and firm until falling prey to either microbial infection or desiccation. In contrast, *Nr* is a dominant mutant characterised by delayed and incomplete ripening more similar to the recessive *alc* mutant than to either *rin* or *nor* (Hobson, 1967; Kopeliovitch *et al.*, 1980). *Nr* fruit mature to an orange-red colour and soften only moderately as compared to their normal counterparts (Hobson, 1967), while fruit harbouring the *alc* mutation ripen slowly to a light red colour (Mutschler, 1984). Additionally, *nor* has been reported to show modest lycopene accumulation in certain genetic backgrounds (Tigchelaar *et al.*, 1978a) and in response to environmental stress (Mizrahi *et al.*, 1982). The fruit of all four mutants produce viable seed.

Attempts to map the *nor* and *alc* mutations genetically relative to RFLP markers have demonstrated that both loci reside in close proximity on tomato chromosome 10 (Kinzer *et al.*, 1990), suggesting the possibility of an allelic relationship between the two mutations. Due to variations in phenotype resulting from environmental influence, and less than complete dominance of the normal allele(s) in heterozygotes, testing for complementation between *nor* and *alc* has proven inconclusive (Kinzer *et al.*, 1990). High resolution genetic mapping of ripening loci (to be detailed later in this chapter) should aid either directly via more accurate localisation on the tomato genetic map, or through confirmation of genotypes during tests for allelism, in the determination of possible allelism between *nor* and *alc*.

It is important to note that partial inducement of ripening has been reported for fruit harbouring either the *rin* or *nor* mutations with various treatments. Exposure to ethylene or propylene of detached *rin* fruit hastens the normal yellowing of *rin* fruit (Tigchelaar *et al.*, 1978a) without discernible effects on lycopene accumulation, softening, ethylene biosynthesis, or polygalacturonase activity (McGlasson *et al.*, 1975; Giovannoni *et al.*, 1989). However, treatment of fruit still attached to the vine with ethylene or 2-chloroethylphosphonic acid (ethephon), a chemical whose degradation products include ethylene, results in 10-20% of lycopene levels observed in normal fruit and an increase in softness (Mizrahi *et al.*, 1975; Buescher, 1977). In addition, similar results have been reported for detached *rin* fruit treated with ethylene or ethephon under high oxygen tension (Frenkel and Garrison, 1976; Tigchelaar *et al.*, 1978b). These results, in conjunction with those described in recent antisense experiments (Hamilton *et al.*, 1990; Penarrubia *et al.*, 1992), demonstrate an emerging realisation that attachment to the vine is involved in promoting normal and complete ripening of tomato fruit. Analysis of developmental regulation of fruit ripening mediated via import of positively acting signals, export of ripening repressors, or some combination of both, is likely a become a focus of ripening research in the near future.

As will be described later, the genes characterised to date which represent the highest rungs yet analysed in the hierarchy of ripening regulation are those involved in the regulation of ethylene biosynthesis (ACC Synthase, ACC Oxidase, E8). The inability of *rin* and *nor* mutant fruit, in particular, to soften, accumulate lycopene, express the fruit polygalacturonase gene, synthesise autocatalytic ethylene, or respond significantly to exogenous ethylene, demonstrates that these mutations represent lesions in the regulatory cascade prior to those points for which genes have been cloned and analysed. Consequently, cloning and characterisation of the normal alleles corresponding to the above described mutants represent a logical next focus for examination of the developmental regulation of fruit ripening. Initial efforts toward the cloning of these genes will be described toward the end of this chapter.

C. Additional Attributes of Tomato as a Model System

Tomato has a number of additional attributes, not necessarily related to ripening, which make it an excellent system for molecular and genetic analysis. The tomato plant itself is easy to grow and maintain and is amenable to clonal propagation. This characteristic, in conjunction with the fact that the cultivated tomato is generally self-pollinating, allows for easy maintenance of experimental materials. In addition, the fact that the crop has served for decades as a model genetic system has resulted in identification of numerous interesting mutants and variants which complement the extensive collection of compatible germplasm available from wild relatives (Rick, 1956, 1986). The net result is availability of an enormous resource of mutant and variant alleles influencing numerous aspects of plant growth, development, physiology and morphology. Tomato is also amenable to *Agrobacterium*-mediated DNA transfer and tissue culture regeneration, allowing for molecular genetic analysis of plant processes, including fruit ripening (McCormick *et al.*, 1986; Fillatti *et al.*, 1987).

Recent generation of a nearly saturated tomato restriction fragment length polymorphism (RFLP) map with an average density of 1.5 cM between markers (Tanksley *et al.*, 1992) demonstrates the continued development of this species as a model genetic system. Markers from this map have been mapped relative to a number of morphological traits including several fruit-ripening loci. In conjunction with the RFLP map, the availability of a tomato yeast artificial chromosome (YAC) library (Martin *et al.*, 1992) should promote the isolation of tomato genes by chromosome walking in the very near future (Rommens *et al.*, 1989; Ganal *et al.*, 1990).

III. ISOLATION OF RIPENING-RELATED GENES

A. Differential Screening

Rattanapanone *et al.* (1977) analysed the relationship between total RNA synthesis and tomato fruit development. The resulting data suggested that peaks in fruit RNA synthesis occurred early during fruit development, at a time corresponding to rapid cell division, and again at the initiation of fruit ripening. Significant declines in RNA synthesis were observed during the intervening time, which was devoted largely to cell expansion, and again during later stages of ripening. In addition, *in vitro* translation of tomato fruit mRNA has demonstrated that ripening fruit display significant changes in the levels of specific mRNAs as compared to fruit analysed just prior to the onset of ripening (Slater *et al.*, 1985; Biggs *et al.*, 1986; Lincoln *et al.*, 1987). Treatment of immature green tomato fruit with ethylene also promoted changes in *in vitro* translation products similar to those observed during fruit ripening (Lincoln *et al.*, 1987). Together, these results suggested the possible utility of differential screening (Grunstein and Hogness, 1975) for the isolation of structural, and possibly regulatory, genes involved in the manifestation of tomato fruit ripening.

Numerous ripening-related and ethylene-inducible genes have been isolated from tomato as the direct result of differential screening strategies (for an extensive list of tomato-ripening related clones, most of which were cloned via differential screens, see

Gray *et al.*, 1992). Slater *et al.* (1985) employed probes synthesised from mature green (mature fruit several days prior to ripening) and ripe tomato fruit to identify clones, harboured within a cDNA library constructed with ripe fruit mRNA, which were preferentially or specifically expressed during ripening. Several clones were shown to correspond to genes expressed only in fruit and only during ripening. Subsequently, Maunders *et al.* (1987) demonstrated that a number of these ripening-related genes were also ethylene inducible. Analysis of three of the clones resulting from these screens (pTOM5, pTOM6 and pTom13) will be described in detail shortly. Most recently, Pear *et al.* (1989) employed differential screening of a ripe fruit cDNA library with probes derived from leaves, immature fruit, and ripening fruit, to isolate a clone, 2A11, specific to fruit (ovary) development extending from pre-anthesis through ripening.

An alternative differential screening strategy employing subtractive hybridisation enrichment (Davis *et al.*, 1984) for the isolation of genes specific to the onset of ripening, yielded additional ripening-related and ethylene-inducible clones (Lincoln *et al.*, 1987). In this instance, cDNAs derived from either ethylene-treated mature green fruit or fruit showing the first signs of ripening, were hybridised to mRNA extracted from non-ethylene-treated mature green fruit, followed by subtraction of hybrid sequences and degradation of residual mRNAs. The remaining ethylene-inducible and/or ripening-specific cDNA sequences were utilised to construct libraries for differential screening with probes derived from either air or ethylene-treated mature green fruit. At least four clones (pE4, pE8, pE17, and J49) were identified which responded positively at the level of mRNA accumulation to the application of exogenous ethylene (Lincoln *et al.*, 1987).

Finally, differential screening has been utilised to isolate sequences from tomato fruit corresponding to genes regulated by low temperature stress (Schaffer and Fischer, 1988). RNA dot blot analysis indicated that two chilling-induced genes (C14 and C17) require continuous exposure to low temperatures to maintain high expression levels. In addition, wild tomato species originating from relatively low and high temperature climates demonstrated significant differences in chilling-induced C14 and C17 expression levels. Subsequent nuclear run-on transcription analysis (Walling *et al.*, 1986) demonstrated that C14 is regulated at the level of transcription by both low and high temperature stress (Schaffer and Fischer, 1990). Comparative sequence analysis demonstrated that C14 is highly homologous to previously identified thiol protease genes (Schaffer and Fischer, 1988).

1. Regulation of ethylene-inducible genes

Maunders *et al.* (1987) demonstrated that several ripening-related genes were also regulated by ethylene at the level of mRNA accumulation, thus suggesting transcriptional control by this hormone. One gene in particular, pTOM13, had been shown previously to encode a 35 kDa polypeptide which accumulated in both fruit and leaves upon exposure to exogenous ethylene (Smith *et al.*, 1988) and more recently, following leaf and fruit wounding (Holdsworth *et al.*, 1988) and during senescence (Davies and Grierson, 1989). Three genomic clones homologous to pTOM13 have been isolated, with at least two demonstrating differential expression in leaves and fruit (Holdsworth *et al.*, 1988). Transcripts from the third pTOM13 homologue were not detected in leaves, green fruit, or ripening fruit which were either wounded or left untreated. Finally, analysis of pTOM5 and pTOM13 mRNA concentrations in the *rin* mutant demonstrated drastic reductions of both messages as compared to levels observed in normal fruit (Knapp *et al.*, 1989).

Increased levels of pE4, pE8, pE17 and J49 mRNA accumulation during ripening, and in response to exogenous ethylene, indicated that the corresponding genes may also be regulated at the level of nuclear transcription. Employing nuclear run-on transcription analysis with nuclei isolated from various stages of fruit development, Lincoln and Fischer (1988a) demonstrated that all four genes are transcriptionally activated in fruit at or before the onset of ripening. Specifically, significant E4 and E8 transcription is not detected until just prior to the onset of ripening, though E8 is transcribed at a low level for several days before. J49 displays a transient increase in transcription preceding the onset of ripening while E17, surprisingly, is transcribed at a fairly consistent rate throughout the mature green stages and into ripening. Upon exposure of mature green fruit to exogenous ethylene, mRNA accumulation from all four genes increases. This increase occurs in parallel with increased transcription for all except E17, whose transcription rates remain essentially the same as in untreated fruit (Lincoln and Fischer, 1988a). These results demonstrate the existence of developmental and/or ethylene-inducible transcription components for accumulation of E4, E8 and J49 mRNAs, while E17 mRNA accumulation is influenced by post-transcriptional processes. It is important to note that the reported mRNA concentrations for E4, E8, E17 and J49 represent polysome-bound messages. Comparisons between polysome-bound and total mRNA preparations must be made before it can be resolved as to whether or not E17 post-transcriptional regulation occurs at the level of mRNA-ribosome loading.

Analysis of E4, E8, E17, and J49 transcription rates in *rin*, *nor*, and *Nr* demonstrated that most of these genes are repressed or completely inhibited in the homozygous ripening mutants (Lincoln and Fischer, 1988b; DellaPenna *et al.*, 1989). The most extreme reduction is for E4, whose transcription is barely detectable in all three mutants, while E17 is the least influenced (DellaPenna *et al.*, 1989). The sole exception to reduced expression is J49 in *nor* fruit, where the corresponding transcription rate and mRNA concentration exceed those of normal fruit by four- and ten-fold, respectively (DellaPenna *et al.*, 1989). Finally, exogenous ethylene has been shown to increase mRNA accumulation from all four genes in homozygous *rin* mutant fruit (Lincoln and Fischer, 1988b). Though transcriptional regulation is likely involved for at least some, the precise level at which ethylene exerts its influence on these genes in the *rin* mutant remains to be demonstrated.

In addition to expression studies, extensive promoter analysis of the E4 and E8 ethylene inducible genes has also been performed. Employing a recombinant lambda DNA 'tag' inserted into the 3' portion of the E8 coding region, Deikman and Fischer (1988) have demonstrated that a 4.4 kb restriction fragment which includes sequences 2 kb 5' and 0.5 kb 3' to the transcribed sequence is sufficient for the observed developmental and ethylene-inducible regulation. In addition, a region approximately 1 kb upstream of the E8 start of transcription was shown via gel shift analysis (Singh *et al.*, 1986) to specifically bind a nuclear factor, C1, associated with ripening. Treatment of mature green fruit with ethylene prior to nuclear protein extraction did not result in binding of C1 to the E8 promoter (Deikman and Fischer, 1988). However, the same factor has been shown to bind specifically to a region of the E4 promoter at a position between -18 and -34 relative to the transcription start via both gel shift and methylation interference footprint analysis (Otwinowski *et al.*, 1988; Cordes *et al.*, 1989). These observations suggest that C1 binding during fruit ripening is developmentally controlled as opposed to being ethylene-mediated. Although sequence analysis around the C1 binding sites of the E4 and E8 5′ regions indicates a conserved region between the two promoters (11 of 17 bases; Cordes *et al.*, 1989), it is curious that a developmentally regulated ripening factor appears to exert its influence on two coordinately expressed genes at such different positions relative to their respective transcription start points.

Finally, Van Haaren and Houck (1991) have performed extensive analysis of sequences both 5′ and 3′ of the fruit-specific 2A11 gene. Utilising numerous GUS reporter constructs (Jefferson, 1987), the authors have defined both positive and negative *cis*-acting regulatory elements regulating chimaeric gene expression in developing fruit. In addition, replacement of sequences 3′ of the 2A11 gene indicated little if any regulatory role for this region. Continued analysis of *cis*-acting sequences and *trans*-acting factors should help elucidate the regulatory components involved in the control of fruit development and ripening. As additional genes are analysed and compared with those already characterised, common developmental and/or inducible regulatory elements and factors are likely to emerge.

B. Ripening-related Genes Cloned on the Basis of Function

1. ACC synthase

As the limiting catalyst in the ethylene biosynthesis pathway, ACC synthase activity, polypeptide and the corresponding gene(s) have proven particularly important targets for analysis by ripening physiologists. ACC synthase activity and resulting ethylene production have been shown to be inducible by numerous environmental and chemical factors, in addition to developmental components such as those active during ripening (Kende, 1989). However, the ACC synthase polypeptide has proven difficult to isolate due to instability and relative low abundance (Bleeker *et al.*, 1986). The first gene encoding ACC synthase was ultimately cloned from zucchini because of the relative high levels of ACC synthase activity and ethylene production yielded by zucchini fruit (Sato and Theologis, 1989). The approach was biochemical and based on production of antiserum from crude extracts containing ACC synthase activity derived from induced tissue. The antiserum was depleted of non-specific antibodies using proteins extracted from uninduced tissues and subsequently used to screen an expression library derived from induced zucchini mRNA. The result was the identification of two putative ACC synthase clones, pACC1 and pACC7. A full length clone (pACC1) was shown to be sufficient for the production of high levels of ACC synthase activity in transformed *Escherichia coli* and yeast (Sato and Theologis, 1989).

Recent isolation and analysis of genomic clones encoding tomato ACC synthase genes revealed the existence of a moderate size gene family (at least five genes) whose members are differentially regulated through development and by various environmental and chemical factors (Olson *et al.*, 1991; Rottmann *et al.*, 1991). At least two genes, LE-ACC2 and LE-ACC4, are developmentally regulated during fruit ripening, while two others, LE-ACC-1A and LE-ACC-1B, are interesting from an evolutionary standpoint in that they are adjacent and convergently transcribed (Rottmann et al., 1991). Also, at least one ACC synthase gene has been shown to be both ripening- and wound-inducible (Olson et al., 1991). The degree of amino acid homology between the ACC synthase family members ranges from 50% to 96% with a region involved in substrate and cofactor (pyridoxal-5'-phosphate) binding being especially well conserved in all five (Rottmann et al., 1991) and among previously characterised aminotransferases (Mehta et al., 1988).

2. Polygalacturonase

Because of its relative abundance during fruit ripening, involvement in pectin degradation, and hypothesised role in ripening-related softening, polygalacturonase has received considerable attention from both pure and applied scientists. Several laboratories have cloned and characterised cDNAs encoding tomato fruit polygalacturonase (DellaPenna et al., 1986; Grierson et al., 1986; Sheehy et al., 1987), and the corresponding developmental, inducible and spatial expression patterns have been analysed extensively in normal and mutant tomato fruits. Also, polygalacturonase was utilised in the first examples of antisense gene repression for examination of gene function in tomato (Sheehy et al., 1988; Smith et al., 1988).

Cloning of tomato fruit polygalacturonase was accomplished by DellaPenna et al. (1986) via a biochemical approach similar to that described above for the isolation of the ACC synthase gene. Comparison of *N*-terminal amino acid sequence from the purified protein with DNA sequence from the cloned cDNA confirmed the isolation of a polygalacturonase clone. A similar strategy was employed by Sheehy et al. (1987). Grierson et al. (1986) also employed comparative amino acid and DNA sequence analysis to demonstrate that a previously isolated ripening-related gene, pTOM6 (Slater et al., 1985), is in fact a polygalacturonase cDNA clone.

Based on genomic restriction analysis, tomato fruit polygalacturonase appears to be encoded by a single gene (Bird et al., 1988; Knapp et al., 1989). Analysis of polygalacturonase mRNA accumulation and transcription patterns indicate that this gene is expressed at the onset of fruit ripening at a time corresponding roughly to autocatalytic ethylene evolution (DellaPenna et al., 1986, 1989; Grierson et al., 1986; Sheehy et al., 1987). Treatment of mature green fruit with exogenous ethylene does result in polygalacturonase mRNA accumulation after several days, but this probably reflects overall inducement of the ripening process as opposed to more direct activation of the polygalacturonase promoter by ethylene (Maunders et al., 1987; Lincoln et al., 1987). In addition, dramatic reductions in polygalacturonase mRNA levels and transcription rates are observed in the *rin*, *nor* and *Nr* ripening mutants (DellaPenna et al., 1986, 1989; Lincoln and Fischer, 1988b; Knapp et al., 1989).

Attempts to localise polygalacturonase protein immunologically within the ripening fruit, via tissue blotting techniques involving the immobilisation of proteins from fruit sections on a membrane surface, demonstrate that the first significant accumulation of polygalacturonase occurs in the collumella region (Tieman and Handa, 1989). This initial accumulation is followed by greatly increased polygalacturonase concentrations throughout the entire ripening pericarp. Substantial polygalacturonase protein was not detected in the locules during any stage of ripening (Tieman and Handa, 1989). *In situ* localisation of polygalacturonase mRNA in tomato fruit showed a marked increase in concentration within cells surrounding pericarp vascular bundles (Pear et al., 1989). Taken together, these results on transcription, mRNA and protein accumulation demonstrate that the timing of polygalacturonase expression is both strictly regulated developmentally (i.e. in ripening fruit) and spatially, to a defined subset of fruit tissues. Furthermore, a polygalacturonase promoter–reporter gene construct expressed in transgenic tomato plants demonstrates that 1.4 kb of DNA sequence upstream of the transcription start is sufficient for fruit and ripening specific expression of this gene (Bird et al., 1988).

Lastly, activity staining of polygalacturonase protein resolved under non-denaturing electrophoresis conditions demonstrates the existence of three distinct isoforms, PG1, PG2A and PG2B (Pressey and Avants, 1973; Tucker et al., 1980; Ali and Brady, 1982). All three isoforms are immunologically related, with PG2A and PG2B differing only at the level of glycosylation (DellaPenna and Bennett, 1988). PG1 comprises approximately one-third of the polygalacturonase protein in ripe fruit and has been shown to be a heterodimer consisting of PG2 isoforms and a 41 kDa polypeptide (Tucker et al., 1981; Moshrefi and Luh, 1983; Pressy, 1984) whose cDNA has been recently cloned and termed the β-subunit of polygalacturonase (D. DellaPenna, personal communication). Recent attempts to define the polygalacturonase isoform(s) involved in ripening-related pectin degradation demonstrate that the majority of in vivo activity results at a time corresponding to the onset of ripening, when PG1 is the prevalent isoform (DellaPenna et al., 1990). The role of PG1 in ripening-related pectin degradation is also supported by the observation that expression of a chimaeric tomato fruit polygalacturonase gene in transgenic tobacco results in accumulation of active (in vitro) PG2A and PG2B isoforms in the cell walls of a variety of tissues, without evidence for significant in vivo pectin degradation (Osteryoung et al., 1990).

3. Pectinmethylesterase

Accessibility of polygalacturonase to cell wall pectins is thought to be influenced by the degree of esterification of the polyuronide substrate (Pressy and Avants, 1982). As a result, pectinmethylesterase, an enzyme involved in the cleavage of methylester groups from cell wall pectins, has been implicated in the regulation of cell wall metabolism during fruit ripening. Pectinmethylesterase activity is detected early in fruit development and reaches maximal levels at a time corresponding roughly to the onset of ripening (Hobson, 1963; Tucker et al., 1982). Tomato fruit pectin esterase polypeptide has been purified to homogeneity, sequenced (Markovic and Jornell, 1986), and the resulting information utilised to generate oligonucleotide primers for the isolation of the corresponding cDNA clone from a ripe fruit library (Ray et al., 1988). An immunological approach, similar to those described above, has also resulted in the cloning of a highly homologous yet distinct cDNA, demonstrating the existence of at least two pectinmethylesterase gene family members which are expressed during fruit development (Harriman et al., 1991).

RNA gel blot hybridisation analysis revealed that pectinmethylesterase mRNA accumulation peaks during early fruit development with the highest levels attained prior to the mature green stage (Ray et al., 1988; Harriman et al., 1991). However, pectinmethylesterase protein concentration increases markedly at the onset of ripening, and continues to increase during later development, even though corresponding mRNA and activity levels decline during this period (Harriman *et al.*, 1991). Examination of the *nor* and *Nr* ripening mutants reveals similar patterns (though reduced levels of expression), while analysis of fruit homozygous for the *rin* mutation demonstrates substantial reductions of pectinmethylesterase mRNA, protein and activity as the fruit matures (Harriman *et al.*, 1991).

IV. DETERMINATION OF THE PHYSIOLOGICAL FUNCTIONS OF RIPENING GENES

The differential screening cloning strategies described above have resulted in a reservoir of genes related to the ripening process via their corresponding expression patterns. Similar strategies have yielded numerous genes related to various aspects of growth and development in a myriad of plant and non-plant species. Nevertheless, for the vast majority of these genes, little is known concerning their respective physiological roles in growth or development of the host organism. Furthermore, only a few of the ripening-related genes (for example polygalacturonase, pectinmethylesterase, ACC synthase and ACC oxidase) have been shown to be responsible for a particular biochemical function.

Form the standpoint of classical genetics, the physiological role of any gene can best be assessed through examination of individuals harbouring mutations at said locus. The advent of reverse genetics (i.e. cloning genes on the basis of their biochemical consequences or patterns of expression) has created the demand for technology to create mutations in specific target genes for analysis of *in vivo* function. Ripening-related gene analysis has proven to be a fertile testing ground for the utilisation of such technologies including antisense gene repression, mutant complementation and co-suppression.

A. Antisense Gene Repression

Antisense gene repression has been demonstrated to operate naturally in a variety of organisms as a system of genetic regulatory control for DNA replication and/or transcription (for review, see Green *et al.*, 1986). Following these initial observations, recombinant DNA technology quickly led to the first examples of engineered antisense RNA activity in plant systems. Ecker and Davis (1986) utilised a tobacco protoplast transient assay system to demonstrate that co-introduction of antisense gene constructs with reciprocal sense orientation chloramphenicol acetyltransferase (CAT) genes resulted in substantial inhibition of extractable CAT activity. van der Krol *et al.* (1988) were the first to report chimaeric antisense gene repression in stably transformed plants. An antisense chalcone synthase gene under the control of the Cauliflower Mosaic Virus 35S promoter (Bevan *et al.*, 1985) resulted in inhibition of endogenous chalcone synthase mRNA and protein accumulation with concomitant changes in floral colour and coloration patterns (van der Krol *et al.*, 1988). Subsequently, a number of ripening-related genes have been assayed for physiological function with the aid of antisense transgenic plants (Table 12.1).

TABLE 12.1. Antisense tomato fruit-ripening genes and corresponding references.

Gene	References
ACC oxidase (pTOM13)	Hamilton *et al.* (1991)
ACC synthase	Oeller *et al.* (1991)
E8 oxidase (E8)	Penarrubia *et al.* (1992)
Pectinmethylesterase	Tieman *et al.* (1992)
Phytoene synthase (pTOM5)	Bird *et al.* (1991)
Polygalacturonase	Smith *et al.* (1988)
	Sheehy *et al.* (1988)

1. Polygalacturonase: pectin hydrolysis but not fruit softening

Employing an inverted 730 bp 5' cDNA fragment under the regulatory control of the CaMV 35S promoter, Smith *et al.* (1988) reported up to 90% inhibition of normal polygalacturonase activity in primary antisense transformants. Similar results were obtained by Sheehy *et al.* (1988) working with a full length antisense cDNA. In addition, antisense gene expression had no significant effect on the transcription rate of the endogenous polygalacturonase gene (Sheehy *et al.*, 1988). Initial phenotypic analysis of ripening parameters in transgenic tomatoes indicated little if any change in lycopene accumulation or fruit texture (Sheehy *et al.*, 1988; Smith *et al.*, 1988). This observation was in contrast to considerable correlative data linking polygalacturonase-mediated pectin hydrolysis with ripening-related softening (Hobson, 1965; Tigchelaar *et al.*, 1978b; Brady *et al.*, 1983; Grierson, 1985). Experiments utilising polygalacturonase antisense constructs represented the first attempts to employ this technology to address questions of ripening-related gene contribution to fruit phenotype.

Because polygalacturonase protein represents approximately 4–5% of total cell wall protein in ripening fruit, 10% of normal polygalacturonase was hypothesised to be sufficient to confer phenotype (Schuch *et al.*, 1989). This hypothesis is supported by the observations cited above indicating the existence of a limiting titratable factor (β-subunit) responsible for *in vivo* polygalacturonase activity. To address this question, Smith *et al.* (1990a) identified original polygalacturonase antisense mutants (T1) which contained single T-DNA insertion events and harvested seed (T2) resulting from their respective self-pollinations. Analysis of *in vitro* polygalacturonase activity from T2 progeny revealed at least one line with three distinct classes of individuals whose ripe fruit contained 100%, 20% and 1% of normal activity, respectively (Fig. 12.1). Individuals expressing 100% polygalacturonase activity had no transgene sequence as determined through gel blot hybridisations, thus indicating they were homozygous non-transgenic. In addition, selfed progeny (T3) from the 1% activity plants bred true for this phenotype, demonstrating homozygosity for the antisense transgene. The remaining progeny, which contained approximately 20% activity, most probably represent hemizygotic transgenic plants (Smith *et al.*, 1990a).

Further analysis of the 0, 1 and 2 transgene copy antisense progeny revealed no significant differences in levels or timing of lycopene or climacteric ethylene biosynthesis, indicating little if any influence by polygalacturonase activity on these

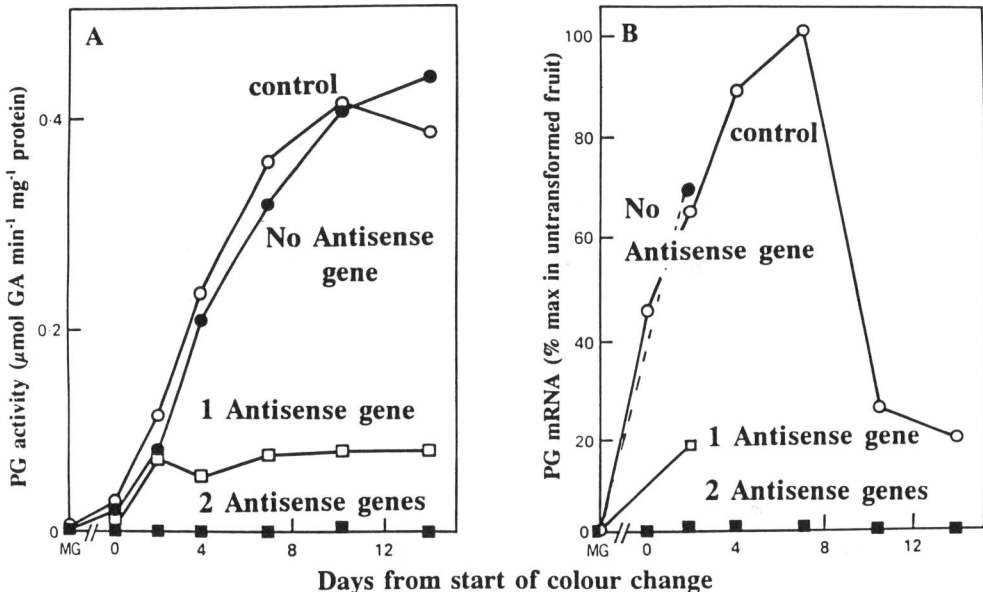

FIG. 12.1. Relationship between polygalacturonase antisense gene dosage and phenotypic effect. Individuals tested included progeny from a F2 population harbouring 0, 1 or 2 copies of the antisense insert and untransformed controls. 'MG' refers to mature green. (A) Polygalacturonase activity measured *in vitro*. (B) Relative accumulation of polygalacturonase mRNA. From Gray *et al.* (1992), with permission.

aspects of tomato ripening physiology. In addition, no significant changes in pectinmethylesterase or invertase activities were observed, even in those fruit retaining less than 1% normal polygalacturonase. Analysis of soluble cell wall pectins from untransformed fruit revealed that the transition from green to ripe resulted in a 49% decrease in average pectin molecular weight. However, this reduction in average soluble pectin size was limited to 40% and 15% in ripening transgenic fruit harbouring 1 and 2 antisense alleles, respectively, with no obvious inhibition of softening (Smith *et al.*, 1990a).

The antisense experiments described here, in conjunction with a mutant complementation experiment to be described shortly, indicate little influence of polygalacturonase activity on ripening characteristics other than pectin hydrolysis. However, reports by Giovannoni *et al.* (1990) and Kramer *et al.* (1990) indicate that polygalacturonase activity may positively influence the fruit-ripening-associated increase of susceptibility to microbial infection. Specifically, field-grown antisense polygalacturonase tomato plants show a significant increase in yield resulting from reduced loss to fruit rot (Kramer *et al.*, 1990), while greenhouse-grown *rin* tomatoes containing an inducible chimaeric polygalacturonase gene construct exhibit increased pathogen susceptibility only in induced fruit (Giovannoni *et al.*, 1990). Although the purpose of ripening-related reductions in pathogen defence are not immediately clear, this phenotype may represent an alternate method of fruit seed release in the event that animal vectors are not available.

2. Pectinmethylesterase

The role of pectinmethylesterase activity in tomato fruit ripening has also been addressed via antisense technology. Tieman *et al.* (1992) employed a CaMV 35S driven reverse orientation partial genomic clone to antisense pectinmethylesterase during tomato fruit development. They focused on an original hemizygotic transformant and its homozygous antisense progeny for subsequent analysis. Pectinmethylesterase mRNA, protein and *in vitro* activity levels were greatly reduced in antisense fruit as compared with untransformed controls (Fig. 12.2). The amount of pectin methylesterification in ripening antisense fruit was 20–40% greater than in untransformed controls, resulting in a corresponding increase in average molecular weight of chelator soluble pectins (Tieman *et al.*, 1992). These results are consistent with reduced pectinmethylesterase activity leading to decreased accessibility of polygalacturonase for its substrate.

An unanticipated consequence of reduced pectinmethylesterase activity was a significant increase in fruit soluble solids. As the authors themselves caution, this result is preliminary and in need of additional scrutiny. Visual examination, in conjunction with measurements of ethylene, lycopene, chlorophyll and polygalacturonase mRNA accumulation in antisense and control fruit, indicated no additional influence of reduced pectinmethylesterase activity on ripening characteristics (Tieman *et al.*, 1992).

3. ACC synthase: regulation of ethylene biosynthesis

As mentioned previously, ACC synthase is the rate-limiting catalyst in ethylene biosynthesis. Oeller *et al.* (1991) utilised antisense gene repression to analyse the mode of action and physiological function of ethylene biosynthesis in ripening tomato fruit. A cDNA (tACC2) corresponding to the most prevalent of 2 ACC synthase mRNAs detected in ripening tomato fruit was employed in a CaMV 35S antisense construct. RNA gel blot analysis of ripening transgenic fruit indicated significant reductions in mRNA accumulation from both ripening-induced ACC synthase genes, LE-ACC2 and LE-ACC4 (Olson *et al.*, 1991; Rottmann *et al.*, 1991; Oeller *et al.*, 1991). Ethylene measurements indicated a corresponding reduction in ripening-related ethylene biosynthesis in a number of the primary transformants.

Homozygous antisense fruit from one particular transformant which demonstrated only 0.5% normal ethylene evolution at a time corresponding to normal ripening was selected for extensive characterisation (Fig. 12.3A). In addition to greatly reduced ethylene biosynthesis, antisense fruit either left on the vine, or harvested and left in air, had a 10–20 day delay in the onset of normal chlorophyll degradation, attained only an orange colour, and failed to soften or develop a 'ripe' aroma. In addition, no significant polygalacturonase protein or activity was detected in antisense fruit. Subsequent treatment with either ethylene or its analogue, propylene, induced climacteric respiration (Fig. 12.3B) and resulted in lycopene accumulation, softening, polygalacturonase activity and aroma identical to the corresponding characteristics displayed by untransformed control fruit (Oeller *et al.*, 1991).

Although the antisense phenotype was reversed by exogenous ethylene, continuous application was required. In fact, 2 days of constant exposure to ethylene was not

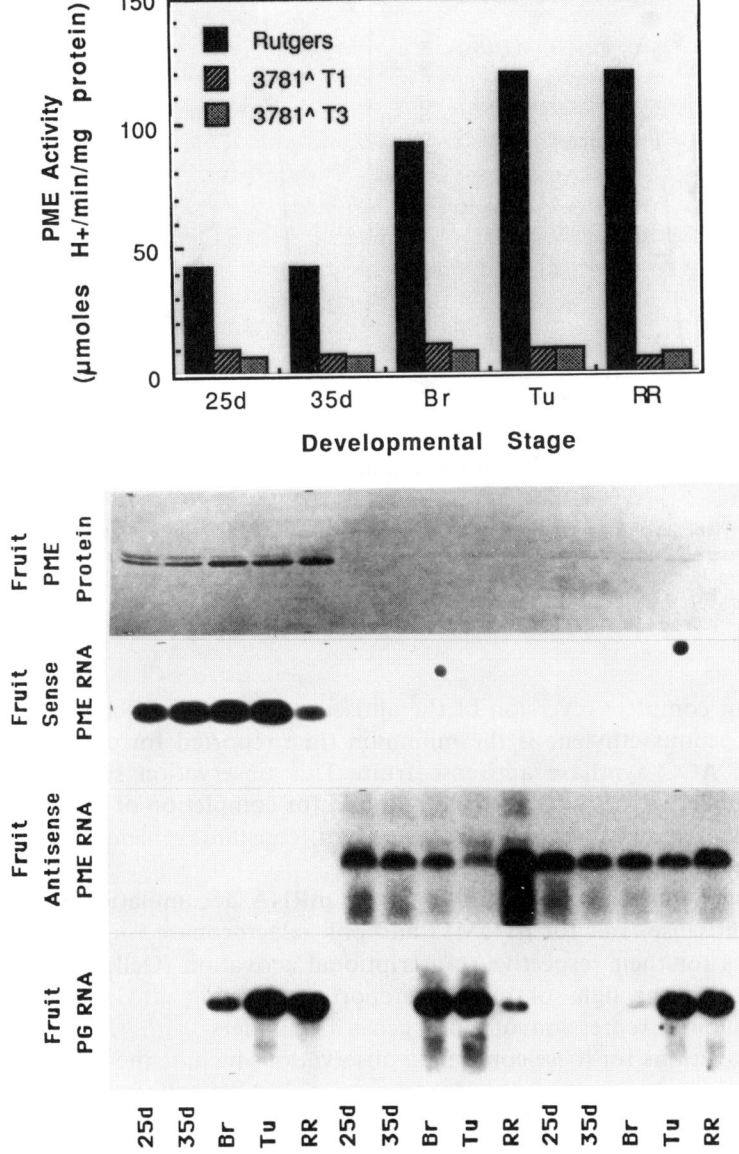

FIG. 12.2. Analysis of antisense pectinmethylesterase tomatoes. *In vitro* pectinmethylesterase activity, immunodetectable protein, sense and antisense pectinmethylesterase mRNA levels, and sense polygalacturonase mRNA levels. Stages of fruit development tested are 25 and 35 days post-anthesis, breaker (Br), turning (Tu), and red ripe (RR). Lines tested contained 0 (Rutgers), 1 (3781 ^ T1), and 2 (3781 ^ T3) copies of the antisense construct. From Tieman *et al.* (1992), with permission.

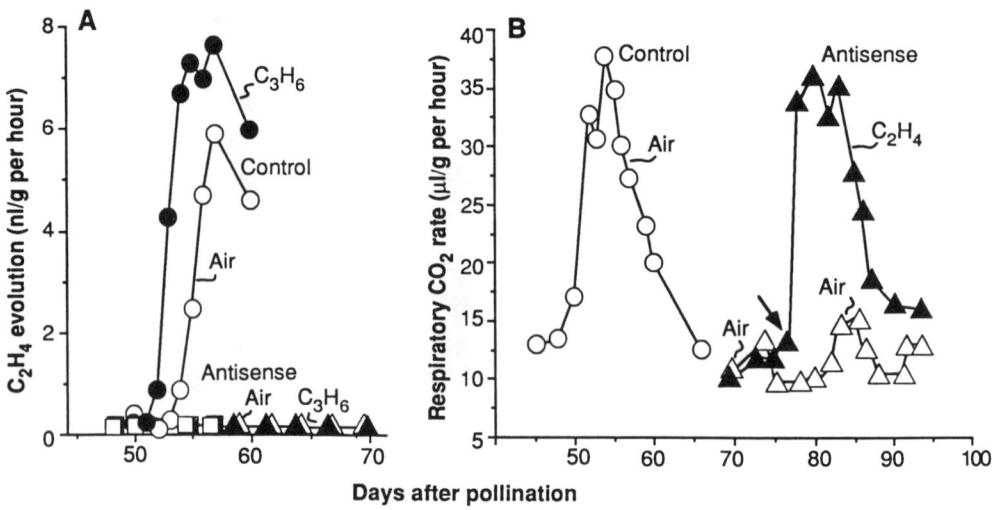

FIG. 12.3. Ethylene evolution and respiration in antisense ACC synthase and untransformed control tomatoes. Antisense and control fruit were harvested, exposed continuously to either air or propylene (C_3H_6), and analysed at the indicated number of days post-pollination for ethylene (C_2H_4) and CO_2 evolution rates. From Oeller et al. (1991), with permission.

sufficient for complete reversion of the antisense phenotype (Oeller et al., 1991). Six days of exogenous ethylene is the minimum time reported for complete phenotypic recovery of ACC synthase antisense fruit. This observation suggests that factors resulting from ethylene induction and required for completion of the normal ripening process are extremely labile and in need of constant replenishment (Theologis, 1992).

Interestingly, normal timing and levels of mRNA accumulation were detected in untreated antisense fruit for pTOM13 and polygalacturonase, suggesting no ethylene requirements for their respective transcriptional activation (Oeller et al., 1991). This is interesting in the light of previous reports suggesting ethylene inducibility and ethylene inhibitor repression of both genes (Maunders et al., 1987; Lincoln et al., 1987). Explanations for these conflicting observations include the possibilities that: (1) 0.5% normal ethylene is sufficient for transcriptional activation of both genes though not post-transcriptional control of polygalacturonase; (2) factors which accumulate prior to ripening and do not require ethylene to promote their effect(s) on polygalacturonase or pTOM13 transcription are themselves regulated by changes in sensitivity to basal ethylene and therefore respond to low levels; and (3) different gene family members respond to different developmental and/or hormonal cues. It is also interesting to note that immunological analysis of antisense fruit indicated polygalacturonase protein accumulation only in the presence of endogenous ethylene, suggesting ethylene-mediated post-transcriptional control (Theologis, 1992). Indeed, ethylene appears to mediate genetic regulation of tomato fruit ripening at a variety of points and through a number of mechanisms!

4. pTOM13: the ethylene-forming enzyme ACC oxidase

Because of correlative evidence linking pTOM13 gene expression with ripening, the wound response and exposure to exogenous ethylene, Hamilton *et al.* (1990) hypothesised that one or more members of this gene family may be involved in ethylene biosynthesis and/or hormone-mediated signal transduction. As a result, a 1.1 kb fragment of the pTOM13 cDNA was utilised in reverse orientation under control of the CaMV 35S promoter to antisense pTOM13 gene expression in transgenic tomatoes (Hamilton *et al.*, 1990). The primary transformants exhibited 32% and 13% of normal ethylene levels in response to wounding and during ripening, respectively. As in the case of polygalacturonase antisense, homozygous antisense individuals resulting from self-crossing of primary transformants showed an even more dramatic phenotype, yielding only 3% of normal ethylene at a time corresponding to ripening of control fruit.

Although the manifestation of ripening characteristics was not as severely repressed as in the ACC synthase antisense mutants described above, pTOM13 antisense fruit did not accumulate normal levels of lycopene and were more resistant to over-ripening and rotting (Hamilton *et al.*, 1990). In addition, wounded homozygous pTOM13 antisense tomato leaves accumulated only 7% of the extractable ACC oxidase (ethylene-forming enzyme) activity recoverable from similarly treated untransformed controls, suggesting that pTOM13 encodes ACC oxidase, a subunit, or a regulator thereof. Subsequent analysis of pTOM13 activity in heterologous systems demonstrated that pTOM13 does in fact encode ACC oxidase (Hamilton *et al.*, 1991; Spanu *et al.*, 1991).

Finally, analysis of polygalacturonase in pTOM13 antisense fruit indicated no effect on the normal accumulation of the corresponding mRNA or extractable activity. This observation suggests that although 0.5% of the normal ethylene level found in ripening tomatoes is not sufficient for accumulation of polygalacturonase protein (Oeller *et al.*, 1991), 3% is (Hamilton *et al.*, 1990).

5. E8: regulation of climacteric ethylene

Lincoln *et al.* (1987) isolated an ethylene-inducible cDNA clone, E8, which showed regions of amino acid sequence homology to pTOM13 yet was expressed only in ripening or ethylene-treated fruit (Deikman and Fischer, 1988). More extensive sequence analysis indicates that E8 is related to dioxegenase enzymes (McGarvey *et al.*, 1992; Penarrubia *et al.*, 1992). To analyse the function of E8 during normal fruit ripening, Penarrubia *et al.* (1992) drove expression of a full length E8 cDNA in reverse orientation under the control of the CaMV 35S promoter. Two independently transformed lines showing the lowest levels of endogenous E8 mRNA accumulation were examined in detail.

Analysis of ethylene evolution rates during fruit development gave no indication of altered ethylene biosynthesis rates prior to the onset of ripening. However, by 2 days post-breaker stage ('breaker' being the first visible signs of lycopene accumulation) the E8 antisense fruit produced up to six-fold more ethylene than their untransformed counterparts (Fig. 12.4). This effect was even more pronounced in transgenic tomatoes

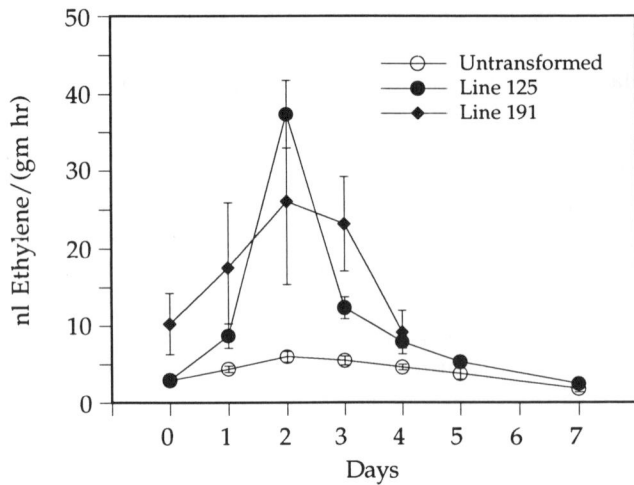

FIG. 12.4. Ethylene evolution rates from antisense E8 and untransformed control fruit harvested at the onset of colour change. 125 and 191 refer to independently transformed lines. Redrawn from Penarrubia *et al.* (1992), with permission.

harvested 4 days before the onset of ripening. In addition, increased ethylene production was transient and limited to ripening fruit, falling to control levels by 7 days post-breaker (Penarrubia *et al.*, 1992).

Immunological detection of E8 protein in a population segregating for the antisense transgene was used to identify progeny with reduced E8 accumulation (Fig. 12.5B). Corresponding ethylene measurements from ripening fruit indicated that the E8 antisense phenotype of ethylene over-production was only observed in those individuals with less than approximately 10% of normal E8 protein levels (Fig. 12.5A: Penarrubia *et al.*, 1992). Together, these results demonstrate that the E8 protein plays a negative regulatory role on climacteric ethylene biosynthesis and that relatively low amounts of said protein are needed to fulfil this function. Based on these observations, it has been postulated that E8 is involved in the perception and associated regulation of autocatalytic ethylene biosynthesis, possibly through interaction with the ethylene receptor (Penarrubia *et al.*, 1992; Theologis, 1992).

6. Phytoene synthase

Until recently, pTOM5 was only characterised as a fruit-specific ripening-related gene of unknown function. However, the availability of sequence data for bacterial carotenoid biosynthetic genes suggested that pTOM5 may be involved in the corresponding pathway in ripening fruit (Armstrong *et al.*, 1990). To ascertain the physiological function of the pTOM5 gene(s), Bird *et al.* (1991) regenerated antisense tomato plants expressing the 5' 794 base pairs of the pTOM5 cDNA in reverse orientation. As in all of the above antisense experiments, regulatory control of the chimaeric gene was mediated by the CaMV 35S promoter.

Visual inspection of the pTOM5 antisense plants revealed some individuals with pale

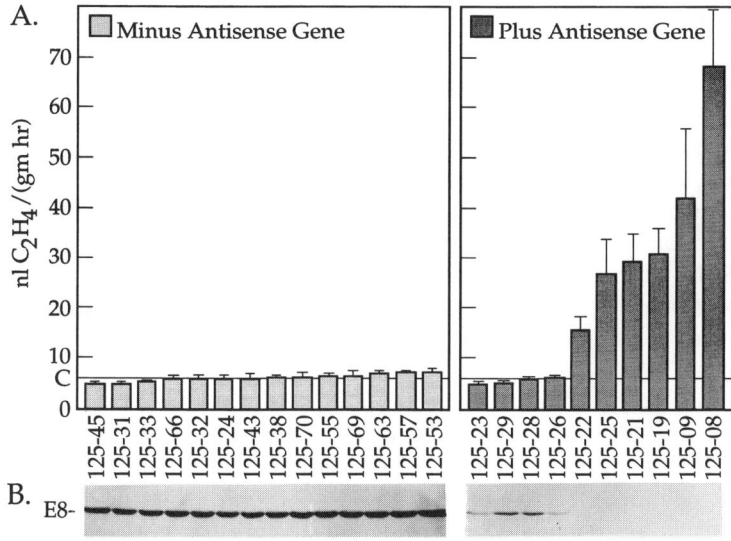

FIG. 12.5. Analysis of ethylene evolution (A) and E8 protein accumulation (B) in an F2 population derived from E8 antisense line 125. From Penarrubia et al. (1992), with permission.

coloured flowers and yellow ripe fruit, suggesting blockage of normal carotenoid biosynthesis in at least these two organs. Analysis of pTOM5 mRNA accumulation demonstrated a corresponding significant reduction in this message in antisense fruit, although additional aspects of floral development and ripening, including polygalacturonase mRNA accumulation, were not significantly influenced. Examination of total carotenoid levels indicated antisense mediated reduction by up to 97% in ripening fruit, with virtually no detectable lycopene in at least one transformant (Bird et al., 1991). Interestingly, no effect on leaf carotenoid biosynthesis was observed in pTOM5 antisense plants, suggesting the existence and activity of additional genes not influenced by expression of the inserted construct.

In addition to elucidating the function of pTOM5 in ripening fruit, the antisense experiments described above may have solved the mystery of the genetic lesion in the *yellowflesh* tomato fruit ripening mutant. Tomato plants homozygous for the mutant *yellowflesh* allele (*r*) display fruit and flower phenotypes very similar to those observed in pTOM5 antisense plants (Darby, 1978; et al., 1991). Also, Kinzer et al. (1990) have RFLP mapped several ripening-related genes and have localised pTOM5 to a region of tomato chromosome 3 known to contain the *r* locus. Comparative analysis of pTOM5 genome organisation and mRNA accumulation patterns between normal and *yellowflesh* mutant tomato plants may help resolve this question. However, complementation of the *yellowflesh* phenotype with a pTOM5 gene construct will likely provide the best proof that pTOM5 corresponds to the *r* locus.

B. Co-suppression of Gene Expression

In addition to the reports of antisense gene repression described above, others have demonstrated reduced mRNA accumulation resulting from expression of full length or partial cDNA sequences in the sense orientation (Napoli *et al.*, 1990; van der Krol *et al.*, 1990; Smith *et al.*, 1990b). Gene repression resulting from homologous sense gene expression is termed co-suppression. With respect to fruit ripening, the same partial cDNA sequence utilised by Smith *et al.* (1988) for antisense repression was successfully employed in a CaMV 35S sense gene construct to inhibit polygalacturonase mRNA accumulation in ripening tomato fruit. The ripening fruit phenotype and level of polygalacturonase mRNA accumulation in co-suppressed transformants was remarkably similar to those observed in corresponding antisense fruit. Polygalacturonase mRNA and protein accumulation was reduced to 1% of normal levels in original transformant progeny homozygous for the sense gene construct, with corresponding reduced pectin degradation and increased resistance to over-ripening (Smith *et al.*, 1990b).

In the case of polygalacturonase, there is no clear superiority of antisense versus co-suppression techniques, or vice versa. The current lack of knowledge regarding mechanisms for either antisense or co-suppression makes it difficult to ascertain which technique will be more effective in reducing expression of a particular target gene. As these mechanisms are elucidated, a set of parameters for both general and specific strategies of gene repression is likely to emerge.

C. Engineering of Pathway Blockage

An alternative method to repressing gene expression for analysing the physiological function of an enzyme is to neutralise the product(s) of the reaction which it catalyses. Although chemical inhibitors have been utilised extensively to block particular biological processes, the ideal scenario would employ highly specific inhibition which is active only at the target in question, thus with zero or only minimal side-effects. Such a strategy was employed by Klee *et al.* (1991) to analyse the roles of ACC synthase and corresponding ethylene production in tomato fruit ripening.

An enzyme, ACC deaminase, capable of catalysing the conversion of ACC to α-ketobutyric acid and ammonia, had been previously identified in *Pseudomonas* soil bacteria (Honma and Shimomura, 1978). Both products represent normally occurring metabolites found in plant cells. Thus, the corresponding bacterial gene was cloned via complementation of *E. coli* and expressed in transgenic tomato plants under the control of the CaMV 35S promoter (Klee *et al.*, 1991). Analysis of ethylene evolution rates from transgenic plants indicated 97% and 90% reductions in endogenous ethylene biosynthesis from leaves and ripening fruit, respectively. Although the timing of the onset of ripening was not affected, the ACC deaminase expressing fruit took approximately three times longer to reach the fully red stage and were firmer and more resistant to over-ripening than their untransformed control counterparts. No obvious perturbations in growth or development were observed in non-fruit tissues (Klee *et al.*, 1991).

D. Complementation

The gene repression and pathway blockage strategies for elucidating gene function described above rely on the ability to reduce a target gene's effective activity to the degree that a change in phenotype is observed. Although most of the examples cited above resulted in 90–99% reductions in normal levels of mRNA and/or corresponding activity, these data must be considered with a wary eye toward the possibility that low expression may be sufficient for manifestation of certain phenotypes. However, it is also possible that complete inhibition of certain genes may result in lethality. Indeed, one might speculate that a tomato plant unable to produce any ethylene whatsoever may have difficulty germinating, much less attain sufficient maturity to yield a fruit for analysis of the role of this hormone in ripening.

An alternative and complementary strategy to functional analysis via gene repression or pathway inhibition involves the addition of gene expression to mutant individuals and/or tissues which do not normally express the gene or activity in question. We have employed such a strategy to test the hypothesised function of polygalacturonase in ripening-related softening of tomato fruit (Giovannoni *et al.*, 1989).

An inducible chimaeric polygalacturonase gene was inserted into the genome of tomato plants homozygous for the *rin* mutation described above. Again, *rin* is pleiotropic in that many aspects of ripening are inhibited including accumulation of polygalacturonase mRNA, protein and activity. Because polygalacturonase is not

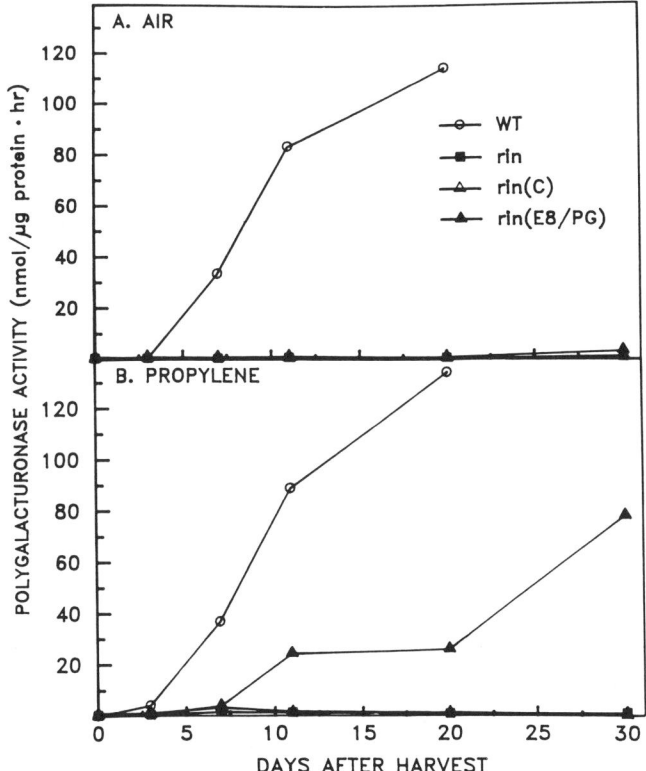

FIG. 12.6. *In vitro* polygalacturonase activity in control and transgenic *rin* tomato fruit. Mature green fruit from normal (WT), untransformed *rin* (rin), and *rin* tomato plants transformed with either the chimaeric polygalacturonase gene (rin(E8/PG)) or the transformation vector alone (rin(C)), were harvested and exposed to either air (A) or propylene (B) for the indicated period of time. From Giovannoni *et al.* (1989).

transcribed in *rin* fruit, E8 regulatory sequences were chosen to drive expression of the chimaeric gene (DellaPenna *et al.*, 1989). Transgenic fruit expressing the inserted gene accumulated high levels of chimaeric mRNA resulting in levels of cell wall extractable polygalacturonase protein and activity similar to those observed in normal ripening tomatoes (Fig. 12.6). Phenotypically, polygalacturonase activity in transgenic *rin* fruit resulted in similar levels of pectin degradation as are observed in normal tomatoes during ripening. However, no discernible effect on fruit softening as measured by compressibility was observed (Giovannoni *et al.*, 1989; DellaPenna *et al.*, 1990). In addition, *rin* fruit expressing polygalacturonase were more susceptible to microbial infection than control *rin* fruit, behaving more like their normal counterparts in this regard (Giovannoni *et al.*, 1990).

The results described here agree well with those observations made with antisense polygalacturonase tomato fruit (Smith *et al.*, 1988, 1990b; Sheehy *et al.*, 1988; Kramer *et al.*, 1990). In addition, they address the question of possible phenotypic consequences attributable to very low levels of polygalacturonase activity in antisense mutants. Finally, in conjunction with the polygalacturonase complementation experiments in transgenic tobacco (Osteryoung *et al.*, 1990), polygalacturonase isozyme analysis in transgenic *rin* fruit supports the role of PG1 as the mediator of *in vivo* activity during tomato fruit ripening (DellaPenna *et al.*, 1990). It appears that in the absence of complete knock-out mutations in a gene of interest (as is the case with polygalacturonase), the best picture of gene function may ultimately be obtained through a combination of experiments employing antisense repression of normal gene expression, and complementation of mutants and/or non-expressing tissues.

V. A MODEL FOR INDUCTION AND REGULATION OF RIPENING

The series of molecular genetic experiments detailed here, in conjunction with previous descriptions of ripening mutants and observations relating to ripening physiology,

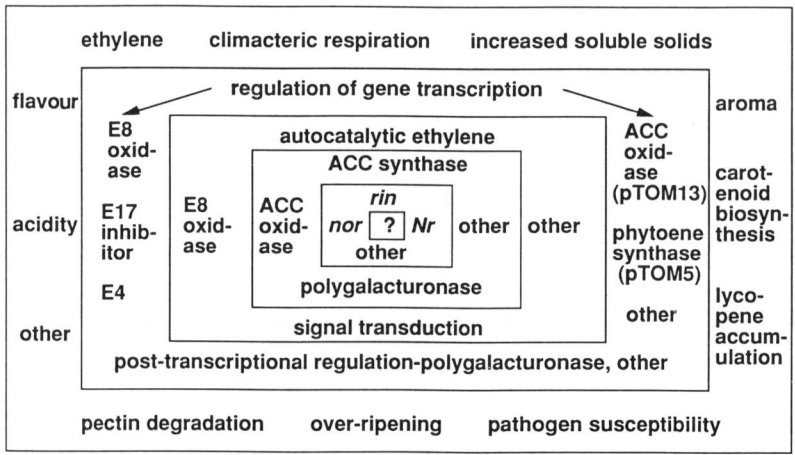

FIG. 12.7. Model for the hierarchy of tomato fruit-ripening regulation.

allow one to make a crude yet useful heirarchial model for the genetic regulation of this final phase of tomato fruit development (Fig. 12.7). The purpose of this model is to focus attention toward questions to be addressed by ripening biologists in the near future.

Antisense gene repression has confirmed the pivotal role of ethylene biosynthesis in coordinating ripening events, and also has shed light on mechanisms of plant hormone action and the regulation of its autocatalytic nature. Through analysis of reduced ethylene evolution in transgenic plants, much has been learned of the sites of action (transcription, post-transcriptional) of this hormone. Contrary to previous hypotheses, polygalacturonase and ACC oxidase gene activity apparently do not require ethylene, though polygalacturonase protein accumulation is ethylene dependent. Tantalising observations of both the need for continuous ethylene, and its excessive concentration as compared to what is minimally required of a ripening fruit, raise interesting new questions concerning the genetic system which responds to this hormone during ripening. Mysteries also remain as to the specific mechanisms of ethylene action, including the nature of the ethylene receptor and that of additional components of the signal transduction machinery. Finally, the function for numerous additional ethylene-induced genes, such as E4, remains to be determined.

Along with insight into the role of ethylene, the regulation of cell wall metabolism and pigment accumulation during ripening have been addressed through antisense technology. Nevertheless, only two plant genes encoding steps in the carotenoid biosynthetic pathway have been isolated. Also, little is known of the regulation of this pathway during ripening. With regards to cell wall metabolism, the function of de-esterification via pectinmethylesterase in modifying net polygalacturonase activity has been shown, while antisense techniques in conjunction with mutant complementation have altered widely held hypotheses on the role of polygalacturonase in ripening-related softening, and focused attention on its possible function in increased pathogen susceptibility.

As the outer tier of Fig. 12.7 shows, genetic components regulating many aspects of the net ripening phenotype remain to be defined, especially those influencing more subtle characters such as flavour and aroma. Many such traits are probably quantitative characters which will only be understood through extensive genetic analysis. The availability of the nearly saturated tomato RFLP map has already allowed the initial genetic characterisation of quantitative characters (Paterson *et al.*, 1988, 1991). Further characterisation will likely be followed by chromosome walking to quantitative trait loci (QTLs).

Finally, the centre-most tiers of Fig. 12.7 point to remaining questions of developmental control of the ripening process. Ethylene certainly coordinates and regulates ripening, but what controls the onset of climacteric ethylene biosynthesis? What changes coordinate polygalacturonase, ACC synthase and ACC oxidase gene transcription, in addition to as yet undefined primary ripening responses? Is there a single genetic 'switch' which initiates the ripening cascade, or is a certain ripening competence achieved through the summative action of individually limited regulators? The answers to at least some of these questions lie in an understanding of the available ripening mutants. *rin* and *nor*, in particular, represent lesions at loci necessary for both the initiation of the ripening climacteric and the transcription of non-ethylene requiring genes such as polygalacturonase. *Nr* delays and attenuates the onset of ripening and virtually all measured ripening characters, indicating that it too represents a primary ripening locus. Together these facts justify placement of *rin*, *nor* and *Nr* at the peak of the hierarchy. Once the corresponding normal and mutant alleles have been isolated and characterised, their regulation can also be addressed. The remainder of this chapter will be devoted to a description of progress toward the isolation of the *rin*, *nor* and *Nr* loci.

VI. STRATEGY AND PROGRESS TOWARD CLONING *rin*, *nor* AND *Nr*

A. Map-based Cloning

Because little is known of the biochemical activities or patterns of expression of any of the three targeted ripening genes, a strategy based solely on the position of these loci on the genetic map was selected. Figure 12.8 summarises the strategy for map-based cloning of fruit-ripening genes, and it should be noted that such a strategy may prove useful for the isolation of any gene which can be scored in a segregating population. In summary, a target system and locus are selected (in this case a tomato-

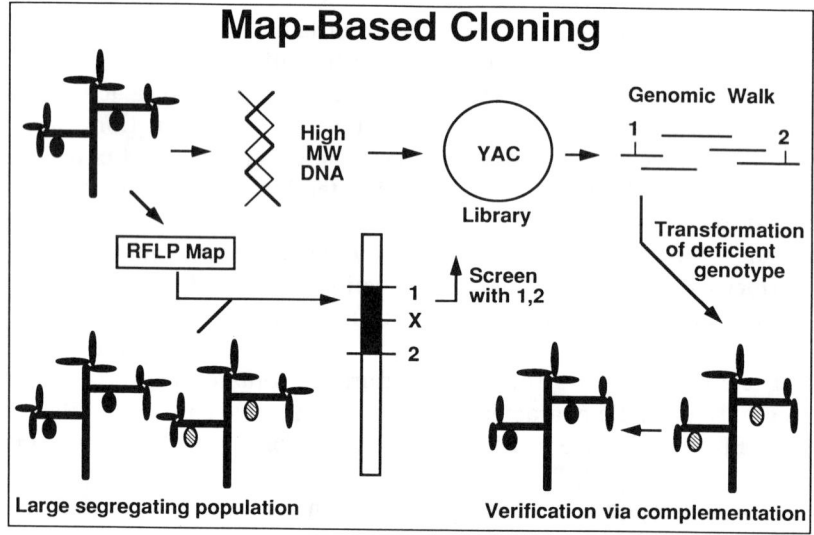

FIG. 12.8. Strategy for map-based cloning of fruit-ripening genes. Segregation analysis of RFLP markers and ripening phenotypes is performed in order to identify single copy markers ('1' and '2') tightly linked to and flanking the target locus ('X'). Flanking markers are then utilised as molecular probes for isolation of high molecular weight genomic DNA fragments from a tomato yeast artificial chromosome (YAC) library. Homologous recombinant clones represent starting points for a genomic walk in which DNA sequences derived from the ends of YAC inserts are used to isolate additional overlapping clones from the tomato YAC library. Ultimately, the genomic region bound by markers '1' and '2' (and containing the target locus 'X') is cloned as overlapping YAC inserts. Verification of the presence of the target locus within the cloned DNA sequences is obtained via transformation and complementation of tomato plants homozygous for the recessive ripening allele. Plants with cross-hatched and filled circles represent those yielding mutant and normal fruit, respectively.

ripening gene) and a population segregating for both RFLP markers and the locus of interest is generated. Gel-blot hybridisation between RFLP probes and genomic DNA isolated from individual progeny is then utilised to identify tightly linked and (preferably) flanking molecular markers. These markers can then be utilised as probes to screen a library of high molecular weight genomic DNA derived from the target organism. At present, yeast artificial chromosomes (YACs) represent the vector system capable of accepting the largest DNA fragments, with inserts over 1 megabase reported (Burke *et al.*, 1987). Under optimal conditions, clones containing both flanking sequences, and thus the intervening target locus, will be recovered. If such is not the case, a 'chromosome walk' must be performed (Rommens *et al.*, 1989; Wallace *et al.*, 1990; Ganal *et al.*, 1990). In a chromosome walk, cloned insert termini are utilised as probes for either genetic mapping relative to the target locus, and/or subsequent screens of the YAC library.

Once a single clone or overlapping clones are isolated, the task becomes one of identifying the locus of interest within the cloned region. Use of an entire YAC clone as a molecular probe for screening a cDNA library, made from tissues likely to express the target gene, has been demonstrated as a viable method for identification of a gene of interest within a YAC (Berger *et al.*, 1992). Alternatively, a target region may be localised within the cloned region through RFLP mapping of subcloned YAC sequences. Whatever the means of localisation, verification of the cloned target locus must ultimately be obtained through complementation of homozygous recessive individuals with the dominant allele.

Based on the above summary, certain prerequisites for a map-based cloning strategy become apparent and include: (1) the ability to score the phenotype of interest in a segregating population; (2) the availability of DNA markers and polymorphic germplasm for RFLP mapping; (3) a representative YAC library which contains the dominant allele of the target locus; and (4) technology for transformation and regeneration of the biological system used. As described previously, all of these requirements are available in tomato and make map-based cloning of the normal *rin* and *nor* ripening genes possible. Because *Nr* is a dominant mutation, this strategy may prove more difficult given that the only tomato YAC library was generated from tomato cultivars with normally ripening fruit (and thus the recessive *nr* allele: Martin *et al.*, 1992).

B. Isolation of Markers Tightly Linked to a Target Locus

Although a nearly saturated RFLP map has been constructed for tomato (Tanksley *et al.*, 1992), one invariably will confront situations in which the closest marker is still too distant from a target locus to reasonably initiate a chromosome walk. One solution to this problem is the use of near isogenic lines (NILs) to detect molecular markers specific to a region of interest. Young *et al.* (1988) used NILs and pools of RFLP probes to detect new markers within the *Tm-2a* region of tomato. Using a similar strategy, Martin *et al.* (1991) were able to use random polymerase chain reaction (PCR) amplification on NILs to isolate new markers near the tomato *Pto* disease resistance locus. Unfortunately, for many potential target intervals, including those containing the *rin*, *nor* and *Nr* loci, useful NILs are not available and are too time-consuming to generate. Fortunately, a technique to target specific chromosomal regions for marker isolation which does not require highly specialised genetic materials arose from efforts to place the *nor* locus on the tomato RFLP map (Giovannoni *et al.*, 1991).

To summarise this technique, individuals are selected from a segregating population (e.g. F2) that are homozygous across a target interval based on known markers. DNA from these individuals is combined into two pools: one homozygous for one parental

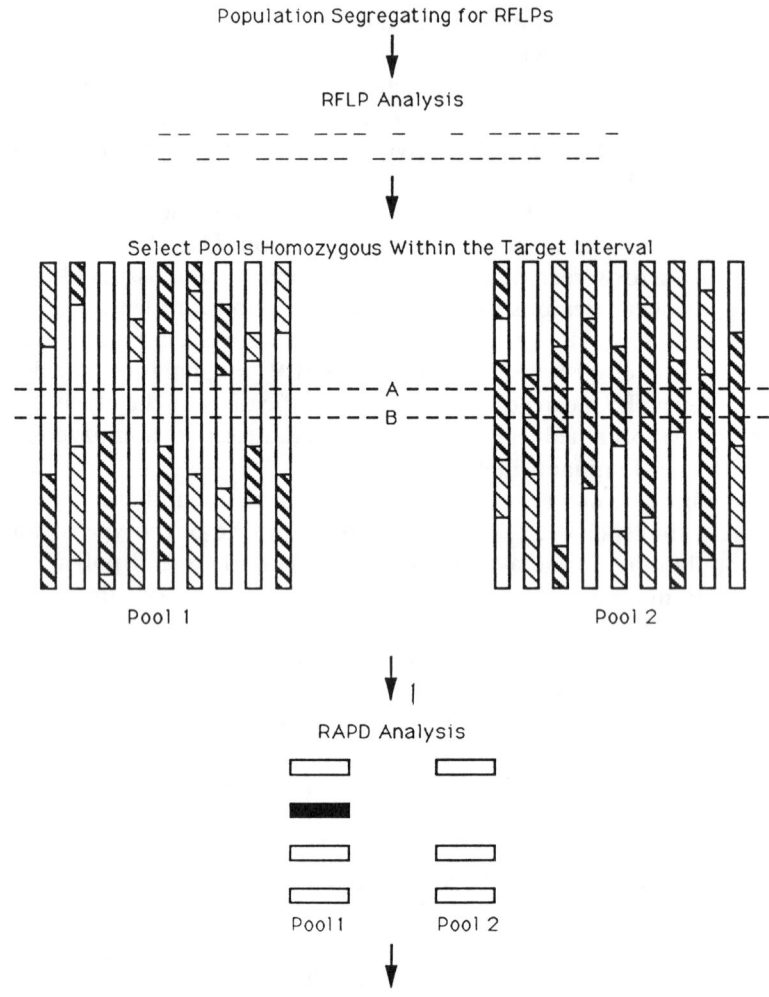

FIG. 12.9. Schematic description of generation and use of isogenic DNA pools. Based on RFLP analysis, individual members of a segregating population are divided into two pools homozygous in the target interval for each of the parental genotypes. Bars represent net chromosomal composition of individual members of a segregating population. Open bars designate chromosomal regions homozygous for one parental genotype, lightly cross-hatched bars designate heterozygous chromosomal regions, and heavily cross-hatched bars designate chromosomal regions heterozygous for the other parental genotype. A and B represent two linked RFLP markers defining the borders of a target interval. Random primer amplification of isogenic DNA pools is subsequently employed to identify molecular markers specific to the targeted region. From Giovannoni *et al.* (1991).

type and one homozygous for the other. The result is two DNA pools homozygous at all loci within and adjacent to the target region. However, the homozygous target region differs between the two pools in parental origin, thus providing the basis for selection of polymorphic markers specific to the targeted region. A sufficiently large number of individuals in each pool assures a high probability for the presence of genomic sequences derived from both parents for all loci except those most tightly linked to and included within the target interval. Pooled DNA samples are subsequently utilised as templates for random primer amplification via PCR (Williams et al., 1990). The amplified products derived from each pool with a specific random primer are then compared by gel electrophoresis for the occurrence of polymorphism between the two pools. Because the pools are essentially homogeneous for all genomic sequences except those within and adjacent to the target interval, polymorphisms should result only if the primer primed within or adjacent to the target interval (Fig. 12.9). Finally, proof of marker localisation within the target interval is obtained through segregation analysis of either the polymorphic PCR product or an RFLP detected via hybridisation of the purified polymorphic DNA band. A similar strategy for linked marker isolation was described by Michelmore et al. (1991).

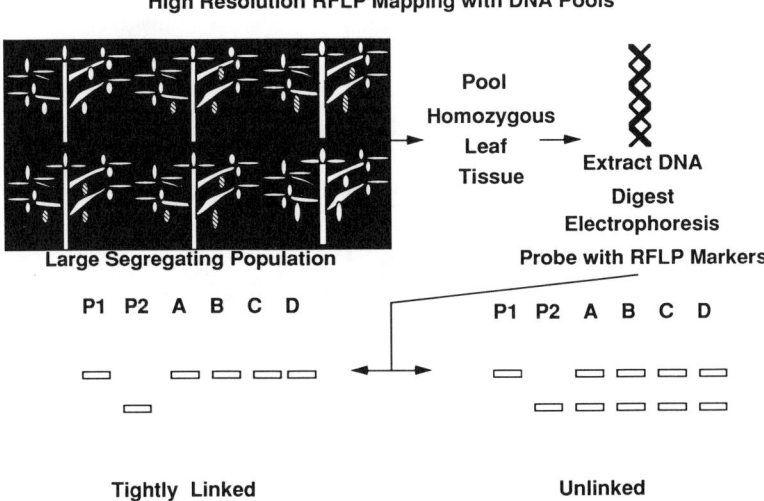

FIG. 12.10. Strategy for high resolution RFLP mapping with pooled DNA samples. A large population segregating for a locus of interest and RFLP markers is generated. Individuals scored as homozygous at the locus of interest are identified and grouped into sets of five whose leaf tissue is pooled. DNA is then extracted from each pool, digested with restriction endonucleases, blotted and probed with RFLP markers. P1 and P2 represent the parents from which the segregating population was derived, while 'A', 'B', 'C' and 'D' each represent DNA pools of five individuals. RFLP markers which are linked to the locus selected to be homozygous will result in hybridisation primarily to the RFLP allele derived from the parent displaying the selected phenotype. However, because selection for members of pools is based solely on homozygosity at the locus of interest, unlinked or distant loci will likely be represented by both parental alleles within most pools. Consequently, RFLP probes which yield predominant gel-blot hybridisation to one parental allele are tightly linked to the locus selected to be homozygous.

C. Pooled Sample Analysis Simplifies Genetic Mapping

Once the availability of markers tightly linked to a target locus becomes less of a limiting factor, the ability to order them efficiently relative to one another and to the gene of interest becomes a primary concern. This is because increasing numbers of segregating individuals must be analysed for recombination between loci as the genetic distances separating the said loci decrease. In an effort to overcome the problem of analysing large populations, a method employing pooled-sample analysis was developed for constructing a high resolution RFLP map for tightly linked markers flanking the *rin* locus (Churchill *et al.*, 1993).

In short, a large population segregating for the locus of interest and RFLP markers is generated. Individuals scored as homozygous at the locus of interest are identified (for example, individuals homozygous for the recessive ripening mutation *rin* can be identified by their non-ripening fruit) and grouped into sets of several homozygous individuals whose leaf tissue is pooled. DNA is then extracted *en mass* from each pool,

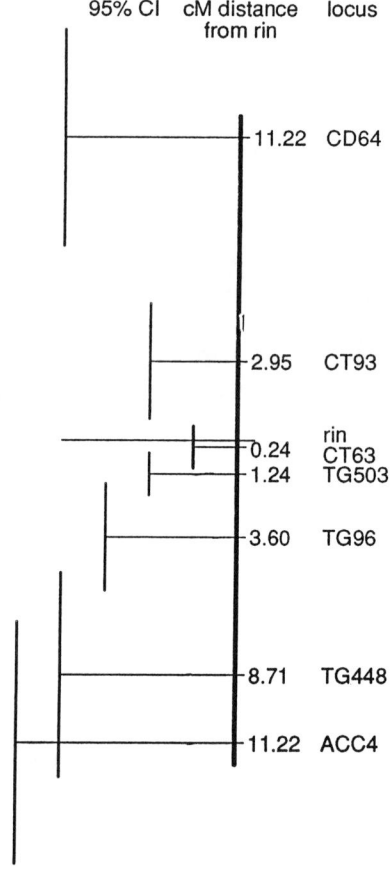

FIG. 12.11. RFLP map of the *rin* locus generated through the use of the pooled sample mapping strategy described in Fig. 12.10. Distances in cM from seven RFLP markers to the *rin* locus are shown. Bars represent regions in which each marker lies with 95% confidence. From Churchill *et al.* (1993), with permission.

digested with restriction endonucleases, blotted and probed with RFLP markers (Fig. 12.10). RFLP markers which are linked to the locus selected to be homozygous will result in hybridisation primarily to the RFLP allele derived from the parent displaying the selected phenotype. Because selection for membership in pools is based solely on homozygosity at the locus of interest, unlinked or distant loci will probably be represented by both parental alleles within most pools. In the example of the recessive *rin* allele, linked RFLP markers show hybridisation primarily to the allele derived from the mutant parent (J. Giovannoni and S. Tanksley, unpublished).

Pooled sample analysis has resulted in the placement of the *rin* locus in the 3.2 cM genomic interval defined by markers CT63 and CT93 (Fig. 12.11: Churchill *et al.*, 1993). In addition, preliminary physical mapping with pulsed field gel electrophoresis indicates that CT63 and CT93 are separated by no more than 800 kb, corresponding to an average maximum of 250 $kb\,cM^{-1}$ in this region of the tomato genome (J. Giovannoni and S. Tanksley, unpublished). Based on the average insert size of 140 kb for the tomato YAC library (Martin *et al.*, 1992), *rin* appears to be within walking distance of CT63.

VIII. SUMMARY

Given the emphasis on the model system of tomato for analysis of fruit development and ripening, it is appropriate to include a note of caution. Although the role of ethylene in climacteric fruit ripening is evident, the precise mechanisms for manifestation of its effects may vary among different species, as do the resulting ripening characteristics. In addition, to date very little has been learned of the genetic basis of ripening in non-climacteric fruit. Because of the obvious importance of non-climacteric fruit in human nutrition and the potential of an alternative system for fruit ripening, the genetic and physiological basis of this developmental programme should not be ignored.

Finally, fruit ripening has proven a fertile field for the analysis of the genetic and physiological basis of plant development. Antisense gene represssion in particular has seen both refinement and utilisation for analysis of gene function during fruit ripening. In the near future, the continued development and use of antisense technology, in addition to emerging strategies such as map-based cloning for the examination of ripening, will surely result in the resolution of old questions and the generation of new ones.

REFERENCES

Abeles, F. (1973). "Ethylene in Plant Biology". Academic Press, New York.
Ali, Z. and Brady, C. (1982). *Aust. J. Plant Physiol.* **9**, 155–169.
Armstrong, G., Alberti, M., Leach, F. and Hearst, J. (1989). *Mol. Gen. Genet.* **216**, 254–268.
Armstrong, G., Alberti, M. and Hearst, J. (1990). *Proc. Natl. Acad. Sci. USA* **87**, 9975–9979.
Baile, J. and Young, R. (1981). *In* "Recent Advances in the Biochemistry of Fruits and Vegetables" (J. Fiend and M. Rhodes, eds), pp. 1–39. Academic Press, London.
Bartley, G. and Skolnik, P. (1989). *J. Biol. Chem.* **264**, 13 109–13 113.
Berger, W., Meindl, A., van de Pol, T., Cremers, F., Ropers, H., Doerner, C., Monaco, A.,

Bergen, A., Lebo, R., Warburg, M., Zergollern, L., Lorenz, B., Gal, A., Bleeker-Wagemakers, E. and Meitinger, T. (1992). *Nature Genetics* **1**, 199–203.
Bevan, M., Mason, S. and Goelet, P. (1985). *EMBO J.* **4**, 1921–1926.
Beyer, E. (1981). In "Recent Advances in the Biochemistry of Fruits and Vegetables". (J. Fiend and M. Rhodes, eds), pp. 107–121. Academic Press, London.
Biggs, M., Harriman, R. and Handa, A. (1986). *Plant Physiol.* **81**, 395–403.
Bird, C., Smith, C., Ray, J., Moureau, P., Bevan, M., Bird, A., Hughes, S. Morris, P., Grierson, D. and Schuch, W. (1980). *Plant Mol. Biol.* **11**, 651–662.
Bird, C., Ray, J., Fletcher, J., Boniwell, J., Bird, A., Teulieres, C., Blain, I., Bramley, P. and Schuch, W. (1991). *Bio/Technology* **9**, 635–639.
Bleecker, A., Kenyon, W., Somerville, S. and Kende, H. (1986). *Proc. Natl. Acad. Sci. USA* **83**, 7755–7759.
Brady, C., Meldrum, S., McGlasson, W. and Ali, M. (1983). *J. Food Biochem.* **7**, 7–14.
Buescher, R. (1977). *HortSci.* **12**, 315–316.
Burg, S. (1962). *Ann. Rev. Plant Physiol.* **13**, 265–302.
Burke, D., Carle, G. and Olson, M. (1987). *Science* **236**, 806–812.
Cass, L., Kirven, K. and Christoffersen, R. (1990). *Mol. Gen. Genet.* **223**, 76–86.
Churchill, G., Giovannoni, J. and Tanksley, S. (1993). *Proc. Natl. Acad. Sci. USA*, **90**, 16–20.
Cordes, S., Deikman, J., Margossian, L. and Fischer, L. (1989). *Plant Cell.* **1**, 1025–1034.
Darby, L. (1978). *Hort. Res.* **18**, 73–84.
Davies, K. and Grierson, D. (1989). *Planta* **179**, 73–80.
Davis, M., Cohen, D., Nielsen, E., Steinmetz, M., Paul, W. and Hood, L. (1984). *Proc. Natl. Acad. Sci. USA* **81**, 2194–2198.
Deikman, J. and Fischer, R. (1988). *EMBO J.* **7**, 3315–3320.
DellaPenna, D. and Bennett, A. (1988). *Plant Physiol.* **86**, 1057–1063.
DellaPenna, D., Alexander, D. and Bennett, A. (1986). *Proc. Natl. Acad. Sci. USA* **83**, 6420–6424.
DellaPenna, D., Lincoln, J., Fischer, R. and Benett, A. (1989). *Plant Physiol.* **90**, 1372–1377.
DellaPenna, D., Lashbrook, C., Toenjes, K., Giovannoni, J., Fischer, R. and Bennett, A. (1990). *Plant Physiol.* **94**, 1882–1886.
Ecker, J. and Davis, R. (1986). *Proc. Natl. Acad. Sci. USA* **83**, 5372–5376.
Ecker, J. and Davis, R. (1987). *Proc. Natl. Acad. Sci. USA* **84**, 5202–5206.
Fillatti, J., Kiser, J., Rose, B. and Comai, L. (1987). In "Tomato Biotechnology". (D. Nevins and R. Jones, eds), pp. 199–210. Alan R. Liss, New York.
Fischer, R. and Bennett, A. (1991). *Ann. Rev. Plant Physiol. Plant Mol. Biol.* **42**, 675–703.
Frenkel, C. and Garrison, S. (1976). *HortSci.* **11**, 20–21.
Ganal, M., Martin, G., Messeuguer, R. and Tanksley, S. (1990). *AgBiotech News and Info.* **2**, 835–840.
Giovannoni, J., DellaPenna, D., Bennett, A. and Fischer, R. (1989). *Plant Cell.* **1**, 53–63.
Giovannoni, J., DellaPenna, D., Lashbrook, C., Bennett. A., and Fischer, R. (1990). In "Horticultural Biotechnology". (A. Bennett, ed.). Alan R. Liss, New York.
Giovannoni, J., Wing, R., Ganal, M. and Tanksley, S. (1991). *Nucl. Acids Res.* **19**, 6553–6558.
Giovannoni, J., DellaPenna, D., Bennett, A. and Fischer, R. (1992). *Hort. Rev.* **13**, 67–103.
Goodwin, T. (1980). "The Biochemistry of the Carotenoids", Vol. 1, 2nd edn. Chapman and Hall, London.
Gray, J., Picton, S., Shabbeer, J., Schuch, W. and Grierson, D. (1992). *Plant Mol. Biol.* **19**, 69–87.
Green, P., Pines, O. and Inouye, M. (1986). *Ann. Rev. Biochem.* **55**, 569–597.
Greve, C. and Labavitch, J. (1991). *Plant Physiol.* **97**, 1456–1461.
Grierson, D. (1985). *CRC Press Crit. Rev. Plant Sci.* **3**, 113–132.
Grierson, D., Tucker, G., Keen, J., Ray, J., Bird, C. and Schuch, W. (1986). *Nucl. Acids Res.* **14**, 8595–8603.
Grierson, D., Purton, M., Knapp, J. and Bathgate, B. (1987). In "Developmental Mutants in Higher Plants". (H. Thomas and D. Grierson, eds), pp. 79–94. Cambridge University Press, Cambridge.
Grunstein, M. and Hogness, D. (1975). *Proc. Natl. Acad. Sci. USA* **72**, 3961–3965.

Hamilton, A., Lycett, G. and Grierson, D. (1990). *Nature* **364**, 284–287.
Hamilton, A., Bouzayen, M. and Grierson, D. (1991). *Proc. Natl. Acad. Sci. USA* **88**, 7434–7437.
Harriman, R., Tieman, D. and Handa, A. (1991). *Plant Physiol.* **97**, 80–87.
Hobson, G. (1963). *Biochem. J.* **86**, 358–365.
Hobson, G. (1965). *J. Hort. Sci.* **40**, 66–72.
Hobson, G. (1967). *Phytochemistry* **6**, 1337–1341.
Holsdworth, M., Schuch, W. and Grierson, D. (1988). *Plant Mol. Biol.* **11**, 81–88.
Honma, M. and Shimomura, T. (1978). *Agric. Biol. Chem.* **42**, 1825–1831.
Jefferson, R. (1987). *Plant Mol. Biol. Rep.* **5**, 387–405.
Kende, H. (1989). *Plant Physiol.* **91**, 1–4.
Kinzer, S., Schwager, S. and Mutschler, M. (1990). *Theor. Appl. Genet.* **79**, 489–496.
Klee, H., Hayford, M., Kretzmer, K., Barry, G. and Kishmore, G. (1991). *Plant Cell* **3**, 1187–1193.
Knapp, J., Moureau, P., Schuch, W. and Grierson, D. (1989). *Plant Mol. Biol.* **12**, 105–116.
Kopeliovitch, E., Mizrahi, Y., Rabinowitch, H. and Kedar, N. (1980). *Physiol. Plant.* **48**, 307–311.
Kramer, M., Sanders, R., Sheehy, R., Melis, M., Kuehn, M. and Hiatt, W. (1990). *In* "Horticultural Biotechnology". (A. Bennett, ed.). Alan R. Liss, New York.
Lackey, G., Gross, K. and Wallner, S. (1980). *Plant Physiol.* **66**, 532–533.
Lincoln, J., Cordes, S., Read, E. and Fischer, R. (1987). *Proc. Natl. Acad. Sci. USA* **84**, 2793–2797.
Lincoln, J. and Fischer, R. (1988a). *Mol. Gen. Genet.* **212**, 71–75.
Lincoln, J. and Fischer, R. (1988b). *Plant Physiol.* **88**, 370–374.
Margossian, L., Federman, A., Giovannoni, J. and Fischer, R. (1988). *Proc. Natl. Acad. Sci. USA* **85**, 8012–8016.
Markovic, O. and Jornell, H. (1986). *Eur. J. Biochem.* **158**, 455–462.
Martin, G., Williams, J. and Tanksley, S. (1991). *Proc. Natl. Acad. Sci. USA* **88**, 2336–2340.
Martin, G., Ganal, M. and Tanksley, S. (1992). *Mol. Gen. Genet.* **233**, 25–32.
Maunders, M., Holdsworth, M., Slater, A., Knapp, J., Bird, C., Schuch, W. and Grierson, D. (1987). *Plant Cell Environ.* **10**, 177–184.
McCormick, S., Neidermeyer, J., Fry, J., Barson, A., Horsch, R. and Fraley, R. (1986). *Plant Cell Rep.* **5**, 81–84.
McGlasson, W. (1985). *HortSci.* **20**, 51–54.
McGlasson, W., Dostol, H. and Tigchelaar, E. (1975). *Plant Physiol.* **55**, 218–222.
McGarvey, D., Sirevag, R. and Christoffersen, R. (1992). *Plant Physiol.* **98**, 554–559.
Mehta, A., Jordan, R., Anderson, J. and Matoo, A. (1988). *Proc. Natl. Acad. Sci. USA* **85**, 8810–8814.
Michelmore, R., Paran, I. and Kesseli, R. (1991). *Proc. Natl. Acad. Sci. USA* **88**, 9828–9832.
Mitcham, E., Gross, K. and Ng, T. (1989). *Plant Physiol.* **89**, 477–481.
Misawa, N., Nakagawa, M., Kobayashi, K., Yamano, S., Izawa, Y., Nakamura, K. and Harashima, K. (1990). *J. Bacteriol.* **172**, 6704–6712.
Mizrahi, Y., Dostol, H. and Cherry, J. (1975). *HortSci.* **10**, 414–415.
Mizrahi, Y., Zohar, R. and Malis-Arad, S. (1982). *Plant Physiol.* **69**, 497–501.
Moshrefi, M. and Luh, B. (1983). *Eur. J. Biochem.* **135**, 511–514.
Mutschler, M. (1984). *J. Am. Soc. Hort. Sci.* **109**, 504–507.
Napoli, C., Lemieux, C. and Jorgensen, R. (1990). *Plant Cell* **2**, 279–289.
Oeller, P., Wong, L., Taylor, L., Pike, D. and Theologis, A. (1991). *Science* **254**, 437–439.
Olson, D., White, J., Edelman, L., Harkins, R. and Kende, H. (1991). *Proc. Natl. Acad. Sci. USA* **88**, 5340–5344.
Osteryoung, K., Toenjes, K., Hall, B., Winkler, V. and Bennett, A. (1990). *Plant Cell* **2**, 1239–1248.
Otwinowski, Z., Schevitz, R., Zhang, R., Lawson, C., Joachimiak, A., Marmorstein, R., Luisi, B. and Sigler, P. (1988). *Nature* **335**, 321–329.
Paterson, A., Lander, E., Hewitt, J., Peterson, S., Lincoln, S. and Tanksley, S. (1988). *Nature* **335**, 721–726.

Paterson, A., Damon, S., Hewitt, J., Zamir, D., Rabinowitch, H., Lincoln, S., Lander, E. and Tanksley, S. (1991). *Genetics* **127**, 181–197.
Pear, J., Ridge, N., Rasmussen, R., Rose, R. and Houck, C. (1989). *Plant Mol. Biol.* **13**, 639–651.
Pecker, I., Chamovitz, D., Linden, H., Sandmann, G. and Hirschberg, J. (1992). *Proc. Natl. Acad. Sci. USA* **89**, 4962–4966.
Penarrubia, D., Aguilar, M., Margossian, L. and Fischer, R. (1992). *Plant Cell* **4**, 681–687.
Percival, F., Cass, L., Bozak, K. and Christoffersen, R. (1991). *Plant Cell Rep.* **10**, 512–516.
Pressy, R. (1984). *Eur. J. Biochem.* **144**, 217–221.
Pressy, R. and Avants, J. (1973). *Biochim. Biophys. Acta* **309**, 363–369.
Pressy, R. and Avants, J. (1982). *J. Food Biochem.* **6**, 57–74.
Ramirez, D. and Tomes, M. (1964). *Bot. Gaz.* **125**, 221–226.
Rattanapanone, N., Grierson, D. and Stein, M. (1977). *Phytochemistry* **16**, 629–633.
Ray, J. Knapp, J., Grierson, D., Bird, C. and Schuch, W. (1988). *Eur. J. Biochem.* **174**, 119–124.
Rick, C. (1956). *Tomato Genet. Coop.* **6**, 22–23.
Rick, C. (1986). *HortSci.* **21**, 918–919.
Robinson, R. and Tomes, M. (1968). *Rep. Tomato Genet. Coop.* **18**, 36–37.
Rommens, J., Iannuzzi, M., Kerem, B., Drumm, M., Melmer, G., Dean, M., Rozmahel, R., Cole, J., Kennedy, D., Hidaka, N., Zsiga, M., Buchwald, M., Riordan, J., Tsui, L. and Collins, F. (1989). *Science* **245**, 1059–1065.
Rottmann, W., Peter, G., Oeller, P., Keller, J., Shen, N., Nagy, B., Taylor, L., Campbell, A. and Theologis, A. (1991). *J. Mol. Biol.* **222**, 937–961.
Sato, T. and Theologis, A. (1989). *Proc. Natl. Acad. Sci. USA* **86**, 621–625.
Schaffer. M. and Fischer, R. (1988). *Plant Physiol.* **87**, 431–436.
Schaffer, M. and Fischer, R. (1990). *Plant Physiol.* **93**, 1486–1491.
Schmidhauser, T., Lauter, F., Russo, V. and Yanofsky, C. (1990). *Mol. Cell Biol.* **10**, 5064–5070.
Schuch, W., Bird, C., Ray, J., Smith, C., Watson, C., Morris, P., Gray, J., Arnold, C., Seymour, G., Tucker, G. and Grierson, D. (1989). *Plant Mol. Biol.* **13**, 303–311.
Sheehy, R., Pearson, J., Brady, C. and Hiatt, W. (1987). *Mol. Gen. Genet.* **208**, 30–36.
Sheehy, R., Kramer, M. and Hiatt, W. (1988). *Proc. Natl. Acad. Sci. USA* **85**, 8805–8809.
Singh, H., Sen, R., Baltinore, D. and Sharp, P. (1986). *Nature* **319**, 154–158.
Slater, A., Maunders, M., Edwards, K., Schuch, W. and Grierson, D. (1985). *Plant Mol. Biol.* **5**, 137–147.
Smith, C., Watson, C., Ray, J., Bird, C., Morris, P., Schuch, W. and Grierson, D. (1988). *Nature* **334**, 724–726.
Smith, C., Watson, C., Morris, P., Bird, C., Seymour, G., Gray, J., Arnold, C., Tucker, G., Schuch, W., Harding, S. and Grierson, D. (1990a). *Plant Mol. Biol.* **14**, 369–379.
Smith, C., Watson, C., Bird, C., Ray, J., Schuch, W. and Grierson, D. (1990b). *Mol. Gen. Genet.* **224**, 477–481.
Spanu, P., Reinhart, D. and Boller, T. (1991). *EMBO J.* **10**, 2007–2013.
Tanksley, S., Ganal, M., Prince, J., de Vicente, M., Bonierbale, M., Broun, P., Fulton, T., Giovannoni, J., Grandillo, S., Martin, G., Messeguer, R., Miller, J., Miller, L., Paterson, A., Puieda, O., Röder, M., Wing, R., Nu, N. and Young, N. (1992). *Genetics*, **132**, 1141–1160.
Theologis, A. (1992). *Cell* **70**, 181–194.
Tieman, D. and Handa, A. (1989). *Plant Physiol.* **90**, 17–20.
Tieman, D. Harriman, R., Ramamohan, G. and Handa, A. (1992). *Plant Cell* **4**, 667–679.
Tigchelaar, E., Tomes, M., Kerr, E. and Barman, R. (1973). *Rep. Tomato Genet. Coop* **23**, 33.
Tigchelaar, E., McGlasson, W. and Buescher, R. (1978a). *HortSci.* **13**, 508–513.
Tigchelaar, E., McGlasson, W. and Franklin, M. (1978b). *Aust. J. Plant Physiol.* **5**, 449–456.
Trewavas, A. (1982). *Physiol. Plant.* **55**, 60–72.
Tucker, G., Robertson, N. and Grierson, D. (1980). *Eur. J. Biochem.* **112**, 119–124.
Tucker, G., Robertson, N. and Grierson, D. (1981). *Eur. J. Biochem.* **115**, 87–90.
Tucker, G., Robertson, N. and Grierson, D. (1982). *J. Sci. Food. Agric.* **33**, 396–400.

van der Krol, A., Lenting, P., Veenstra, J., van der Meer, I., Koes, R., Gerats, A., Mol, J. and Stuitje, A. (1988). *Nature* **333**, 866-869.
van der Krol, A., Mur, L., Beld, M., Mol, J. and Stuitje, A. (1990). *Plant Cell* **2**, 291-299.
Van Haaren, M. and Houck, C. (1991). *Plant Mol. Biol.* **17**, 615-630.
Wallace, M., Marchuk, D., Andersen, L., Letcher, R., Odeh, H., Saulino, A., Fountain, J., Brereton, A., Nicholson, J., Mitchell, A., Brownstein, B. and Collins, F. (1990). *Science* **249**, 181-186.
Walling, L., Drews, G. and Goldberg, R. (1986). *Proc. Natl. Acad. Sci. USA* **83**, 2123-2127.
Williams, J., Kubelik, A., Livak, K., Rafalski, J. and Tingey, S. (1990). *Nucl. Acids Res.* **18**, 6531-6535.
Yang, S. (1980). *HortSci.* **15**, 238-243.
Yang, S. (1985). *HortSci.* **20**, 41-45.
Yang, S. and Hoffman, N. (1984). *Ann. Rev. Plant Physiol.* **35**, 155-189.
Young, N., Zamir, D., Ganal, M. and Tanksley, S. (1988). *Genetics* **120**, 579-585.
Yu, Y., Adams, D. and Yang, S. (1979). *Arch. Biochem. Biophys.* **198**, 280-286.

Index

Note: Figures and tables are indicated by *italic page numbers*

A

A9 (anther-specific) cDNA, 58, *59*
  temporal expression pattern, 59, *60*
ABA-responsive element (ABRE), 244
Abscisic acid (ABA)
  effects, 237, 242–243, 245, 246
  inhibition of biosynthesis, 244
ACC deaminase, 272
ACC oxidase
  gene encoding, 269
  role in ethylene biosynthesis, 253–254
ACC synthase
  genes encoding, 260–261
  role in ethylene biosynthesis, 253, 266, 268
Adaptor molecules, cDNA cloning, 66
Affinity chromatography, 22
African clawed toad *see Xenopus*
Agarose gel electrophoresis, 26, 27, 29, 87, 91
Albino mutant protoplasts, somatic hybrid selection, 198–199
Alkaline phosphatase immunoreaction system, 109, 152–153, 169
1-Aminocyclopropane-1-carboxylic acid (ACC), 253
Anhydrobiosis, 240
Annealing temperature, determination, 131, *132*
Anther development, 60–61
Anther-specific cDNAs, autoradiographic identification, 58, *59*
Anther-specific genes, cloning of, 58
Antibodies
  detection of proteins using, 107–108
  recommended protocol, 109
  purification by immunoaffinity blotting, 111
  types used in ICC, 171
Antibody precipitation, analysis of mRNA translation products using, 48–49
Antibody screening, cDNA, 77, 226
Antifade mounting medium, 175
  preparation, 183
Antigens
  diversity in plant cells, 160
  immunolocalisation techniques for, 160–183
Antimetabolites, somatic hybrids selected using, 199
Antisense gene repression, fruit-ripening, 263–271, 275
Arabidopsis thaliana (thale cress)
  embryonic development, 221–223
  embryo-specific mutants, 231–235
  protoplast fusion example, 202
  as representative of dicotyledonous plants, 221, 230
Arabidopsis tissue
  DNA fingerprinting, 125
  RNA isolation method, 9–10
ATP
  chloroplast protein import affected by, 215
  protein synthesis affected by, 38
Atropa belladona (deadly nightshade), somatic hybrids, 201
Autoradiography
  analysis of mRNA translation products using, 47

Autoradiography (*cont.*)
　anther-specific cDNA studied, 58, *59*
　co-localisation with DIG-detection, 154–155
　procedure, 154
Autotrophy, 199
Avena sativa (oat)
　RNA isolation method, 14–15, *19*
　see also Oat
Avian reverse transcriptase (AMV-RT), 63
　action, 63
Avocado fruit, RNA isolation method, *19*

B

Bacteriophage vectors, cDNA cloning using, 66–68
Barley
　embryogenesis, 227, *228*
　　polypeptides accumulating, 242
　Northern analysis of root-tip RNA, *86*
　RNA isolation method, 12–13, *19*
　see also *Hordeum* spp.
Biochemical inhibitors, somatic hybrids selected using, 199
Biotin labelled probes, 142, 167
　compared with DIG-labelled probes, 142
Biotin–streptavidin immunodetection system, *167*, 167–168
　protocol for use, 178
Blocking proteins, 84, 87, 106–107, 165
Blotted proteins, microanalysis, 111–113
Blotting membranes, 82, 87, 102
Blotting techniques see Northern . . . ; Southern . . . ; Southwestern . . . ; Western blotting
Bovine lacto transfer technique optimiser (BLOTTO), 107
Brassica napus (rape)
　cDNA libraries, 58
　nucleotide sequence, 76
Brassica spp.
　DNA fingerprinting, 125
　protoplast fusion examples, *189*, 201, 202
Bread wheat
　chromosome mapping, 94, *95*
　mutant line studied, 98
　Southern blot analysis applied, *92*, *95*, *96*, *98*

C

Caesium chloride centrifugation method (in RNA isolation), 14, 15

Capillary blotting method, 82, 103
　see also Southern blotting
Capsella bursa-pastoris (shepherd's purse), embryonic development, 221–223
Carotenoid biosynthesis, tomato fruit ripening, 254, 270–271
Carrot
　embryonic development, 226
　RNA isolation method, 10–11, *18*
Cauliflower Mosaic Virus (CaMV) 35S promoter, 244, 263
cDNA see Complementary DNA
Cell wall glycoproteins, 229
Cereals
　desiccation tolerance, 238–239
　see also Barley; Maize; Oat; Wheat
Cereal seed proteins, 238
　identification
　　after gel electrophoresis, 109, *110*
　　multigene family members, 96–97
Cetyltrimethyl ammonium bromide (CTAB)
　RNA isolation using, 16–17, *19*, 20
　as RNase inhibitor, *3*
Chaotropic agents
　RNA isolation using, 13–16, *19*
　see also Guanidine hydrochloride; . . . thiocyanate; Sodium perchlorate
Chemical fixatives, 144, 162
Chemically induced protoplast fusion
　compared with electrofusion, 189, 194–195
　first reported, 188
　large-scale techniques, 190–194
　small-scale techniques, 197
Cherry tissues, RNA isolation method, 10–11, *18*
Chicken embryo tissue, RNA isolation method, 7–8
Chlorophyll *a/b* preproteins, analysis method, *48*, 49
Chlorophyll-deficient mutants, somatic hybrid selection, 198–199
Chlorophyll degradation, ripening of tomato, 254
Chloroplast proteins
　biogenesis, *208*
　import into intact chloroplasts, 213–215
　　assay procedure, 213–215
　　exogenous ATP used to drive import, 215
　import into isolated thylakoids, 215–218
　　assay procedure, 217–218
　　processes involved, 215–216
　　requirements, *216*
　　two-phase model, 216
　in vitro synthesis, 211–213

INDEX

Chloroplasts
 biogenesis, 207–209
 import of proteins, 213–215
 isolation, 209–211
  growth conditions necessary, 209
  procedures, 209–211
Chromosome mapping, Southern blotting used, 94, *95*
Citropsis gilletiana, protoplast fusion example, *189*
Citrus spp., protoplast fusion examples, *189*
Climacteric ripening, 252–253
 ethylene biosynthesis, 253
  gene affecting, 255, 269–270
Cloning
 cDNA, 65–72, 126
 PCR used, 126–129
Coagulating fixatives, 144, 162
Cocoa, DNA fingerprinting, 125
Codon usage, 133
Cold-plaques, 61
Cold-plaque screening, 61, 74–76
 diagnostic Northern dot blots used, 75–76
 preparation of probe stocks, 75
Colloidal gold, immunolocalisation using, 170, 179–181
Complementary DNA (cDNA)
 applications, 61
 cDNAs encoding specific peptides, 76–78
 cloning, 65–72
  adaptors used prior to, 66
  linkers used prior to, 65–66
  PCR used, 126
  preparation of cDNA, 65–66
  size selection prior to, 65
  subtraction protocols, 65
  tailing of cDNA, 65
  vectors used, 66–72
 first strand synthesis, 63, 126
  primers used, 62–63
 identification methods, 49–50
 library
  amplification procedures, 68, *81*, 123–124
  construction, 58, 62–64
  screening, 61–62, 72–78, *81*, 226
  titre measurement, 68
  tomato, 260
 screening techniques, 61–62, 72–78
  cold-plaque screening, 61, 74–76
  differential screening, 70–71, 72–74
  heterologous DNA probes used, 76
  PCR used, *81*, 123–124
 second strand synthesis, 63–64, 126–127
 synthesis, 62–64, 126–127
  RNA quality required, 62

Conjugated enzyme-labelled protein-detection methods, 107–108
 disadvantage, 108
 recommended protocol using alkaline phosphatase, 109
Co-suppression (of gene expression)
 meaning of term, 272
 tomato fruit ripening, 271–272
Cotton tissues, RNA isolation method, 8–10, *18*
Craterostigma (resurrection plant), desiccation tolerance, 240–241
Crosslinking fixatives, 144, 162
Cryosectioning techniques, 143, 163–164, 174–175
 limitations, 143, 163–164
Cybridisation, 203
Cysteine, labelled, incorporation during RNA translation, 38, 45
Cytoplasmic genomes, effect of protoplast fusion, 202–203

D

Datura innoxia, somatic hybrids, 199, 201
Daucus spp., somatic hybrids, 199
Dehydrins, 241
Denaturing fixatives, 162
Denaturing sucrose gradient method, RNA fractionation using, 25–26
Desiccation tolerance
 cereals, social impact, 238
 seeds, 238–239
  factors affecting enhancement/ acquisition, 239–243
Detergent (RNA isolation) method, 8–10, *18*
Dextran sulphate, nucleic acid hybridisation affected by, 87, 150
Diagnostic Northern dot blots, 75–76
Diazobenzyloxymethyl (DBM) paper, binding of cDNA to, 50
Dicotyledonous plants, embryonic development, 221–223
Diethylpyrocarbonate (DEPC)
 precautions required, 143
 as RNase inhibitor, *3*, 88–89
Differential growth, somatic hybrids selection using, 200
Differential hybridisation, 58
Differential precipitation (RNA isolation) method, 10–11, *18*
Differential screening
 cDNA, 58–61, 72–74
  limitations, 61
  nucleic acid hybridisation methods, 73–74

Differential screening, cDNA (*cont.*)
   preparation of first strand cDNA probe stocks, 72–73
   preparation of replicate lifts, 70–71, 72
   purification of first strand cDNA probes, 73
   fruit ripening-related genes isolation, 257–260
   regulation of ethylene-inducible genes, 258–260
Diffusion transfer method, 103
Digoxigenin (DIG) labelled probes, 142
   co-localisation with autoradiography, 154–155
     advantages, 156–157
   detection of, 148, 152–153
   factors affecting performance, 148
   protococol for preparation, 147–148
Dimethylsulphoxide, denaturation using, 26, 51
Dithiothreitol (DTT), *3*, 150, 165
DNA binding proteins, detection, 113–114
DNA fingerprinting, 124–125
DNA sequences
   alteration/manipulation by PCR, 129
   screening for, 61–62, 72–78
     cold-plaque screening, 61, 74–76
     differential screening, 70–71, 72–74
     heterologous DNA probes used, 76
     PCR used, *81*, 123–124
DNA, *see also* Complementary DNA (cDNA)
Dog pancreatic microsomal membranes, *in vitro* processing using, 52–53
Doppelwurzel type mutants (*Arabidopsis thaliana*), 233
Double-mutants technique (for somatic hybrids selection), 200
Drosophila melanogaster (fruit fly), genetic analysis, 230
Durum wheat, chromosome mapping, 94, 95

E

Electroblotting, 104–105
   apparatus commercially available, 105
   typical protocol, 105–106
   see also Western blotting
Electrofusion (protoplast fusion), 194
   compared with chemically induced fusion, 189, 194–195
   first reported, 188
   large-scale techniques, 194–196
   small-scale techniques, 197–198
Electrophoresis *see* Gel electrophoresis

Embedding techniques, 145, 146, 163, 164–165, 173–174, 177
Embryogenesis
   genetic analysis, 230–237
   molecular analysis, 225–230
Embryonic development, 220–225
   dicotyledonous plants, 221–223, 225
   molecular biochemical approach, 225–230
   monocotyledonous plants, 223–225
Embryonic maturation, 237–246
Em proteins, 241–242
Enhanced chemiluminescence (ECL) Western blotting system, 108
Enhancer trapping, 235
Enzyme-labelled antibodies
   immunolocalisation using, 169
   protein detection using, 107–108
   recommended protocol, 109
Escherichia coli
   bacteriophage lambda vectors, 68–69
   mRNA translation using, 34, 35, 36–39
Ethylene
   biosynthesis, 253–254
     rate-limiting catalyst, 253, 266, 268
   as ripening hormone, 253–254
Ethylene-inducible genes
   antisense gene expression studies, 269–270
   regulation in tomato, 258–260
Eukaryotic mRNA translation, systems used, 39, 41

F

Fackel type mutants (*Arabidopsis thaliana*), 232
Ferritin, 169
   immunodetection using, 170
Fingerprinting (DNA/genome), PCR used, 124–125
Fixatives
   immunocytochemistry, 162–163, 172, 176, 180
   RNA hybridisation, 144–145
Flow cytometry, selection of protoplasts, 201
Fluorescein isothiocyanate (FITC), 107, 168
Fluorescent dye labelled antibodies, 168
   problems encountered, 168
   protein detection using, 107
   protocol for use, 178
Formaldehyde solution, preparation, 144, 182
Formalin, 144
Formalin–acetic acid–ethanol (FAA) fixative, 145

Formamide/formaldehyde method, RNA denaturation using, 28–29
Fruit development/ripening, 251–281
 climacteric vs non-climacteric ripening, 252–253
 commonality among species, 252
 tomato as model system, 254–257

G

Gel blot techniques
 A9 gene studied, 59–60, *60*
 see also Northern . . . ; Southern . . . ; Southwestern . . . ; Western blotting
Gel electrophoresis
 fractionation using
  denaturing conditions, 26–29, 45–47
  non-denaturing conditions, 26
Gene copy number reconstruction experiments, 91, *92*, 93
Gene mapping
 chromosomal location, 94, *95*
 molecular location, 93–94
Genetic aberrations (in cell cultures), 227
Genetic analysis
 classical studies, 230
 dicotyledonous plants, 231–235
 monocotyledonous plants, 235–237
Genome fingerprinting, 124–125
β-Glucuronidase (GUS) gene, 235, 244
Glutaraldehyde, as fixative, 144, 145, 162, 176
Glyoxal/dimethylsulphoxide method, RNA denaturation using, 26–27
Gnom type mutants (*Arabidopsis thaliana*), 232
Gold labelling, immunolocalisation using, 170, 179–181
Grape tissues, RNA isolation methods, 9–10, 15–16, *18*, *19*
Guanidine hydrochloride, as RNase inhibitor, *3*
Guanidine thiocyanate
 RNA isolation using, 14–15, *19*
 as RNase inhibitor, *3*
Gurke type mutants (*Arabidopsis thaliana*), 232

H

Heterokaryons, 188
 manual isolation, 200–201
 production, 198
 selection, 198–200
Heterologous DNA probes, cDNA screening by, 76

HistoResin, 165, 177
Homopolymer tailing (for cDNA cloning), 65
Hordeum spp.
 protoplast fusion example, *189*
 see also Barley
Hot phenol (RNA isolation) method, 7–8, *18*
Hyacinth root tips, immunolocalisation of microtubules, 172–176
Hybaid plastic bags, 84, 88
Hybond-N membrane, 70, 82
Hybrid arrest method, mRNA translation products analysed using, 49
Hybridomas, 171
Hybrid release (selection) method, mRNA translation products analysed using, 49, 50–52
Hypophysis, 223

I

Immunoaffinity blotting, purification of antibodies using, 111
Immunocytochemistry (ICC), 160–183
 antibody production/storage, 170–171
 cryosectioning techniques, 163–164, 174–175
 detection systems, 166–170
  alkaline phosphatase used, 169
  colloidal gold used, 170
  controls required, 170
  direct methods, 166
  ferritin used, 169–170
  fluorescent labels used, 168
  indirect methods, 166–168
  peroxidase used, 168–169
  protein A used, 169
 embedding media/techniques used, 163–165, 173–174, 177, 180
 fixatives used, 162–163, 172, 176, 180
 general considerations, 161
 photomicrography, 172, 176
 protocols
  immunogold localisation of proteins, 179–181
  localisation of proteins in resin-embedded tissue, 176–179
  visualisation of microtubules, 172–176
 recipes for reagents, 181–183
 sectioning techniques, 163–164, 174–175, 178, 180
 staining considerations, 165–166
 tissue selection, 161–162
Immunodetection
 DIG-labelled probes, 152–153, 155

INDEX

Immunodetection (cont.)
  direct methods, 166, *167*
  indirect methods, 166-168
Immunoprecipitation, analysis of mRNA translation products using, 48-49, 62
Immunoreaction, protein detection using, 107-108, *110*
Immunostaining, 165-166, 175-176, 178-179
Insertional mutagenesis, 235
In situ (nucleic acid) hybridisation technology, 142-157
  advantages, 156
  DIG-labelled transcripts
    detection, 148, 152-153
    factors affecting performance, 148
    preparation protocol, 147-148
  embedding of tissue, 145
  hybridisation, 150
    protocol, 151
  immunodetection, 152-153
    protocol, 153-154
  microtechnique, 143-145
    protocol, 145-146
  posthybridisation treatment, 151-152
    protocol, 152
  precautions required, 143
  prehybridisation, 149
    protocol, 149-150
  probe preparation, 146
  protocols
    DIG-labelled transcript preparation, 147-148
    dot blots preparation, 148-149
    hybridisation, 151
    immunodetection/staining, 153-154
    linearisation of vector, 147
    microtechnique, 145-146
    posthybridisation treatment, 152
    prehybridisation, 149-150
  sectioning/mounting of tissue, 145
  tissue selecion, 143
In vitro processing, 52-53
In vitro translation systems (for mRNA)
  analysis of products, 45-52
  chloroplast protein synthesis, 211-213
  preparation/use, 36-45
  selecting, 35, 211-212

K

Keule type mutants (*Arabidopsis thaliana*), 232-233
Knolle type mutants (*Arabidopsis thaliana*), 232-233
Knopf type mutants (*Arabidopsis thaliana*), 233

L

Labelled amino acids, incorporation during RNA translation, 38, 45
Lactuca spp., protoplast fusion example, *189*
Lambda (phage-based) vectors, 66-67
  advantages, 68
  host cells infection, 68-69
  ligation, 67
  packaging *in vitro*, 67-68
  preparation, 67
  use in cDNA cloning, 66-67
  use in cDNA library screening, 68-72
    identification of plaques, 71-72
    master plates, 69
    orientations marks for plaque lifts, 71
    preparation/infection of host cells, 68-69
    preparation of replicate lifts, 70-71
    purification of plaques, 72
    transfer to support membranes, 70
Lambda ZAP-II vector, 66, 67
Late embryogenic abundant (LEA) proteins, 240, 241
Ligase chain reaction (LCR), 121
Light microscopy, immunolocalisation protocols, 172-179
Light-sensitive mutants, somatic hybrid selection, 199
Linker molecules, cDNA cloning, 65-66
Lotus spp., somatic hybrids, 200
LR White resin, 165, 177, 180
Lycopene (tomato pigment), 254
Lycopersicon esculentum (tomato)
  fruit development/ripening, 251-281
  in vitro translation systems used, *43*
  RNA isolation method, 10-11
  see also Tomato

M

Maize (Indian corn)
  DNA fingerprinting, 125
  embryonic development, 223-224
  embryo-specific mutants, 235-237
  endosperm development, 224
  see also Zea mays
Male sterility
  advantages, 202
  transfer, 203
Map-based cloning, fruit-ripening genes, 276-277
Medicago spp., somatic hybrids, *189*, 199

Melting temperature
   annealing temperature determined, 131, 132
   calculation, 89, 131–132
   effect of nucleic acid hybridisation, 89–90
Membranes
   blocking agents for, 84, 87, 106–107
   nucleic acid blotting, 70, 82
   protein blotting, 102
β-Mercaptoethanol, as RNase inhibitor, *3*, 16
Messenger RNA (mRNA)
   abundance, 2
   in vitro translation, 33–53
      analysis of products, 45–53
      components required, 34, *35*
      Escherichia coli system, 34, 35, 36–39
      rabbit reticulocyte lysate system, 34, 35, 41–42
      selecting appropriate system, 35
      wheat germ system, 34, 35, 39–41, 211, 213
      Xenopus oocyte system, 34, 42
Mickey type mutants (*Arabidopsis thaliana*), 233
Microsequencing methods, blotted proteins, 111–113
Microsomal membranes, *in vitro* processing using, 52–53
Monoclonal antibodies, 171
Monocotyledonous plants
   embryonic development, 223–225
   import of proteins, 209
Monopteros type mutants (*Arabidopsis thaliana*), 232
Moricandia arvensis, protoplast fusion example, *189*
mRNA *see* Messenger RNA
Multigene family members, identification, 94, 95–97
Murine reverse transcriptase (M-MLV RT), 63
Mutants
   analysis by Northern/Southern blotting, 97–98
   tomato ripening-impaired, 255–256
Mutations
   detection with PCR, 125
   introduction into sequences by PCR, 129

N

Near-isogenic lines (NILs), tomato, 277
Nick columns (Sephadex G50), 73
Nicotiana spp., somatic hybrids, 199

Nicotiana tabacum (flowering tobacco)
   anther-specific cDNAs, 58, 61
   protoplast fusion examples, 189, 198, 199
   see also Tobacco . . .
Nitrate reductase deficiency, somatic hybrids selected using, 199
Nitroblue tetrazolium (NBT), 108, 153, 169
Nitrocellulose (NC) membranes, 102, 104
Non-radioactive labelled probes, 142
   double labelling protocol with radiolabelled probes, 154–155
      advantages, 156–157
   see also Digoxigenin (DIG) labelled probes
Norflurazon (herbicide), effects, 244
Northern blotting, 88–89
   A9 gene studied, *60*
   barley root-tip RNA studied, *86*
   compared with Southern blotting, 89
   diagnostic dot blots, 75–76
Nuclear genomes, effect of protoplast fusion, 201–202
Nucleic acid hybridisation
   bottles used, 84, 88
   in cDNA screening, 73–74
   factors affecting, 89, 90
   in situ hybridisation technology, 142–157
   in Southern blotting, 84
      comments on method, 87–88
   stringency conditions
      effect of variation, 95–96
      estimation, 90
Nylon membranes, 82, 102
   see also Hybond-N membrane

O

Oat seedlings, RNA isolation method, 14–15, *19*
Octant (embryonic tissue), 221
Oilseed rape
   storage protein synthesis, 243–244
   see also Brassica napus
Oligo(dT), first strand cDNA synthesis using, 62, 63, 72, 126
Oligo(dT)-cellulose, poly(A)$^+$ RNA isolation using, 24
Osmotic stress respones, 242–243

P

Paraffin embedding, 145, 164
Paraformaldehyde
   as fixative, 144, 162, 172, 176, 180
   preparation of solution, 144, 182

pBluescript, 66, 67
Pea, isolation of chloroplasts, 209–211
Peach fruit, RNA isolation method, 12–13, *19*
Pear fruit, RNA isolation method, 10–11, *19*
Pearl chains, 194
  electrofusion frequency affected by, 195
Pectinmethylesterase, 262–263
  role in tomato fruit ripening, 266
Peptide synthesis, *in vitro* technique, 52
Peroxidase-antiperoxidase (PAP) immunodetection method, 168–169
Peroxidases
  action, 228
  in immunodetection, 168
Petunia spp., protoplast fusion procedures, 190–193
Phagemid, 66
Phage vectors, cDNA cloning using, 66–68
Phytoene synthase, 254, 270–271
Pisum sativum (pea), isolation of chloroplasts, 209–211
Plaque lift (in cDNA screening), 70
  orientation marks for, 71
Plasmid vectors
  cDNA cloning using, 66
  compared with viral vectors, 68, 70
Plastic resins, embedding techniques, 164–165, 177
Poly(A)$^+$ RNA
  cDNA synthesised, 72–73, 126
  preparation, 22–24, 72
Polyacrylamide gel electrophoresis (PAGE)
  gel preparation procedure, 46–47
  mRNA translation products analysed, 45–47
  proteins transferred from, 102–103
  RNA fractionation using, 26
Polyclonal antibodies, 171
Polyethylene glycol (PEG)
  cytotoxic effects, 193
  embedding wax, 164
  nucleic acid hybridisation affected by, 87
  protoplast fusion using, *189*, 191–194
Polygalacturonase
  gene encoding, 261–262
  role in fruit ripening, 255, 264–265, 273–274
Polymerase chain reaction (PCR), 117–136
  applications, 123–130
    cloning experiments, 126–129
    detection of mutations, 125
    detection of specific DNA sequences, 123–124
    DNA fingerprinting, 124–125
    sequencing reactions, 129–130
  cloning of PCR products, 127–128
  contamination problems, 124, 135
  controls needed, 134
  costs, 122–123
  enzymes used, 121
    cost, 123
  equipment used, 122, 135–136
    cost, 122–123
    machines commercially available, 122, 135–136
  negative results, 134
  primers, 131–133
    annealing temperature calculation, 131–132
    cost, 123
    general features, 131
    heterologous primers, 132–133
  principle, 118–119
  publications covering, 118, 136
  quantification methods, 130
  reagents used, 121–122
  template preparation, 120
  troubleshooting hints, 133–136
  typical protocol, 119–120
Polysomes
  translation of, 42–44
    S-30 system, 43–44
    S-100 system, 44
Polyvinyldifluoride (PVDF) membranes, 102, 111
Ponceau S, as protein stain, 106, 108
Pooled-sample analysis, 278–279
  tomato genetic mapping, 280–281
Potato
  protoplast fusion examples, 189
  RNA isolation method, 10–11, *19*
Precipitating fixatives, 162
Prehybridisation, 149
  procedures, 73, 149–150, 154–155
Primer-dimers (PCR products), 127, 131
  removal procedure, 127
Probe labelling, 80, 82–83, 87
Prokaryotic mRNA translation, system used, 36
Protein A, immunodetection using, 169
Proteinase K
  protein removal (from RNA) using, 21
  RNA isolation method using, 11–13, *19*
  as RNase inhibitor, *3*, 12
Proteinase inhibitors, 44–45
Protein blotting, 101–114
  detection methods used, 107–108
    recommended protocol, 108–109
  double staining technique, 106, 109–111
  membrane blocking agents used, 106–107

membranes used, 102
microanalysis of blotted proteins, 111–113
protocols recommended, 108–109
stains used, 106
 recommended protocol, 109
transfer methods, 102–106
 capillary action, 103
 diffusion method, 103
 electroblotting, 104–105
 vacuum blotting, 103–104
see also Western blotting
Protein stains, 106
Protein synthesis, *in vitro* technique, 33
Protoderm, 223
Protoplast fusion, 187–203
 chemically induced fusion
  first reported, 188
  large-scale techniques, 190–194
  number of protoplasts required, 190
  small-scale techniques, 197
 effect on cytoplasmic genome, 202–203
 effect on nuclear genome, 201–202
 electrofusion
  first reported, 188
  large-scale techniques, 194–196
  number of protoplasts required, 190
  small-scale techniques, 197–198
 first observed, 188
 large-scale techniques, 190–196
  number of protoplasts required, 190
 plant-regeneration from protoplast-derived tissues, 200
 selection of somatic hybrids after, 198–200
 small-scale techniques, 196–198
  number of protoplasts required, 190
Prunus spp., somatic hybrids, 196
Pyrus spp., somatic hybrids, 196

R

Rabbit reticulocyte lysate
 mRNA translation using, 34, 35, 41–42
 RNase inhibitor used, 44
 storage procedure, 42
Radiolabelled antibodies, protein detection using, 107
Radiolabelled probes
 disadvantages, 142
 double labelling protocol with non-radioactive labelled probes, 154–155
  advantages, 156–157
Randomly Amplified Polymorphic DNA Sequences (RAPDS), 124–125
Random primed labelling, 82, 87

Replacement synthesis, double-stranded cDNA, 63–64, *64*
Replicate plaque lifts, 70–71
Resistance markers, somatic hybrids selected using, 199
Restriction fragment length polymorphisms (RFLPs)
 gene mapping based on, 92–93, 189, 198
 tomato, 257, 277
Resurrection plants, 239, 240–241
Reverse genetics, 263
Reverse transcriptases, 63
Reversible embedment cytochemistry (REC), 165
Ribonucleases (RNases), 2
 removal procedures, 3
 sources, 2, 3, 44
 see also RNase inhibitors
Rice tissues
 immunolocalisation of proteins, 176–181
 RNA isolation method, *19*
Ripening (of fruit)
 climacteric vs non-climacteric ripening, 252–253
 hormone responsible, 253–254
 model for induction/regulation, 274–276
Ripening-related genes
 cloning
  on basis of function, 260–263
  isolation of markers, 277–279
  map-based cloning, 276–277
  pooled sample analysis, 280–281
 determination of physiological functions, 263–274
  antisense gene repression, 263–271
  complementation, 272–274
  co-suppression of gene expresion, 271–272
  pathway blockage strategy, 272
 isolation, 257–263
  differential screening used, 257–260
RNA fractionation, 24–29
 gel electrophoresis
  denaturing conditions, 26–29
  non-denaturing conditions, 26
 sucrose gradient methods
  denaturing conditions, 25–26
  non-denaturing conditions, 25
RNA gel, transfer to membrane, 29
RNA *in vitro* translation systems, 33–53
 see also Messenger RNA, *in vitro* translation . . .
RNA isolation, 2–20
 chaotropic agents used, 13–16, *19*
 CTAB method, 16–17, *19*, 20
 detergent method, 8–10, *18*

RNA isolation (cont.)
   differential precipitation method, 10–11, *18*
   guanidinium thiocyanate method, 13–15, *19*
   hot phenol method, 7–8, *18*
   proteinase K method, 11–13, *19*
   SDS/phenol method, 3–7, *18*
      tomato fruit tissues, 6–7
      wheat leaf chloroplasts, 5–6
      wheat leaf extracts, 4–5
      yield data, *18*
   sodium perchlorate method, 15–16, *19*
   storage of samples, 3
   yields listed for various methods, 13, 15, *18–19*
RNA purification, 21–24
   carbohydrate-removal method, 23
   DNA-removal methods, 5, 21–22
   poly(A)$^+$ RNA isolation, 22–24, 72
   protein-removal methods, 6, 21
RNA purity, measurement, 21
RNA, see also Messenger RNA
RNase inhibitors, *3*, 12, 44, 88
Rough endoplasmic reticulum (RER), peptide processing involved, 52
Rudbekia spp.
   protoplast fusion method, 195–196
   somatic hybrids regeneration, 200

S

S-30 fraction, meaning of term, 36
S-100 extracts, preparation method, 44
Seed development, 219–246
   molecular biochemical approach, 225–230
   see also Embryonic development
Seed reserve substances, 219–220
   factors affecting, 220
Sephadex G50 columns, 73, 83
   first strand cDNA probes purified, 73
Sequencing reactions, PCR used, 129–130
Sinapsis turgida, somatic hybrids, 200
Slab gel drier, vacuum blotting system developed from, 104
Sodium dodecyl sulphate (SDS), as RNase inhibitor, 3
Sodium dodecyl sulphate (SDS)/phenol RNA isolation method, 3–7
   examples of use, 4–7
   yield data, *18*
Sodium perchlorate
   RNA isolation using, 15–16, *19*
   as RNase inhibitor, *3*
Somaclonal variation, 227
Somatic embryogenesis studies, 226–227
Somatic hybridisation, 189, 196
   selection techniques used, 198–200

see also Protoplast fusion
Somatic hybrids
   examples, *189*, 196, 198, 199, 200, 201
   selection methods, 198–200
Southern blotting
   applications, 91–98
      analysis of mutants, 97–98
      gene copy number, 91, 93
      gene location, 93–94
      identification of multigene family members, 94–97
   compared with Northern blotting, 89
   first described, 80, 103
   method, 81–85
      comments on method, 85–90
      DNA transfer, 81–82, 85–86
      prehybridisation/hybridisation, 84
      probe labelling, 80, 82–83, 87
      probe removal, 85
      washing procedure, 84–85, 88
   modification for Northern analysis, 88–89
   principle, 80
Southwestern blotting, 113
   DNA binding proteins detection, 113–114
Spermine, effects on RNA translation, 41
Stains
   immunodetection, 153, 169
   protein-visualisation, 106
Steedman's wax, 164
Stomatal apertures, closure, 242–243
Strawberry tissues, RNA isolation method, 10–11, *18*
Streptavidin-biotin immunodetection system, *167*, 167–168
Sucrose gradient methods
   RNA fractionation using
      denaturing conditions, 25–26
      non-denaturing conditions, 25
Sweet potato, RNA isolation method, *19*

T

Taq polymerase, 121
Thylakoids, *208*
   import of proteins, 215–218
Tissue cultures, genetic aberrations occurring, 227
Tobacco
   RNA isolation method, 9–10
   see also *Nicotiana* ...
Tobacco mosaic virus RNA, translation procedure, 41
Tomato
   advantages as model system, 257
   DNA fingerprinting, 125

fruit development/ripening, 251–281
  characteristic changes, 254–255
  cloning of fruit-ripening genes, 276–281
  determination of physiological functions of ripening genes, 263–274
  isolation of ripening-related genes, 257–263
  model for induction/regulation, 274–276
  ripening-impaired mutants, 255–256
  in vitro translation systems used, *43*
  as model system for fruit development/ripening, 254–257
  ripening-impaired mutants, 255–256
  RNA isolation method, 6–7, 10–11, *19*
Tomato Spotted Wild Virus (TSWV), transfer in *Nicotiana* spp., 189
Toro type mutants (*Arabidopsis thaliana*), 233
Transcription factors, 245
Transfer methods, 70, 82
  factors affecting, 104
Transformed cell lines, somatic hybrids selected using, 200
Transmission electron microscopy, immunolocalisation protocols, 179–181
Trehalose, 240
Triticum . . . see Bread . . . ; Durum wheat; Wheat

V

Vacuum blotting, 103–104
Vectors
  cDNA cloning
    plasmid vectors, 66
    tomato fruit-ripening genes, 268
    viral vectors compared with plasmid vectors, 68, 70
Viviparous (*vp*) mutants, 238
  stress-responsive gene expression, 244–245
VP-1 gene (maize), role in embryogenesis, 245–246

W

Western blotting
  meaning of term, 102, 104
  see also Protein blotting
Wheat, Southern blot analysis used, *92, 95, 96, 98*
Wheat Em protein, 241
Wheat germ
  mRNA translation using, 34, 35, 39–41, 211, 213
    disadvantage, 211
    RNase inhibitor used, 44
  sources, 39
Wheat leaf
  analysis of mRNA translation products, *48*, 49, 50
  RNA isolation methods, 4–6
Wheat storage proteins, identification after gel electrophoresis, 109, *110*
White spruce, DNA fingerprinting, 125

X

Xenopus oocytes
  difficulty in maintaining supply, 42
  in vitro processing using, 52
  mRNA translation using, 34, 42

Y

Yeast artificial chromosomes (YACs), as vectors for tomato, 259, 279

Z

Zea mays (Indian corn/maize)
  embryonic development, 223–224
  embryo-specific mutants, 235–237
  as representative of monocotyledonous plants, 221, 223–224
  see also Maize